# Technical Writing

# Technical Writing

**Fifth Edition**

## John M. Lannon

Southeastern Massachusetts University

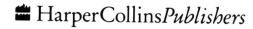 HarperCollins*Publishers*

Sponsoring Editor: Constance Rajala
Development Editor: Vicky Anderson Schiff
Project Coordination and Text Design: The Wheetley Company, Inc.
Cover Design: Jaye Zimet
Production: Michael Weinstein
Compositor: Pam Frye Typesetting, Inc.
Printer and Binder: R.R. Donnelley & Sons Company
Cover Printer: The Lehigh Press, Inc.

Technical Writing, Fifth Edition

Library of Congress Cataloging-in-Publication Data

Lannon, John M.
    Technical writing / John M. Lannon. — 5th ed.
        p.    cm.
    Includes index.
    1. Technical writing.     I. Title.
  T11.L24    1990
  808'.0666–dc20                                          90-24538
                                                          CIP


        ISBN 0-673-52095-1 (student edition)
        ISBN 0-673-52153-2 (teacher edition)

91 92 93 9 8 7 6 5 4 3 2 1

# Brief Contents

# Contents

## Chapter 7   Summarizing Information     140

## Chapter 8   Defining Your Terms     154

## PART III  SEQUENCE, SHAPE, AND STYLE IN YOUR DOCUMENT     173

### Chapter 9  Outlining     174

### Chapter 10  Shaping the Paragraphs     186

## Chapter 11    Revising for a Readable Style    198

## PART IV    GRAPHIC AND DESIGN ELEMENTS    239

### Chapter 12    Creating Visual Aids    240

## Chapter 17   Memos and Short Reports    389

## Chapter 18   Supplements to Documents    407

## Chapter 19    **Proposals**    **419**

## Chapter 20    **Analytical Reports**    **453**

## Chapter 21  **Oral Reports**    483

## Appendix A    Review of Grammar, Usage, and Mechanics    492

## Appendix B    Writers and Audiences on the Job    526

# Preface

*Technical Writing,* Fifth Edition is a comprehensive and flexible introduction to technical and professional communication. Designed for classes in which students from a variety of majors are enrolled, it addresses a wide range of interests. Rhetorical principles are explained, illustrated, and applied to an array of assignments, from brief memos and summaries to formal reports and proposals. Exercises parallel the writing demands students will face both in college and on the job. Throughout the book, students learn how to focus on their purpose and connect with their audience while working through the planning, drafting, and revising activities of the writing process. Beyond a traditional focus on the *informative* dimension of workplace writing, this new edition treats in detail the *persuasive* and *ethical* dimensions as well, emphasizing interpersonal considerations and accountability.

## ORGANIZATION

**In Part I,** "Communication in the Workplace," job-related writing is presented as a process of problem solving and critical thinking. The first four chapters analyze writers' essential decisions in solving problems regarding information, persuasion, and ethics in their communications. Chapter 5 synthesizes these decisions by illustrating the entire process in a workplace writing situation.

**In Part II,** "Information Retrieval and Synthesis," readers will find strategies used routinely by technical communicators. Students learn how to retrieve information from primary and secondary sources, how to shape their material for readers' needs, and how to specify their exact meaning.

**In Part III,** "Sequence, Shape, and Style in Your Document," the material demonstrates how to organize and express a message for specific readers' expectations and understanding. Students learn how to control their material and how to develop a style that connects with readers.

**In Part IV,** "Graphic and Design Elements," there is an exploration of the rhetorical implications of graphics and page design. The part shows how to enhance a document's access, appeal, and visual impact.

**In Part V,** "Specific Documents and Applications," the concepts and strategies introduced earlier are adapted to the composing of actual documents. Various letters, memos, reports, and proposals offer a balance of examples from the workplace and from student writing. Each sample document has been chosen so that students can emulate it easily.

Finally, the **appendixes** contain a brief handbook of grammar, usage, and mechanics; the full text of interviews with four writers on the job; and a sample proposal, progress report, and final report for an actual workplace project.

## THE FOUNDATIONS OF *TECHNICAL WRITING*

1.  Although it follows no single, predictable sequence, the writing process is no set of random actions; rather it is a series of decisions in problem solving. Beyond studying model documents, students need to understand that any effective writing is the product of good thinking.

2.  Writers who lack rhetorical awareness often ignore initial decisions that determine the effectiveness of documents. Only by sizing up writing situations, defining the communication problems, and asking important questions can writers find the answers they need. Workplace writing always requires critical thinking.

3.  The most vital workplace writing involves struggling with persuasion problems such as readers' egos or resistant attitudes, and with ethics problems such as groupthink or misplaced loyalties.

4.  All students can learn to incorporate in their writing the essential rhetorical features: worthwhile content, sensible organization, readable style, and inviting and accessible design.

5.  Because most technical writing courses contain an assortment of students with varied backgrounds, explanations that are thorough, examples and models that are broadly engaging and intelligible, and goals that are rigorous but collectively achievable are called for.

6.  Rather than reiterating information found in the course textbook, classroom workshops can apply textbook principles by focusing on the students' writing. (Suggestions for workshop design are found in the Instructor's Manual.)

## NEW TO THIS EDITION

- More emphasis on problem solving and critical thinking helps students better understand the writing process.

- A new chapter on persuasion, based on current research in rhetoric and social science, communication, and workplace practices, enables students to anticipate and assess the interpersonal challenges, constraints, and pitfalls of any persuasive effort.

- A new chapter on communication ethics offers a realistic and practical introduction to the long-overlooked notion of *accountability,* and serves as a reference point for ethical considerations throughout the text.

- Coverage of legal implications in workplace writing introduces students to laws governing privacy, product liability, copyrights, and specifications.

- Ample new examples, models, and exercises are inspired by current workplace issues ranging from energy conservation to employee monitoring to the space shuttle *Challenger* disaster.

- More team projects throughout offer practice in the kinds of collaborative writing often done in the workplace.

- Treatment of visual communication is based on the latest computer technology and on recent research in rhetoric and document design.

- More visuals and an improved design enhance the book's accessibility and appeal.

- A new appendix presents a unified set of documents (proposal, progress report, and analytical report—fully annotated and discussed) prepared by one student in response to an actual workplace situation.

- A comprehensive educational package includes an instructor's manual with test bank and chapter quizzes, acetate transparencies, and interactive software for style editing. The software (an automated version of Chapter 11) offers five editing lessons: on clarity, conciseness, fluency, exactness, and tone. (A demonstration disk is available from the publisher.)

Much of the improvement in this edition was inspired by generous and invaluable help from the following reviewers: Gerald Alred, University of Wisconsin, Milwaukee; Carolyn Brown, Auburn University; Pamela Gardner, Utah Technical College; Peter Hager, University of Texas at El Paso; Judith Kaufman, Eastern Washington University; Nancy Mackenzie, Mankato State; Fiore Pugliano, University of Pittsburgh; Mark Rollins, Ohio University; Edith Weinstein, University of Akron. Thank you all. Thanks also to Jack Bordan for expert advice about many technical details.

At Southeastern Massachusetts University, Raymond Dumont provided, as always, countless ideas and suggestions. Louise Habicht responded patiently and lucidly to my wild ideas. Many other colleagues provided encouragement and helpful critiques. And my students offered feedback, examples, and inspiration.

Thanks to Constance Rajala and Vicky Anderson-Schiff for enabling this whole project to go forward with care and quality, and to Elizabeth Gabbard for a thoroughly professional job of production.

For the ones who keep me going: Chega, Daniel, Sarah, Patrick, and Max.

John M. Lannon

# Communication in the Workplace

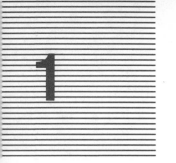

# Introduction to Technical Writing

Technical Writing Serves Practical Needs

Writing Is Part of Most Careers

Writers Today Need Better Skills Than Ever

Advanced Writing Technology Still Requires Mental Labor

Technical Writers Face Three Related Problems

Problem Solving Requires Critical Thinking

In technical writing, you communicate and interpret specialized information for your readers' use. Readers may need your information to perform a task, answer a question, solve a problem, or make a decision. Whether you write a memo, letter, report, or manual, the document must present material your readers find useful.

## TECHNICAL WRITING SERVES PRACTICAL NEEDS

Unlike poetry or fiction, which appeal primarily to our *imagination,* technical documents appeal to our *understanding.* Technical writing therefore rarely seeks to entertain, create suspense, move emotions, or invite differing interpretations. Those of you who have written any type of lab or research report already know that technical writing has little room for ambiguity.

To serve practical needs in the workplace, technical documents must be reader oriented and efficient.

### Technical Documents Are Reader Oriented

Instead of focusing on the writer's desire for self-expression, a technical document addresses the reader's desire for information. This doesn't mean that your writing should sound like something produced by a robot, without any personality (or *voice*). Your document may in fact reveal a lot about you (your competence, knowledge, integrity, etc.), but it should not focus on you personally. Readers are interested in *what you have done* or in *what you recommend,* but they have only a professional interest in *who you are* (your feelings, hopes, dreams, visions). A personal essay, then, would not be technical writing. Consider this essay fragment:

> Computers are not a particularly forgiving breed. The wrong key struck or the wrong command typed is almost sure to avenge itself on the inattentive user by banishing the document to some electronic trash can.

Focuses on the writer's feelings

This view of computer reality communicates a good deal about the writer's resentment and anxiety but very little about computers themselves.

The following example can be called technical writing because it focuses (in italics) on the subject, on what the writer has done, and on what the reader should do:

> On VR 320 terminals, *the BREAK key* is adjacent to keys used for text editing and special functions. Too often, users inadvertently strike the BREAK key, causing the program to quit prematurely. To prevent the problem, *we have* modified all database management terminals: to quit a program, *you must* now strike BREAK twice successively.

Focuses on the subject

This next example can be called technical writing because it focuses on what the writer recommends:

> I recommend that our VAX 780 hardware be upgraded by a maximum addition to main memory, a disk input/output control unit, and an additional disk storage unit. This expansion will (1) increase the number of simultaneous terminal users from 60 to 80, (2) increase the system's responsiveness, and (3) provide sorely needed disk storage for word processing and company data bases.

Focuses on the writer's recommendation

And so your document never makes you "disappear," but it does focus on that which is most important to the reader.

## Technical Documents Strive for Efficiency

Professors read to see what we know; colleagues, customers, and bosses read to see what they can use. Workplace readers hate waste and demand efficiency. They want only as much as they need: "When it comes to memos, letters, proposals, and reports, there's no extra credit for extra words. And no praise for elegant prose. Bosses want employees to get to the point—quickly, clearly, and concisely" (Spruell 32). Efficient documents save time and energy.

In any system, efficiency is the ratio of useful output to input. For the product that comes out, how much energy goes in?

ENERGY $\longrightarrow$ | SYSTEM | $\longrightarrow$ PRODUCT
(input)                                            (output)

When a system is efficient, the output nearly equals the input.

Similarly, a document's efficiency can be measured by how hard readers have to work with the document to get the message they need. Is the product worth the reader's effort?

| READER SPENDS<br>ENERGY<br>(input) | $\longrightarrow$ | DOCUMENT | $\longrightarrow$ | READER GETS<br>THE MESSAGE<br>(output) |
|---|---|---|---|---|

No reader should have to spend ten minutes deciphering a message worth only five minutes. Consider this wordy message:

> At this point in time, we are presently awaiting an on-site inspection by vendor representatives relative to electrical utilization adaptations necessary for the new computer installation. Meanwhile, all staff are asked to respect the off-limits designation of said location, as requested, due to liability insurance provisions requiring the on-line status of the computer.

Inefficient documents drain the readers' energy; they are too easily misinterpreted; they waste time and money. Notice how hard we had to work with the message above to extract information that could be expressed this simply:

> Hardware consultants soon will inspect our new computer room in order to recommend appropriate wiring. Because our insurance covers only an *operational* computer, this room must remain off limits until the computer is fully installed.

Instead of locking us out, efficient documents invite us in.

Readers have a sense of how hard they will have to work to understand a document. When readers sense they are working too hard, they tune out the message – or they stop reading altogether.

Inefficient documents have varied origins. Even though information in a document may be accurate, errors like the following make readers work too hard:

- more (or less) information than readers need
- irrelevant or uninterpreted information
- no discernible organization
- fancier or less precise words than readers need
- more words than readers need
- uninviting appearance or confusing layout
- no visual aids when readers need them

Far more than a *list* of information, an efficient document sorts, organizes, and interprets that information to suit the audience's needs, abilities, and interests.

Instead of merely "happening," an efficient document is carefully designed to include these elements:

- *content* that makes the document worth reading
- *organization* that reveals the line of thinking and provides emphasis
- *style* that is economical and clear

- *visuals* (graphs, diagrams) that depict concepts and relationships
- *format* (layout, typeface) that is accessible and appealing

Reader orientation and efficiency are more than abstract "rules": All writers are accountable for their documents. In questions of liability, faulty writing is considered no different from any other faulty product. If your inaccurate or unclear or incomplete information leads to injury or damage or loss, *you* can be held legally responsible.

## WRITING IS PART OF MOST CAREERS

Although you will not likely join the workforce primarily as a "writer," your skills will be tested repeatedly in situations like these:

- proposing various projects to upper management or to clients
- writing progress reports to the boss
- contributing articles to the employee newsletter
- describing a product to employees or customers
- writing procedures and instructions for employees or customers
- justifying to management a request for funding or personnel
- editing and reviewing documents written by colleagues

Like most professionals in any field, you can expect to spend a good percentage of your day as a part-time technical writer.

Your value to any organization will depend on how well you communicate what you know. Many working professionals spend at least 40 percent of their time writing or dealing with someone else's writing. The higher their position, the more they write (Barnum and Fisher 9–11). Here is a corporate vice president's description of some audiences you can expect to face:

> The technical graduate entering industry today will, in all probability, spend a portion of his or her career explaining technology to lawyers—some friendly and some not—to consumers, to legislators or judges, to bureaucrats, to environmentalists and to representatives of the press (Florman 23).

And these audiences, among countless others, expect to read efficient documents.

Here is what two top managers for a major auto maker have to say about the effect a document can have on the organization *and* on the writer:

> A written report is often the only record which is made of results that have come out of years of thought and effort. It is used to judge the value of the *person's* work and serves as the foundation for all future action on the project. If it is written clearly and precisely, it is accepted as the result of sound reasoning and careful observation. If it is poorly written, the results presented in it are placed in a bad light and are often dismissed as the work of a careless or incompetent worker (Richards and Richards 6).

Instead of consciously judging your writing, the audience is most likely to see it as a measure of your *thinking*. Good writing gives you and your ideas *visibility* and *authority* within your organization. Bad writing, on the other hand, is not only useless to readers and politically damaging to the writer; it is also expensive:

> The biggest untapped source of net profits for American business lies in the sprawling, edgeless area of written communication, where waste cries out for management action. Daily, this waste arises from the incredible amount of dull, difficult, obscure, and wordy writing that infests plants and offices (Keyes 105).

The estimated cost of communication in American business and industry is more than $75 billion yearly. And roughly 60 percent of the writing produced is inefficient: unclear, misleading, irrelevant, or otherwise wasteful of time and money (Max 5–6).

Although automated office systems (word processing, electronic mail, teleconferencing, and so on) speed the flow of information, no computerized device can convert bad writing to good. When you struggle through that first draft, it makes no difference whether you use quill and ink or the most sophisticated word processor; if the thinking is shabby, the writing will be useless. Moreover, because today's writers produce the *finished document* on word processors, they no longer can rely on secretaries to "fix up" their writing.

Whatever your career plans, you probably will have to work as a part-time technical writer. Employers first judge your writing by your application letter and résumé. In a large organization, your future may be decided by executives you've never met. And one concrete measure of your job performance will be your letters, memos, and reports. As you advance professionally, skill in communication becomes more important than technical background. The higher your goals, the better you need to communicate.

## WRITERS TODAY NEED BETTER SKILLS THAN EVER

Computer technology has brought us the Information Age. No longer merely the knowledge stored in our brains or in printed documents, *information* today includes all the computerized data to which we have access. Information has become our ultimate product—a product whose volume more than doubles in each decade.

Today's professionals are information workers, increasingly reliant on computers for communication. In 1986, 15 million American workers—one in eight—were using a computer ("Calming Words" 99). By 1988, computer keyboards in the American workplace outnumbered white-collar workers (International Data Corporation 84). In the 1990s, you can expect to spend at least part of your workday communicating by computer.

Every two or three years, computers double their capacity to process information. But even though automation is increasing the *speed* and *volume* of communication, the information itself still needs to be *written*. A recent survey of 17,000 office workers revealed that managers spend nearly half their workday on writing and clerical tasks such as filling in forms and typing (International Data Corporation 86). Today's writers still struggle for control over the massive volume of information in the workplace. As one executive observes, "Productivity in the Information Age is not measured by the mile of text produced" ("Gassee" 121). Information becomes useful knowledge only when it has been sorted, interpreted, and organized to lead to a conclusion.

Beyond grappling with an increasing volume of information, today's writers must address many readers who have no expertise in technology, but who must *use* the technology in their jobs and lives. The manager using a teleconferencing network, the data-entry clerk using a computer terminal, or the bank teller using a check-verification system—these are among the countless "nontechnical" workers who rely on technology to do their jobs. As sophisticated equipment becomes more easily available and affordable, writers need to explain to laypeople how to use software, or how to operate computer hardware, medical equipment, telephone systems, precision instruments, and a growing array of automated products, from videocassette recorders to microwave ovens.

Information in too great a volume actually can *prevent* communication, unless you (1) impose some order on your material; (2) select, from everything you know, only that which is *useful* to your readers; and (3) interpret for your readers. With so much information available, no writer can afford merely to "let the facts speak for themselves."

---

## ADVANCED WRITING TECHNOLOGY STILL REQUIRES MENTAL LABOR

Besides being the frequent subject of your writing, the new technology can serve as a multipurpose writing tool as well. Here is how computers are changing the way we write.

### Word Processing

Basically a typewriter with a memory, a word processor reduces the drudgery of producing a typed or printed page. Writing and editing directly on the computer screen, you design as many versions of a document as you wish without retyping the whole thing. A basic word processing package lets you insert, delete, or move blocks of text; other software allows you to change formats, integrate visuals, or search the document to change a word or phrase, or to have your document examined automatically for correct spelling, accurate word choice, and readable style. By streamlining the writing process, word processors leave you more time to think, to experiment, and to refine your document.

### Research and Reference

Document preparation often involves research. Instead of thumbing through newspapers, journals, reference books, or filing cabinets, you can do much of your research at the computer terminal. A phone call to an on-line database will provide the latest facts and figures on the stock market, the environment, nuclear energy—virtually any major topic. More recently, information systems on compact disks (CDs) offer a growing array of information products, ranging from electronic encyclopedias (twenty volumes on a five-inch compact disk) to reports filed with the Securities and Exchange Commission to the latest edition of *Books in Print*. Now in development are systems that will store all documents in electronic file cabinets for instantaneous retrieval (Ilkovics 9).

### Data Transmission

Using electronic mail, you can distribute copies of your document to a network of computer terminals, whether within your own company or worldwide. Your audience can in turn provide feedback on your message. Through the same electronic network, you might work as a coauthor with colleagues who are preparing, editing, or reviewing different parts of a document. Using a FAX network, you can send and receive printed documents of all kinds.

### Text Management

After composing on the computer, you can file your document electronically, cataloging it for easy record keeping and rapid retrieval. When you have documents or parts of documents that you use often (such as sales letters or proposal parts), you can retrieve them rapidly, alter them as needed, and insert them directly into some other document.

### Desktop Publishing

Any writer with a personal computer, a printer, and page design software can become a desktop publisher. Beyond the actual writing, desktop publishing includes page design, graphics production, typesetting, printing, and distribution of the published document to its audience.

Desktop publishing software offers such design options as dividing pages into columns, controlling the size and appearance of type, adjusting the size and shape of graphics, inserting predrawn images, and printing headlines. These options enable you to produce newsletters, brochures, pamphlets, manuals, and other publications that traditionally have required graphic artists and print shops.

The appearance of a document affects readers' acceptance. Because of the volume of paper generated by computers, any document needs to be designed for visual impact, "simply to get . . . noticed in that sea of documents both in and out of the office" (Gubernat 72).

## Tomorrow's Writing

In development are powerful multiprocessors that will function as "smart" computers, able to simulate a degree of intelligence. Such computers will recognize spoken commands, thus eliminating the need for typing—but not for careful editing. Machines will also be able to "read" images into memory—a text page, say, or a graphic figure placed before the computer screen. A Massachusetts company has developed a "smart" computer that can recognize up to a thousand spoken words and phrases, and a machine with a five- to fifteen-thousand-word recognition is expected soon (Reade 20).

You might one day be a "telecommuter," performing most of your research, writing, and other work on a home terminal, which is then linked to a company or even a world-wide information network.

Figure 1.1 gives an overview of the high-tech writing tools that are steadily increasing our writing efficiency.

By streamlining the physical labor of writing, advanced technology gives writers more time to *think* about what they are writing. But no computer can

**FIGURE 1.1** How Writers Use the New Technology

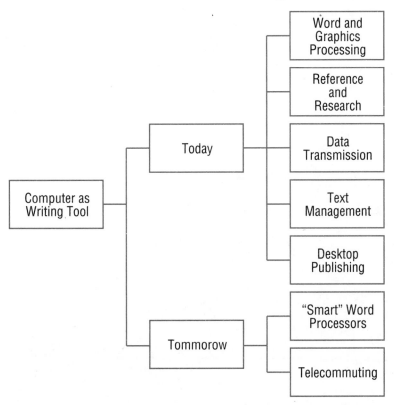

do this thinking for you. No computer can transform shabby thinking into effective writing. Beautifully formatted documents that say nothing are useless.

Even if voice-activated computers do eventually eliminate typing, the writer's central task—thinking—will continue to be every bit as essential. Whether you set out to write that first draft with quill and ink or with a state-of-the-art computer, you face identical problems. Useful writing is a product of mental labor, of critical thinking and problem solving, of hard work and constant revision. No matter how "smart" your computer, it cannot replace your ability to write—that is, to think, formulate, and evaluate.

## TECHNICAL WRITERS FACE THREE RELATED PROBLEMS

No matter how sophisticated the technology, automated writing systems remain incapable of *thinking* for the writer. This section examines how writers must continue to solve three kinds of problems:

- the *information problem,* because different readers have different information needs in different situations
- the *persuasion problem,* because people often disagree about what the information means and about what should be done
- the *ethics problem,* because any workplace writing can have consequences for other people

Figure 1.2 helps us visualize how these three problems relate.

**FIGURE 1.2**   Writers Face Three Related Problems

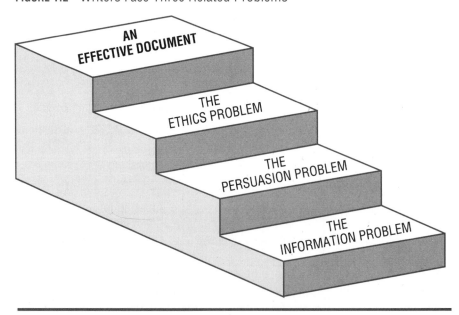

The following scenarios illustrate how a typical professional faces the triple problem of communicating in the workplace.

## The Information Problem

Sarah Burnes has worked two months as a chemical engineer for Millisun, a leading maker of cameras, multipurpose film, and photographic equipment. Sarah's first major assignment is to evaluate the quality of the plant's incoming and outgoing water. (Contaminants in incoming water can damage film during production, and production itself can pollute outgoing water.) Upper management wants an answer to this question: How often should we change water filters? Sarah will study endless printouts of chemical analysis, review current research, and do some testing of her own. When she finally decides on the most accurate answer, Sarah will prepare a recommendation report for her bosses. Later, for other employees, she will write an instruction manual on how to check and change water filters.

The initial task here seems straightforward: reporting the facts, drawing conclusions from these facts, and making recommendations. But Sarah still has to solve the problem of giving readers the information they need: How much data will I need to support my recommendations? How much explaining should I do? How will I organize? Do I need visuals? And so on.

In other situations, Sarah will face a persuasion problem, for instance, when decisions must be made or actions taken on the basis of incomplete or inconclusive facts or conflicting interpretations (Hauser 72). In these instances, Sarah will need to win reader acceptance for *her* view; her writing will have to be persuasive as well as informative.

## The Persuasion Problem

Millisun and other electronics producers are located on the shores of a small harbor, the port for a major fishing fleet. For almost twenty years before 1980, each company discharged into the harbor effluents containing metal compounds, PCB's, and other toxins. Sarah is on a team from these companies, assigned to work with the Environmental Protection Agency to clean up the harbor.

Enraged local citizens are demanding immediate action, and the companies themselves are anxious to put this public-relations nightmare behind them. But in analyzing the problem, Sarah's team discovers that any type of cleanup would stir up harbor sediment, possibly dispersing the pollution into surrounding waters and the atmosphere. (Many of the contaminants can be airborne.) In short, premature action might actually *increase* danger.

Sarah's communication here takes on a persuasive dimension: She will have to justify the delays to her bosses and the public-relations office. She will have to make readers understand the dangers as well as she understands them.

In this situation, the facts are neither complete nor conclusive, and views about what these facts mean will conflict. Sarah's documents will have to balance the various "political" pressures and make a case for *her* interpretation of the facts. Some elements of Sarah's persuasion problem: What do these facts mean? Are other interpretations or conclusions possible? What, if anything, should be done? Is there a better way? And so on.

In some different situation, Sarah might have to reckon with the ethical dimensions of her writing, over the issue of "doing the right thing." For instance, if she ended up with an unethical employer or supervisor, she might feel pressured to overlook or sugarcoat or suppress facts that would be costly or embarrassing to the company. Or sometimes the best technical solution to a problem might be a poor solution in human terms (as when a heavy industry decreases local pollution by building a smokestack tall enough to carry emissions hundreds of miles away).

---

**The Ethics Problem**
Millisun's production manager asks Sarah to test the air purification system in the chemical division. After finding the filters hopelessly clogged, she decides to test the air quality, and finds dangerous levels of benzene (a potent carcinogen). She reports these findings in a memo to the manager, with an urgent recommendation that all employees in the chemical division be tested for benzene poisoning. The manager phones and tells Sarah to "have the filters replaced, and forget about it." Sarah now has to decide what to do next: whether to bury the memo in some file cabinet or to defy her boss and forward copies to other readers who might take action.

---

In situations where truth and fairness call for a particular stand, Sarah will face the hardest choices of all: whether to shut up and look the other way or to speak out and risk retaliation for "bucking the system." Some elements of Sarah's ethics problem: Is it fair? Who might benefit or suffer? What other consequences will this have? Should I reconsider? And so on.

Technical writing involves more than merely throwing facts at readers. For Sarah or any other professional, writing means problem solving and critical thinking.

## PROBLEM SOLVING REQUIRES CRITICAL THINKING

The situations faced by Sarah Burnes illustrate how effective writing involves a lot more than cooking up some information, serving it, and expecting readers to digest it whole. Because writing is nothing more or less than *thinking* on paper, it is never merely a "by-the-numbers" exercise. But even though we can't boil writing down to a formula, we can improve our chances for success through a deliberate strategy known as *critical thinking*.

**FIGURE 1.3** A Critical-Thinking Model for Technical Writing

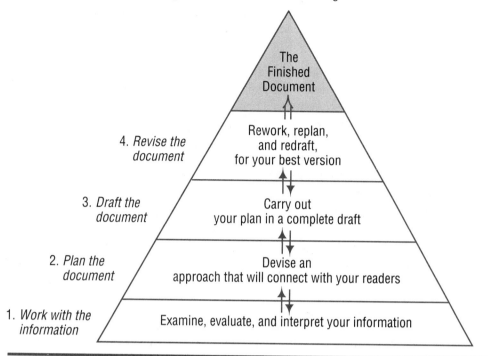

Critical thinking is a way to test the worth of information or the strength of an idea. Instead of accepting the idea at face value, we examine, evaluate, verify, analyze, weigh alternatives, and consider consequences—at every stage of that idea's development. In short, we use critical thinking to solve problems and to test the effectiveness of our solutions.

When we approach writing problems as critical thinkers, we engage in the *writing process,* a process with four distinct and related stages:

1. We work with the information.
2. We plan the document.
3. We draft the document.
4. We revise the document.

Each stage calls for deliberate decisions on the writer's part. And it is during this process (Figure 1.3) that our best ideas often occur (Dumont 14).

One engineering professional describes how critical thinking enriches every stage of the writing process:

> Good writing is a process of thinking, writing, revising, thinking, and revising, until the idea is fully developed. An engineer can develop better perspectives and

even new technical concepts when writing a report of a project. Many an engineer, at the completion of a laboratory project, senses a new interpretation or sees a defect in the results and goes back to the laboratory for additional data, a more thorough analysis, or a modified design (Franke 13).

Writing is a process of *discovering* what you want to say, "a way to end up thinking something you couldn't have started out thinking" (Elbow 15). Figure 1.3 depicts critical thinking during the writing process. As the arrows indicate, no one stage of the process is complete until *all* stages are complete. Like the exposed tip of an iceberg, the finished document ultimately provides the only visible evidence of the quality of your writing labor.

In the workplace you often may need to complete the stages in Figure 1.3 rapidly, under deadline pressures—nonetheless, they must be completed every time you set out to produce a document that does the job. Figure 1.4 lists typical questions you need to answer in moving through these stages. For practicing such decisions in various situations, versions of these questions appear in later chapters. Chapter 5 shows one writer's critical thinking as he prepares a sensitive report for his bosses.

## EXERCISES

**1.** Locate a brief example of a technical document (or section of one) in your library or elsewhere. Make a photocopy, and bring it to class. Prepare to explain why your selection can be called technical writing.

**2.** Research the writing skills you will need in your career. (Begin by looking at the *Dictionary of Occupational Titles* in your library's reference section.) You might interview a successful person in your profession. Why and for whom will you write on the job? Explain in a memo to your instructor.

**3.** In a memo to your instructor, describe the specific skills you hope to acquire in your technical writing course. How, exactly, will you apply these skills in your career?

**4.** Find an example of a poorly written memo or letter. In a memo to your instructor, explain why and how the message fails. Attach a copy of the document to your memo. (Check around campus or look through unsolicited sales letters you receive. Or ask a business acquaintance for a sample.) Now do the same for a well-written document, explaining instead how it succeeds.

**5.** Assume that a friend in your major thinks that writing skills no longer are needed. This person argues that secretaries or word processors can "fix up" any piece of writing, and that "smart" computers soon will eliminate the need for most writing. Write your friend a letter, detailing why you think these assumptions about word processors and secretaries are mistaken, and why you think skilled writing will help your friend become a more effective communicator. Use examples to support your position.

**6.** Describe one of your past writing experiences, what you wrote and for whom, and the rewards and frustrations you experienced in writing. What specific problems did you encounter, and how did you try to overcome them?

**7.** Write a memo to your boss, justifying reimbursement for this course. Describe how the course will help you become more effective in the workplace.

**FIGURE 1.4** Critical Thinking in the Writing Process

### 1. Work with the information:

- Is the information accurate, reliable, and unbiased?
- Can it be verified?
- How much of it is useful?
- Do I need more information?
- What do these facts mean?
- Do the facts conflict?
- Are other interpretations or conclusions possible?
- What, if anything, should be done?
- Is it fair?
- Is there a better way?
- Who might benefit or suffer?
- What other consequences will this have?
- Should I reconsider?

### 2. Plan your document:

- What do I want it to do?
- Who is my audience, and why will they read it?
- What do they need to know?
- What are the "political realities" (feelings, egos, etc.)?
- How will I organize?
- What format and visuals should I use?

### 3. Draft your document:

- How do I begin, and what comes next?
- How much is enough?
- What can I leave out?
- Am I forgetting anything?
- How will I end?

### 4. Revise your document:

- How does this draft measure up?
- Does it do what I want it to do?
- Is the content worthwhile?
- Is the organization sensible?
- Is the style readable?
- Is everything easy to find?
- Does everything look good?
- Is everything correct?

## TEAM PROJECT

### Introducing a Classmate

Throughout this semester, members of your class will be working together to evaluate and edit each other's documents. So that everyone can become acquainted quickly, one of your first tasks in this course will be to introduce to the class the student seated next to you. (And that person, in turn, will introduce you.) In preparing your introduction, please follow this procedure:

*a.* Share with your neighbor any information about yourself you think will be useful to the class. (Your neighbor will share the same kind of information with you.) In five minutes or less, provide *relevant* details about your background, major, career goals, course goals, or anything else you think the class should know about you. Then allow five minutes for your neighbor's information.

b. Record your neighbor's information accurately, asking any questions you feel are necessary and appropriate.

c. Take your collected material home and review it carefully, keeping only that which would be useful to the classroom audience.

d. Organize your selected material in a memorandum to your classmates. (See pages 392–393 for memo format.) Your completed memo should be no longer than one page, and should provide a clear picture of *who this person is* — without details that your audience would consider needless.

e. Ask your neighbor to review your document for accuracy.

f. At the next class meeting, present an oral paraphrase of your memo and submit a copy to your instructor.

# Writing for Readers: The Information Problem

- Assess Readers' Information Needs
- Maintain an Appropriate Level of Technicality
- Focus on Needs and Attitudes
- Brainstorm for a Useful Message
- Sample Writing Situations

All technical writing is for readers who will use and react to your information. You might write to *define* something—as to insurance customers who want to know what *variable annuity* means. You might write to *describe* something—as to an architectural client who wants to know what a new addition to her home will look like. You might write to *explain* something—as to a stereo technician who wants to know how to eliminate bass flutter in your company's new line of speakers. Whatever your purpose, you write not for yourself but to inform readers. As depicted in Figure 2.1, your information problem is in enabling readers to understand exactly what you mean.

Most readers would prefer not to have to read your document at all. They are not interested in how smart or eloquent you are, but they *do* want to find what they need, as quickly and as easily as possible.

**FIGURE 2.1** One Problem Confronted by Writers

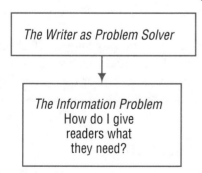

## ASSESS READERS' INFORMATION NEEDS

Different readers have different backgrounds. Therefore, a specialized concept that makes sense to you might confuse someone else, unless it is explained at that reader's level.

Good writing connects with its readers by recognizing their differences in background, their specific needs, and their preferences. In some cases, a single message may appear in several versions for several audiences. An article about a new cancer treatment might appear in the *Journal of the American Medical Association,* read by doctors and nurses. The same article may later be rewritten less technically for a medical textbook, read by medical and nursing students. Or a nontechnical version might appear in *Good Housekeeping.* All three versions treat the same subject, but each is adapted to the needs of its audience.

Technical writing is intended to be *used.* You become the teacher and the reader becomes the student. Because your readers know less than you do, you can expect them to ask questions:

- *What is the purpose of this document?*
- *Who should read it?*
- *What is being described or explained?*
- *What does it look like?*
- *How do I do it?*
- *How did you do it?*
- *Why did it happen?*
- *When will it happen?*
- *Why should we do it?*
- *How much will it cost?*
- *What are the dangers?*

The documents you write should answer these kinds of questions. And each specific question is a version of the one overriding question: What, exactly, do you mean? You always write to show a specific audience what you mean; and during the work of making your meaning clear, you often discover new meanings for yourself.

Think of your readers' information needs as you think of your own. If Chapter 1 of your introductory math textbook covered differential equations (advanced calculus), the author would be ignoring your needs. For your background and purposes, the chapter would be useless. The message must connect with the reader's level of understanding.

## MAINTAIN AN APPROPRIATE LEVEL OF TECHNICALITY

When you write for a close acquaintance (friend, fellow computer buff, psychology classmate, fellow technician, engineering colleague, chemistry professor who reads your lab reports, or your immediate boss), you know a good deal about your reader's background. And so you automatically adapt your report to that reader's knowledge, interests, and needs. But sometimes you have to write for less clearly defined audiences, particularly when the audience is large (when you are writing for a magazine or professional journal, or asked to explain your company's stock-options program to members of many departments). If you have only a general notion about your audience's background, you must decide whether your message should be *highly technical, semitechnical,* or *nontechnical.*

## The Highly Technical Message

Readers at your specialized level expect technical data, without long explanations. The following report of treatment given a heart attack victim is highly technical. The writer, a physician on call in the emergency room, is reporting to the patient's doctor. This reader needs an exact record of his patient's symptoms and treatment.

Mr. X was brought to the emergency room by ambulance at 1:00 A.M., September 27, 1987. The patient complained of severe chest pains, shortness of breath, and dizziness. Auscultation and electrocardiogram revealed a massive cardiac infarction and pulmonary edema marked by pronounced cyanosis. Vital signs were as follows: blood pressure, 80/40; pulse, 140/min; respiration, 35/min. Lab tests recorded a wbc count of 20,000, an elevated serum transaminase, and a urea nitrogen level of 60 mg%. Urinalysis showed 4+ protein and 4+ granular casts/field, suggesting acute renal failure secondary to the hypotension.

**Readers here need merely the facts**

The patient was given 10 mg of morphine stat, subcutaneously, followed by nasal oxygen and a 5% D & W intravenously. At 1:25 A.M. the cardiac monitor recorded an irregular sinus rhythm, suggesting left ventricular fibrillation. The patient was defibrillated stat and given a 50 mg bolus of Xylocaine intravenously. A Xylocaine drip was started, and sodium bicarbonate was administered until a normal heartbeat was established. By 3:00 A.M., the oscilloscope was recording a normal sinus rhythm.

As the heartbeat stabilized and cyanosis diminished, the patient was given 5 cc of Heparin intravenously, to be repeated every six hours. By 5:00 A.M. the BUN had fallen to 20 mg% and the vital signs had stabilized as follows: blood pressure, 110/60; pulse, 105/min; respiration, 22/min. The patient was now conscious and responsive.

Written at the highest level of technicality, this report is clear only to the medical expert. The writer correctly assumes that her reader has the background to understand the message. She therefore defines no technical terms (pulmonary edema, sinus rhythm). Nor does she interpret lab findings (4+ protein, elevated serum transaminase). She uses abbreviations that she knows her reader understands (wbc, BUN, 5% D & W). Because her reader knows the reasons for specific treatments and medications (defibrillation, Xylocaine drip), she includes no theoretical background. Her report is designed to answer concisely the main questions she can anticipate from her reader: What happened? What course of treatment was followed? What were the results?

## The Semitechnical Message

Less specialized readers expect technical data to be explained and interpreted. Semitechnical messages may address a broad array of readers. First-year medical students have some technical background, but not as much as second-, third-, and fourth-year students. Yet students in all four groups could be considered semitechnical readers. When you write for a semitechnical audience, identify the *lowest* level of understanding in the group, and write to that level. You are wiser to give too much explanation than too little.

Here is a partial version of the earlier medical report. Written at a semitechnical level, it might appear in a textbook for first-year medical or nursing students, in a report for a medical social worker, in a patient's history for the medical technology department, or in a monthly report for the hospital administration.

**Readers here need the facts explained and interpreted**

Examination by stethoscope and electrocardiogram revealed a massive failure of the heart muscle along with fluid buildup in the lungs, which produced a cyanotic **discoloration of the lips and fingertips from lack of oxygen.**

The patient's blood pressure at 80 mm Hg (systolic)/40 mm Hg (diastolic) was **dangerously below its normal measure of 130/70.** A pulse rate of 140/minute was **almost twice the normal rate of 60–80.** Respiration at 35/minute was more than **twice the normal rate of 12–16.**

Laboratory blood tests yielded a white blood cell count of 20,000/cu mm (normal value: 5,000–10,000), **indicating a severe inflammatory response by the heart muscle.** The elevated serum transaminase enzymes (produced in quantity only when the heart muscle fails) confirmed the earlier diagnosis. A blood urea nitrogen level of 60 mg% (normal value: 12–16 mg%) **indicated that the kidneys had ceased to filter out metabolic waste products.** The 4+ protein and casts reported from the urinalysis (normal value: 0) **revealed that the kidney tubules were degenerating as a result of the lowered blood pressure.**

The patient was immediately given **morphine to ease the chest pain,** followed by **oxygen to relieve strain on the cardiopulmonary system,** and an intravenous solution of **dextrose and water to prevent shock.**

This version explains (in boldface) the raw data. Exact dosages are not significant to these readers because they are not treating the patient; dosages therefore are not mentioned. Normal values of lab tests and vital signs, however, make interpretation easier. (Highly technical readers would know these values, as well as the significance of specific dosages.) Knowing what medications the patient received would be especially important to the lab technician, because some medications affect blood test results. For a nontechnical audience, however, the message needs even further translation.

## The Nontechnical Message

Readers with little or no training in your field expect technical data to be translated into their simplest version. Nontechnical readers are impatient with abstract theories but want enough background to help them make the right decision or take the right action. They are bored by long explanations but frustrated by bare facts not explained or interpreted. They want to understand a message without having to become students of the discipline. They expect a report that is clear on first reading, not one that requires review or study.

Now for a nontechnical version of the earlier medical report. The physician might write this version for the patient's spouse who is on a business trip overseas, or as part of a script for a documentary film about emergency room treatment.

Both heart sounds and electrical impulses were abnormal, **indicating a massive heart attack caused by failure of a large part of the heart muscle.** The lungs were swollen with fluid and the lips and fingertips showed **a bluish discoloration from lack of oxygen.**

Readers here need everything translated into simplest terms

Blood pressure was **dangerously low, creating the risk of shock.** Pulse and respiration were **almost twice the normal rate, indicating that the heart and lungs were being overworked** in keeping oxygenated blood circulating freely.

**Blood tests** confirmed the heart attack diagnosis and **indicated that waste products usually filtered out by the kidneys were building up in the bloodstream. Urine tests showed that the kidneys were failing as a result of the lowered blood pressure.**

The patient was given **medication to ease his chest pain, oxygen to ease the strain on his heart and lungs, and intravenous solution to prevent his blood vessels from collapsing and causing irreversible shock.**

Nearly all interpretation, this nontechnical version translates (in boldface) all specialized information into simple language. No medications, lab tests, or normal values are mentioned, because these would mean nothing to the nontechnical reader. The writer merely summarizes the events and explains the causes of the crisis and the reasons for the particular treatment.

In some other situation, however (say, in a jury trial for malpractice), the nontechnical audience might need information about specific medication and treatment. Such a report would be much longer, with detailed explanations to simplify technical concepts—a mini-course in emergency coronary treatment.

Each version of the medical report is useful *only* to readers at a specific level of understanding. Doctors and nurses have no need for the explanations in the two latter versions, but they do need the specialized data in the first. Beginning medical students and paramedics might be confused by the first version and bored by the third. Nontechnical readers would find both the first and second versions meaningless.

## Primary Versus Secondary Readers

When you write the same basic message for readers at different levels of technicality, classify your readers as *primary* or *secondary*. The primary readers usually are those who have requested the document and who probably will use it as a basis for decisions or actions. The secondary readers are those who will carry out the project, who will advise the primary readers about their decision, or who will somehow be affected by this decision. They will read your report (or perhaps only part of it) for information that will help them to get the job done, to provide educated advice, or to stay abreast of new developments.

As often as not, these two audiences will differ in technical background. Primary readers may require highly technical messages, and secondary readers may need semitechnical or nontechnical messages—or vice versa. When you must write for audiences at different levels, follow these guidelines:

1. If your document is short (a letter, memo, or anything less than two pages), rewrite it at various levels for various readers.

2. If your document is longer than two pages, maintain a level of technicality that connects with your primary readers. Then supplement the report with appendixes addressed to the secondary readers (technical appendixes when secondary readers are technical, or vice versa). Letters of transmittal, informative abstracts, and glossaries are other supplements that help nonspecialized audiences understand a highly technical report.

The next scenario shows how some documents must be tailored for both primary and secondary readers.

---

Different readers have different information needs

**Tailoring a Document for Different Readers**

You are a metallurgical engineer in a Detroit consulting firm. Your supervisor has asked that you test the fractured rear axle of a 1987 Delphi pickup truck recently involved in a fatal accident. Your job is to determine whether the fractured axle was the cause or a result of the accident.

After testing the hardness and chemical composition of the metal and examining microscopic photographs of the fractured surfaces (fractographs), you conclude that the fracture resulted from stress that developed *during* the accident. Now you must report your procedure and your findings to a variety of readers.

Because your report may serve as evidence in a court case, you must explain your findings in meticulous detail. But your primary readers (the decision makers) will be nonspecialists (the attorneys who have requested the report, insurance representatives, possibly a judge and a jury), and so you will have to translate your report, explaining the principles behind the various tests, defining specialized terms such as "chevron marks," "shrinkage cavities," and "dimpled core," and showing the significance of these features as evidence.

Secondary readers will include your supervisor and outside consulting engineers who will be evaluating your test procedures and assessing the validity of your findings. Consultants will be focusing on various parts of your report, to verify that your procedure has been exact and faultless. For these readers, you will have to include appendixes spelling out the technical details of your analysis: *how* hardness testing of the axle's case and core indicated that the axle had been properly carburized; *how* chemical analysis ruled out the possibility that the manufacturer had used inferior alloys; *how* light-microscope fractographs revealed that the origin of the fracture, its direction of propagation, and the point of final rupture indicated a ductile fast fracture, not one caused by torsional fatigue.

---

In this situation, the primary readers need to know *what your findings mean*, whereas the secondary readers need to know *how you arrived at your conclusions*. Unless you serve the needs of each group independently, your information will be worthless.

## FOCUS ON NEEDS AND ATTITUDES

Deciding on the best level of technicality is one requirement, but you face other decisions as well. Your choices of *what* you write (content) and *how* you write it (format, arrangement, style) are determined by your writing situation. Identify your purpose, and learn all you can about your audience and situation *before* you write.

When you write for a particular reader or a small group of readers, you can focus sharply on your audience by asking specific questions:

1. Who wants the document? Who else will read it?
2. Why do they want the document? How will they use it? What purpose do I want to achieve?
3. What is the technical background of the primary audience? Of the secondary audience?
4. How much does the audience already know about the subject? What material will have informative value?
5. What exactly does the audience need to know, and in what format? How much is enough?
6. When is the document due?

Follow the suggestions on the next few pages for answering these questions.

### Reader Identification

Identify the primary readers by name, job title, and speciality. (Martha Jones, Director of Quality Control, B.S. and M.S. in mechanical engineering). Are they superiors, colleagues, or subordinates? Are they inside or outside your organization? What is their attitude toward your subject likely to be? Are they apt to agree or disagree with your conclusions and recommendations? Will your report be taken as good or bad news?

Although your document should satisfy the primary readers, you should not ignore secondary readers. Identify those additional readers who are likely to be interested in or affected by your document, or who will affect the primary reader's perception or use of your document.

### Purpose of the Request

Find out why your readers want the document and how they will use it. Ask them directly. Do they merely want a record of your activities or progress? Are you expected to supply only raw data, or conclusions and recommendations as well? Will your readers take immediate action based on your report? Do they need step-by-step instructions? Are they merely collecting information for later use? Will the document be read and discarded, filed, or published? In your audience's view, *what* is most important? From what you know of your audience, what purpose do you want your document to achieve?

### Readers' Technical Background

Colleagues on your project speak your technical language and can understand raw data. Supervisors responsible for several technical areas may want interpretations and recommendations. Managers may have only vague technical knowledge and thus will expect definitions and explanations. Clients with no technical background will expect simplified versions spelling out what the data mean to *them* (to their health, pocketbook, business prospects). However, none of these generalizations may apply to *your* situation. Find out for yourself.

By assessing your readers' technical background, you can avoid insulting their intelligence or writing over their heads. Remember, it's easier to make the mistake of being too technical than too simple. In any mixed audience, aim for the lowest level of technicality.

### Readers' Knowledge of the Subject

Do not waste time rehashing information readers already have. If your audience has read preliminary reports on your subject, a brief summary of their major points will do.

Readers expect something *new* and *significant* from your message. Writing has informative value[1] when it (1) conveys knowledge that is new *and* worthwhile to the intended audience; (2) reminds the audience of something they know but ignore; or (3) offers fresh insight about something familiar, causing readers to see things in a new way.

[1] Adapted from James L. Kinneavy's assertion that discourse ought to be unpredictable, in *A Theory of Discourse* (Englewood Cliffs: Prentice, 1971).

The informative value of any message is measured by its relevancy to the writer's purpose and the audience's needs. Assume, for instance, that you have no computer background (only general knowledge), but you are trying to decide whether to take a computer course. And my purpose is to provide facts that will help you decide. In this situation, which of these bits of information would you find useful?

1. Interest in computers has grown immensely in the last decade.
2. Roughly 70 percent of businesses are computer-dependent.
3. The first digital computer was built by Howard Aiken.
4. Information can be transmitted rapidly by computer.

The information in 2 may be new to many novices, and is relevant to your needs and my purpose (stated above). Because 1 merely repeats common knowledge, it has no informative value here. Although 3 offers an unfamiliar fact, it is not relevant to my purpose and your needs. (But 3 could be relevant in another situation: if you were already in a computer course.) Everyone would agree with 4, and so it offers you nothing worthwhile.

A message also has informative value if it offers fresh insight on a familiar fact. As this book's audience, for instance, you expect to learn about technical writing, and my purpose is to help you do so. In this situation, which of these statements would you find useful?

1. Technical writing is hard work.
2. Technical writing is a process of making deliberate decisions in response to a specific situation. In this process, you discover important meanings in your topic, and give your readers the information they need to understand your meanings.[2]

Statement 1 is no news to anyone who has ever picked up a pencil, and so it has no informative value for you. But 2 offers a new way of seeing something familiar. Even if you've done much writing, and struggled through decisions about punctuation, organization, and so on, you probably have not viewed writing as entailing the many kinds of decisions discussed in this book (and illustrated in Chapter 5). Because 2 provides new insight into a familiar activity, you can say it has informative value. A writer *selects* only material that has informative value for a given audience.

The more nonessential information readers are given, the more likely they are to overlook or misinterpret the important parts of your document. Always take the time to carefully determine what your readers need, and give them just that.

---

[2]My thanks to Major Robert M. Hogge, U.S. Air Force Academy, for this definition.

## Appropriate Details and Format

In the earlier medical reports we saw that dosages, drug names, and medical terms are significant to some readers but not others. The amount and kind of detail in your report (How much is enough?) will depend on what you have learned about your readers and their needs. Were you asked to "keep it short" or to "be comprehensive"? Can you summarize certain material or does it all need spelling out? How deeply do they need to know this material? What length will your readers tolerate? Are your primary readers most interested in conclusions and recommendations, or do they want a full description of your investigation as well? Have they requested a letter, a memo, a short report, or a long, formal report with supplements (title page, table of contents, appendixes, and so on)? What kinds of visuals (charts, graphs, drawings, photographs) make your material more accessible? What level of technicality will connect with your primary readers?

**High technicality**    The diesel engine generates 10 BTUs per gallon of fuel, as opposed to the conventional gas engine's 8 BTUs.

**Low technicality**    The diesel engine yields 25 percent better fuel mileage than its gas-burning counterpart.

Every professional in the Information Age has to keep abreast of technology, in order to use it effectively. What one has to know changes quickly and often. A senior engineer (and supervisor) might know less about VLSI circuits (very-large-scale integrated circuits—the heart of microprocessors) than a recent graduate.

If you cannot identify all audience members, write for the least specialized. You are safer to risk boring some experts than to write over other people's heads.

## Due Date

Does your report have a deadline? Find out. Give yourself plenty of time to collect data and to write and revise the report. If possible, ask your primary readers to review an early draft and to suggest improvements. A report dashed off at the last minute impresses no one.

## BRAINSTORM FOR A USEFUL MESSAGE

When you begin working with an idea, content is raw material: ideas, insights, statistics, facts, or examples—anything which advances your meaning, which helps you answer this question: How can I find something worthwhile to say, something that will convey my meaning?

Outlines for discovering material about specific writing tasks are covered in later chapters. But the technique of *brainstorming* is one good way to find worthwhile content. The aim is to get as much raw material as possible down *on paper*. Business and technical people often use group brainstorming sessions to get ideas for new projects, marketing campaigns, and problem solving.

Some people brainstorm as a very first writing step, even before deciding about their purpose and their audience's needs, just as a way of getting started. Others save brainstorming until they have written a rough draft. Regardless of the sequence, good writers almost always brainstorm at some stage in the writing process. They do so to ensure that they will discover *all* the material readers might find useful.

The brainstorming procedure is simple: simply concentrate on your writing situation, and jot down *every thought.* Don't stop to judge relevance or worth, and don't worry about complete sentences or spelling. Just write everything that comes to mind. The more ideas, the better. Trust your imagination; even the wildest idea might lead to a valuable insight.

Brainstorming helps you achieve the concentration needed for coming up with *all* the details you might want to include. This technique invariably produces more information than you can use. After deleting nonessential ideas, you should still have plenty of material. And chances are you will discover more while writing your letter, memo, or report. From this broad inventory, you select only the useful material: namely, worthwhile content.

In Chapter 5, you will see how a working professional brainstorms while preparing an important memo to his superiors.

## SAMPLE WRITING SITUATIONS

In many jobs you will have to ask questions daily about your audience's needs. Here are two scenarios in which specialists who also must be "part-time" technical writers assess their audiences' needs in different situations.

### A Day with a Police Officer
You are a police officer in the burglary division of an urban precinct.

- A local community college has asked you to lecture on the fields of police work during its career week. Your audience will be interested laypersons, some considering a police career. You know they've been captive audiences at lectures for years, and so you decide to liven things up with visuals. The police artist agrees to draw up a brightly colored flip chart and you use graphics software on the office microcomputer to generate graphs and flowcharts illustrating salary levels and career paths.
- Tomorrow in court you will testify for the prosecution of a felon you caught red-handed last month. His lawyer is a sly character who is known for making police testimony look foolish, and so you are busy reviewing your notes and getting your report in final form. Your audience will consist of legal experts (judge, lawyers) as well as laypersons (jury). The key to a convincing testimony will be *clarity, accuracy,* and *precision,* with impartial and factual descriptions of *what* happened, as well as *when* and *where.*

- You are drafting an article about new fingerprinting techniques for a law enforcement journal. Your readers will be expert (veteran officers) and informed (junior officers). Because they understand shoptalk and know the theory and practice of fingerprinting, they won't need extensive background or definitions of specialized words. Instead, they will read your article to learn about a new *procedure*. You will have to convince them that your way is better. You will later write a manual for using this technique.

- Your chief asks you to write a manual on investigative procedures for junior officers in your division. The manual will be bound in pocket size and carried by all junior officers on duty. Your audience is informed, but not expert. They have studied the theory at the police academy, but now they need the "how-to." They will want rapid access to instructions for handling problems, with each step spelled out. They probably will use your instructions as a guide to *immediate* actions and responses. Therefore, you will have to label warnings and cautions *before* each step. Clarity throughout will be imperative.

- Next month you will speak before the chamber of commerce on "Protecting Your Business Against Burglary." The audience will be highly interested laypersons. Because burglary protection can be costly (alarm systems, guard dogs), the audience is apt to be most interested in the less expensive precautions they can take. They will want to remember your advice, and so you decide to supplement your talk with visuals: specifically, a checklist for business owners that you will discuss item-by-item, using an overhead projector.

---

### A Day with a Civil Engineer

As a civil engineer in a materials-testing lab, you are in charge of friction studies on road surfaces.

- Your firm is competing for a contract to study the safety of state bridge surfaces. As engineer in charge, you are responsible for the quality of this proposal. Thus you are carefully reviewing and editing, on disk, proposal sections written by employees. Your audience consists of experts (state engineers), informed readers (officers in the highway department), and laypersons (members of the state legislature). The legislature ultimately will decide which firm gets the contract, but they will act on the advice of engineers and the highway department. You decide therefore to keep your proposal at a level of technicality that will connect with the specialized audience. The legislative committee will read only the informative abstract, conclusions, and cost data.

- Checking your electronic mail, you find a message from the vice president asking you to write safety instructions for paving crews on the new bridge. Your instructions will be read by all crew members, from project engineer to laborer, but when it comes to safety on this kind of job, even the specialized personnel may be considered laypersons. You ask the graphics department to design a brochure for your instructions, including sketches and photographs of the hazardous areas and situations.

- At next month's convention in Dallas you are scheduled to deliver a paper describing a new road-footing technique that reduces frost heave damage in northern climates. (This paper will soon be published in an engineering journal.) Your audience will be civil engineering colleagues who want to know what it is and how it works—without lengthy background. You decide to compress your data with conventional visuals (cross-sectional drawings, charts, graphs, maps, and formulas), as well as computer graphics.
- You draft a letter to a colleague you've never met, to ask about her new technique for increasing the durability and elasticity of rubberized asphalt surfaces. Your expert reader is a busy professional with little time to read someone's life story, but she will need *some* background about your own work, along with clear, specific, and precise questions that she can answer quickly. As a gesture of goodwill, you close your letter with an offer to share *your* findings with her.
- On your microcomputer, you draft and revise a memo justifying your request to the vice president for an additional lab technician. You know that the present lab technician feels threatened by the prospect of someone invading his space, and so you have to be sensitive, diplomatic, and persuasive—and to make sure that Joe (the lab technician) receives a copy of your memo.

---

In these situations, the writers addressed their audiences' needs for clear, appropriate, and useful information. Anything they could learn about each audience beforehand guided their choice of material.

Much of your writing in this course—and much of it on the job—will be aimed at general readers, but even when your readers are informed or expert, they will know less than you do about the subject. Whereas in college you write for an audience (the professor) who knows more than you, and who is *testing* your knowledge, on the job you write for an audience who knows less, and who is *using* your knowledge.

## EXERCISES

**1.** In a short report, discuss the kinds of writing and speaking assignments you expect in your career. (If necessary, interview someone in your field for information.) For whom will you be writing (colleagues, supervisors, clients)? Will you need to address both primary and secondary audiences? How will your readers use your information? For which audience are you likely to write most often? Do you envision any political problems?

**2.** Locate a short article from your field. (Or select part of a long article or a section from one of your textbooks for an advanced course.) Choose a piece written at the highest level of technicality you can understand and then translate the piece for a layperson. (Be sure to use the simplest language, as in the example on page 21.) Exchange translations with a classmate from a different major. Read your neighbor's translation and write a paragraph evaluating its level of technicality. Submit to your instructor a copy of the original, your translated version, and your evaluation of your neighbor's translation.

**3.** Assume that a new employee is taking over your job (part time or full time) because you have been promoted. Identify a specific problem in your old job that could cause difficulty for the new employee. Write for the employee instructions for avoiding or dealing with the problem. Before writing, perform an audience analysis by answering (on paper) the questions on page 23. Then brainstorm for details. Submit to your instructor your audience analysis, your brainstorming list, and your instructions.

**4.** Assume you live in the Northeast, and citizens in your state are voting on a solar energy referendum that would channel millions of tax dollars toward solar technology. These two paragraphs are versions of a message designed to help you, as a voter, make an educated decision. Do both messages have informative value, or does one merely repeat common information? Explain.

Solar power offers a realistic solution to the Northeast's energy problems. In recent years, the cost of fossil fuels (oil, coal, and natural gas) has risen sharply while supplies continue to decline. High prices and short supply will prolong a worsening energy crisis. Because solar energy comes directly from the sun, it is an inexhaustible resource. Using this energy to heat and air-condition our buildings, as well as to provide electricity, we could decrease substantially our consumption of fossil fuels. In turn, we would be less dependent on the unstable Middle East for our oil supplies. Clearly, solar power is a good alternative to conventional sources of energy.

Solar power offers a realistic solution to the Northeast's energy problems. To begin with, solar power is efficient. Solar collectors installed on fewer than 30 percent of roofs in the Northeast would provide more than 70 percent of the area's heating and air-conditioning needs. Moreover, solar heat collectors are economical, operating for up to twenty years with little or no maintenance. These savings recoup the initial cost of installation within only ten years. Most important, solar power is safe. It can be transformed into electricity through photovoltaic cells (a type of storage battery) in a noiseless process that produces no air pollution— unlike coal, oil, and wood combustion. In sharp contrast to its nuclear counterpart, solar power produces no toxic waste and poses no catastrophic danger of meltdown. Thus, massive conversion to solar power would ensure abundant energy and a safe, clean environment for future generations.

## TEAM PROJECT

Form teams according to major (electrical engineering, biology, etc.), and respond to the following situation.

Assume your team has received the following assignment from your major department's chairperson: An increasing number of freshmen are dropping out of the major during their first year because of low grades or stress or inability to keep up with the work load. Your task is to prepare a "Survival Guide," for distribution to incoming freshmen. This one or two page memo should focus on the challenges and the pitfalls and should include a brief motivational section, along with whatever else your team feels readers need.

## Analyze Your Audience

Use the following questions as a guide for developing your audience-and-use profile. (One set of possible responses to these questions is shown.)

- *Who is my audience?* Incoming freshmen in the major (and faculty).
- *How will readers use the information?* To develop a clear sense of what to expect and how to proceed (say, in managing workloads or meeting deadlines).
- *How much is the audience likely to know about this topic?* Very little. They will need everything spelled out.
- *What else does the audience need to know?* They need answers to questions like these: How big is the problem? How can it affect me? What are the department's expectations? How much homework will I need to do? How should I budget my time? Can I squeeze in a part-time job? Are there any skills I should try to acquire beforehand (say, word processing or graphics and basic design skills)? Where do most freshmen make their big mistakes? (Based on your own experience, can your team anticipate any other questions?)
- *What attitudes or misconceptions about this topic is the audience likely to have?*
  Any who are overly optimistic ("No problem!") about their chances will need to visualize the real challenges ahead.
  Any who are overly pessimistic ("I'm dead for sure!") will need encouragement, along with the facts.
  Any who are indifferent ("Who cares?") will need some motivation, along with the facts.
  Whatever combination of attitudes the audience holds, we have to address each attitude—as well as we can identify it.
- *What probable attitude does the audience have toward the writers?* (Are we seen as trustworthy, sincere, threatening, arrogant, or what?) Since we are all students, readers will likely identify with and trust us to a decent extent. They'll probably realize we're on their side.
- *Who will be affected by this document?* Mostly the incoming students (primary audience), and possibly the department.
- *In this situation, how can we characterize the audience's temperament and probable reaction?* Most readers should be eager for this information and should take it seriously.
- *Do we risk alienating anyone?* Gifted students who don't know the meaning of failure might feel patronized or offended. And some faculty might resent any suggestions that courses are too demanding, and so we don't want to editorialize. The purpose of this piece is informative and advisory—not evaluative.
- *How did this document originate, and how long should it be?* Because it was requested by the department and not by the primary audience, we can't expect students to tolerate more than a page or two.
- *What material will be most important to this audience?* They will want clear advice about what and what not to do.

- *What arrangement would be most effective for this audience and purpose?* We should provide brief background on the dropout problem, discuss its causes, suggest ways to survive, and end on a positive note of encouragement and motivation.
- *What tone would this audience expect?* We are all students; a friendly, relaxed and positive (to avoid panic), but serious tone seems best.
- *What is the document's intended effect on its audience?* If it manages to connect it will, we hope, cut down the dropout rate.

Follow the model on page 33 for designing a profile sheet to record your audience-and-use analysis. (Feel free to improve on the design and content of this model.)

## Devise a Plan for Achieving Your Goal

From the audience traits you have identified, develop a plan for getting your information across. Express your goal and plan in a statement of purpose.

> We need to give incoming freshmen a clear sense of the challenges and pitfalls of the first year in our major. We will show how dropouts have increased, discuss what seems to go wrong, give advice on avoiding some common mistakes, and emphasize the benefits of staying in the program.

## Plan, Draft, and Revise Your Document

Brainstorm for worthwhile content (for this exercise, make up some reasons for the dropout rate if you need to), do any research that may be needed, write a workable draft, and revise until it represents your team's best work.

Appoint a team member, to present the finished document (along with a complete audience-and-use analysis) for class evaluation, comparison, and response.

## Alternative Projects

a. In a one or two page memo to *all* incoming freshman, develop a "Freshman Survival Guide." Spell out the least any freshman should know in order to get through the first year.
b. In one or two pages, describe the job outlook in your field (prospects for the coming decade, salaries, subspecialties, promotional opportunities, etc.). Write for high school seniors interested in your major. Your team's description will be included in the career handbook published by your college.
c. Identify an area or situation on campus that is dangerous or inconvenient or in need of improvement (endless cafeteria lines, poorly lit intersections or parking lots, noisy library, speeding drivers, inadequate dorm security, etc.). Observe the situation as a group during a peak-use period. Spell out the problem in a letter to a specified decision maker (dean, campus police chief, head of food service) who presumably will use your information as a basis for action.

# AUDIENCE-AND-USE PROFILE SHEET

## Audience Identity and Needs

Primary audience: _____ *(name, title)*

Other potential reader(s): _____

Relationship: _____ *(client, employer, other)*

Intended use of document: _____ *(perform a task, solve a problem, other)*

Prior knowledge about this topic: ____ *(knows nothing, a few details, other)*

Additional information needed: _____ *(background, only bare facts, other)*

Probable questions: _____?

_____?

_____?

_____?

_____?

## Audience's Probable Attitude and Personality

Attitude toward topic: _____ *(indifferent, skeptical, other)*

Probable objections: _____ *(cost, time, none, other)*

Probable attitude toward this writer: ___ *(intimidated, hostile, receptive, other)*

Persons most affected by document: _____

Temperament: _____ *(cautious, impatient, other)*

Probable reaction to document: ___ *(resistance, approval, anger, guilt, other)*

Risk of alienating anyone: _____

## Audience Expectations About the Document

Reason document originated: _____ *(audience request, my idea, other)*

Acceptable length: _____ *(comprehensive, concise, other)*

Material important to this audience: _____ *(interpretations, costs,*
*conclusions, other)*

Most useful arrangement: _____ *(problem-causes-solutions, other)*

Tone: _____ *(businesslike, apologetic, enthusiastic, other)*

Intended effect on this audience: ____ *(win support, change behavior, other)*

# Writing for Readers: The Persuasion Problem

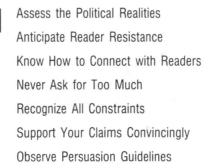

Assess the Political Realities

Anticipate Reader Resistance

Know How to Connect with Readers

Never Ask for Too Much

Recognize All Constraints

Support Your Claims Convincingly

Observe Persuasion Guidelines

Chapter 2 showed how writers face the *information problem:* How can I make readers understand exactly what I mean? But workplace writers face a frequent *persuasion problem* as well, as outlined in Figure 3.1. Persuasion means trying to influence people's thinking or win their cooperation. And the size of your persuasion problem depends on who your readers are, how you want them to respond, and how strongly they identify with their position.

You face a persuasion problem whenever you express a viewpoint that readers might dispute. Viewpoints ordinarily are expressed in a thesis or a *claim* (a statement of the point you are trying to prove). For instance, you might want readers to *recognize* facts they've ignored:

> The O-rings in the space shuttle's booster rockets have a serious defect that could have disastrous consequences.

Or you might want to influence how they *evaluate* the facts:

> Taking time to redesign and test the O-rings is better than taking unacceptable risks to keep the shuttle program on schedule.

Or you might want readers to *take immediate action:*

> We should call for a delay of tomorrow's shuttle launch because the risks simply are too great.

**FIGURE 3.1** Two Problems Confronted by Writers

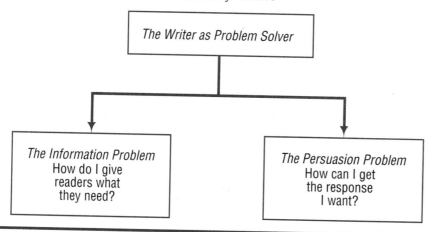

Whenever your audience disagrees about what things mean or what is better or worse or what should be done, you face a persuasion problem.

Many of your own letters, memos, and reports will be asking readers to accept and act on claims like these:[1]

- We cannot meet this production deadline without sacrificing quality.
- We're doing all we can to correct your software problem.
- This policy is discriminatory.
- Our software is superior to the competing brand.
- We should begin this project immediately.
- I deserve a raise.

Your goal might be to change the readers' way of thinking, to reinforce a way of thinking, or to create a new way of thinking about an issue. In any event, you need to make the best possible case for seeing things *your* way.

## ASSESS THE POLITICAL REALITIES

Besides their varied backgrounds, different readers have different attitudes. On one level, your readers are users of information; on another they are *people*, who react on the basis of their personality and feelings, and who create their own meanings for what they have just read (Littlejohn and Jabusch 5).

Any document can evoke different reactions in each reader—according to individual temperaments, preferences, interests, fears, biases, misconceptions, ambitions, or anything else that affects a person's attitude. Depending on their attitude, readers might approach your document with questions like these:

[1]This list of sample claims was inspired by Jeanette W. Gilsdorf. The full citation appears in Works Cited, pp. 560–564.

- *Says who?*
- *So what?*
- *Who cares?*
- *Why should I?*
- *Why rock the boat?*
- *What's in it for me?*
- *What's in it for you?*
- *What does this* really *mean?*

Some readers might be impressed and pleased by your suggestions for increasing productivity; some might feel offended or threatened; and others might perceive your document as an attempt to make yourself "look good" or to make them look bad. People can read much more between the lines than what is actually on the page. Such are the political realities of writing in any organization.

If you have worked with other employees, you have experience with office politics: how some people seek favor, status, or power; how some resent, envy, or intimidate others; how some are easily threatened. Author, teacher, and writing consultant Robert Hays sums up any writer's political situation:[2] "A writer must labor under political pressures from boss, peers, and subordinates. Any conclusion affecting other people can arouse resistance" (19). Some readers might resist your suggestion for shortening lunch breaks, cutting expenses, or automating the assembly line. Or in your document you might disagree with company policy or with your boss's approach. Or you might have to "blow the whistle" on a project or product that is unsafe, inefficient, or worthless.

No one wants bad news; some people even prefer to ignore it (as the O-ring flaw on the shuttle *Challenger's* booster rockets made all too clear). If you know that something is wrong, you have to decide whether "to try to change company plans; to keep silent; to 'blow the whistle'; or to quit" (Hays 19). Does your organization encourage or discourage outspokenness and constructive criticism? Find out—preferably before you accept the job offer.

Whichever combination of needs and attitudes your audience has, you must do your best to satisfy each reader's major interests. By ignoring the political realities in your situation, you might write an otherwise informative and useful report that offends someone and hurts your career.

## ANTICIPATE READER RESISTANCE

Before most people change their minds, they need good reasons. And sometimes, no matter how good your reasons, people will refuse to budge; the O-ring claims on page 34 were made—and supported—by engineers before the Shuttle

---

[2]I am indebted to Professor Hays's article for excellent suggestions about analyzing and addressing political realities faced by writers.

*Challenger* exploded on January 28, 1986. That such claims were ignored by decision makers sadly illustrates how audiences sometimes resist even the most compelling arguments.

Whenever you question someone's stand on an issue, or try to influence people's behavior, you can expect your audience to react defensively. Winning support and cooperation from coworkers is especially challenging. One researcher offers this explanation of why persuasion is so difficult:

> By its nature, informing "works" more often than persuading does. While most people do not mind taking in some new facts, many people do resist efforts to change their opinions, attitudes, or behaviors (Gilsdorf 61).

The bigger the readers' stake in the issue, the more personal their involvement, the more resistance you can expect.

To overcome reader resistance, persuasion requires more than mere "giving in" from its audience, especially if you want to keep your audience's goodwill. When people do "yield" to persuasion, they yield either grudgingly, willingly, or enthusiastically. Researchers categorize these responses as *compliance, identification,* or *internalization* (Kelman 51–60):

- *Compliance:* "I'm yielding to your demand in order to get a reward or to avoid punishment. I really don't accept it, but I feel pressured and so I'll go along to get along."

- *Identification:* "I'm yielding to your appeal because I like and respect you, I believe what you say, and I want you to like me."

- *Internalization:* "I'm yielding because what you're saying makes good sense and it fits my goals and values."

Although achieving compliance is sometimes necessary (as in military orders or workplace safety regulations) nobody likes being "pushed." Effective persuasion instead usually elicits feelings of identification or internalization. If readers merely comply or feel coerced, then you probably have lost their loyalty and goodwill—and as soon as the threat or reward disappears, you will lose their compliance as well.

**FIGURE 3.2** Ways People Yield to Persuasion

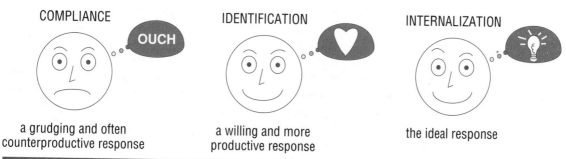

COMPLIANCE — OUCH — a grudging and often counterproductive response

IDENTIFICATION — a willing and more productive response

INTERNALIZATION — the ideal response

## KNOW HOW TO CONNECT WITH READERS

Persuasive people know when to merely declare what they want, when to reach out and create a relationship, when to appeal to reason—or when to use some combination of these strategies. These three strategies for connecting have been categorized as *hard, soft,* and *rational* (Kipnis and Schmidt, 40–46). Let's call them the *power connection,* the *relationship connection,* and the *rational connection,* as shown in Figure 3.3.

**FIGURE 3.3** Ways of Connecting

POWER CONNECTION

"Do it because I said so!"

RELATIONSHIP CONNECTION

"Do it because I'd appreciate it."

RATIONAL CONNECTION

"Do it because it makes sense."

To envision how these different connections work, imagine yourself in the following situation: You are a junior member of XYZ, a large engineering firm which has recently instituted a companywide fitness campaign, based on findings that healthy employees do better work and cost less in lost time and insurance rates. The campaign includes clinics for smoking, stress reduction, and weight loss, along with various exercise programs. In your second month on the job you receive this notice through Electronic-mail:

To: All Employees                                           5/20/90
From: George Maximus, Human Resources Director
Subject: <u>Physical Fitness</u>

*Orders readers to show up*

On Monday, June 10, all employees will report to the company gymnasium at 8:00 A.M. for the purpose of choosing a walking or jogging group. Each group will meet 30 minutes three times weekly during lunch hour.

How would you feel about this memo—and about the person who wrote it? Here the writer seeks nothing more than compliance. Although he speaks of "choosing," you are given no choice about attending; you simply are ordered to show up. When this *power connection* is used, it is typically used by bosses and others in power. And while it may or may not achieve its goal, the power connection almost surely will alienate its audience.

Now assume instead that you have received the following memo on the same topic. How would you feel about this message and its writer?

To: All Employees                                                    5/20/90
From: George Maximus, Human Resources Director
Subject: <u>An Invitation to Physical Fitness</u>

I realize most of you spend lunch hour playing cards, reading, or just enjoying    **Invites readers to**
a bit of well-earned relaxation in the middle of a hectic day. But I'd like to invite    **participate**
you to join our lunchtime walking/jogging club.

We're starting this club in hopes that it will be a great way for us all to feel better.    **Leaves the choice to**
Why not give it a try?    **the reader**

Here the writer seeks to instill in readers a sense of identification, of shared
feelings and goals. Instead of being ordered to participate, readers are being
invited—they are given a real choice. This *relationship connection* seeks goodwill between writer and reader.

An audience's perception of the writer is often the biggest factor in persuasion. Audiences invariably are more receptive to people they like, trust and
respect[3]. Of course, you would be unethical in appealing to the relationship
or faking the relationship merely to hide the fact that you had no evidence to
support your claim (Ross 28). And while this personal emphasis has emotional
appeal, the relationship connection might strike some readers as too "chummy"
to carry any real authority—and so the request might be ignored.

Here is a third version of the walking/jogging memo. As you read it, think
about the ways the author makes a persuasive case.

To: All Employees                                                    5/20/90
From: George Maximus, Human Resources Director
Subject: <u>Invitation to Join One of Our Jogging or Walking Groups</u>

I want to share a recent study from the *New England Journal of Medicine,* which    **Presents authoritative**
reports that adults who walk two miles a day could increase their life expectancy    **evidence**
by three years.

Other research repeatedly shows that 30 minutes of moderate aerobic exercise,
at least three times weekly, has a significant and long-term effect in reducing stress,
lowering blood pressure, and improving job performance.

As a first step in our exercise program, XYZ Corporation is offering a variety    **Offers alternatives**
of daily jogging groups: The One-Milers, Three-Milers, and Five-Milers. All groups
will meet at designated times on our brand new, quarter-mile, rubberized clay track.

For beginners or skeptics, we're offering daily two-mile walking groups. And for
the truly resistant, we offer the option of a Monday-Wednesday-Friday two-mile
walk.

Coffee and lunch breaks can be rearranged to accommodate whichever group you    **Offers a compromise**
select.

---

[3]No matter how good the relationship, audiences also need to (a) find the claim believable ("Exercise will
help me feel better.") and (b) feel the claim is relevant ("I personally need this kind of exercise").

Leaves the choice up
to the reader

Offers incentives

Why not take advantage of our hot new track? As small incentives, XYZ will reimburse anyone who signs up as much as $100 for running or walking shoes, and will even throw in an extra fifteen minutes for lunch breaks. And with a consistent turnout of 90 percent or better, our company insurer may be able to eliminate everyone's $200 yearly deductible in medical costs.

Here the writer appeals primarily to reason and offers a compromise ("If you do this, I'll do that"). This *rational connection* communicates respect for the reader's intelligence *and* for the relationship by presenting good reasons, a variety of alternatives, and attractive incentives—all framed as an invitation. Whenever an audience is willing to listen to reason, the rational connection stands the best chance of succeeding.

Keep in mind that each kind of connection (or some combination) can work in particular situations. But no cookbook formula exists. Each situation calls for careful decisions on the writer's part.

## NEVER ASK FOR TOO MUCH

Not even the most rational connection can persuade people to accept something they consider unreasonable. And definitions of *reasonable* vary with the individual. Employees at XYZ, for example, will differ widely as to which walking/jogging option they might ultimately accept.

**Options offered
by XYZ**

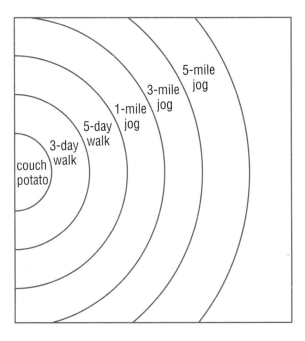

To the jock writing the memo, a daily 5-mile jog might seem perfectly reasonable. But to many other employees, the 5-mile option would be unreasonable. And so XYZ's exercise campaign has to offer something most of its audience (except, say, couch potatoes and those in poor health) accept as reasonable. Any request that exceeds its audience's range of acceptance (Sherif 30–59) is doomed from the start.

## RECOGNIZE ALL CONSTRAINTS

Effective communicators observe certain limits or restrictions imposed by the particular situation. These are the *constraints,* and they govern what should or should not be said, who should say it and to whom, when and how it should be said, and through which medium (printed document, computer screen, telephone, face to face, and so on). In the workplace you need to account for constraints like those in Figure 3.4.

**FIGURE 3.4** Every Writing Situation Poses Its Own Constraints

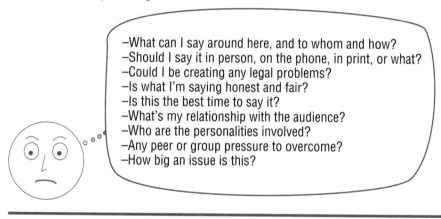

–What can I say around here, and to whom and how?
–Should I say it in person, on the phone, in print, or what?
–Could I be creating any legal problems?
–Is what I'm saying honest and fair?
–Is this the best time to say it?
–What's my relationship with the audience?
–Who are the personalities involved?
–Any peer or group pressure to overcome?
–How big an issue is this?

### Organizational Constraints

Communicators in organizations often face "official" constraints: for instance, schedules, deadlines, budget limitations, writing style, the way a document is organized and formatted, and its chain of distribution throughout the organization.

But beyond these "official" matters, communicators face a whole set of unofficial constraints:

> Most organizations have clear rules for interpreting and acting on (or responding to) statements made by colleagues. Even if the rules are unstated, we know who can initiate interaction, who can be approached, who can propose a delay, what topics can or cannot be discussed, who can interrupt or be interrupted, who can order or be ordered, who can terminate interaction, and how long interaction should last (Littlejohn and Jabusch 143).

The exact rules vary from organization to organization, depending on whether the communication channels are open and flexible or closed and rigid. Does the organization encourage or discourage employee responses and participation in decision making? Any of you who have had jobs already have firsthand experience with organizational constraints, both official and unofficial—including organizational grapevines.

These rules of the game usually are unspoken. But anyone who ignores them invites disaster. For example, a junior employee would first request the immediate supervisor's approval before addressing work-related requests or suggestions to a higher authority.

Airing even the most legitimate gripe in the wrong way through the wrong medium to the wrong person can be fatal to your work relationships and your career. The following memo, for instance, is likely to be seen by the executive officer as petty and whining behavior, and by the maintenance director as a public attack.

Dear Chief Executive Officer:

<table>
<tr><td>Wrong way to the<br>wrong person</td><td>Please ask the Maintenance Director to get his people to do their job for a change. I realize we're all short-staffed, but I've gotten fifty complaints this week about the filthy restrooms and overflowing wastebaskets in my department.</td></tr>
</table>

cc. Maintenance Director

Instead, why not address the memo directly to the key person—or better yet—phone the person instead?

A better way to the
right person

Dear Maintenance Director:

I'd like to meet with you because I have some ideas about how we might work together to provide mutual support during these staff shortages.

Can you identify some the unspoken rules where you have worked? What happens when such rules are ignored?

## Legal Constraints

Sometimes what you are allowed to say is limited by contractual obligations or by laws protecting confidentiality or customers' rights, or laws affecting product liability. For example, in a collection letter for an overdue payment, you can threaten legal action but you are not allowed to threaten any kind of violent action or to publicize the debtor's refusal to pay or to misrepresent yourself as an attorney (Varner and Varner 31–40). Similarly, if someone requests information on one of your employees, you face certain legal constraints about what you can say. And when writing sales literature or user manuals, you and your company face potential liability for incorrect information that leads to injury or damage.

Always be sure to know how the law applies to any document you prepare. Suppose, for instance, an employee drops dead while participating in the new jogging program you've marketed so persuasively. Could you and your company be held liable? Perhaps you should require physical exams and stress tests (at company expense) for all participants.

## Ethical Constraints

While legal constraints are defined by federal and state laws, ethical constraints are defined by good conscience, honesty, and fair play. For example, you may be acting within the law when you promote a new pesticide by emphasizing its effectiveness while downplaying its possible carcinogenic effects; whether your action is *ethical,* however, is another issue entirely. To earn the trust of clients and colleagues, you will find that "saying the right thing" involves more than legal considerations. Chapter 4 is devoted to various ethics problems in communication.

## Time Constraints

Persuasion often is a matter of good timing. Should you wait to say your piece, say it immediately—or what? Let's assume you're trying to "bring out the vote" among members of your professional society on some hotly debated issue: say, whether to refuse work on any project related to biological warfare. On the one hand, you might want to wait until you've gathered all the information you need or until you've analyzed the situation and planned a strategy. But you don't want to delay so long that rumors, misinformation, or paranoia cause people to harden their position *and* their resistance to your appeals. Your timing, then, is a matter of close judgment: if further delay might place the situation beyond your control, you just might have to act sooner than you would like.

## Social and Psychological Constraints

Too often, what we say can be misunderstood or misinterpreted. Here are just a few of the "human" constraints routinely encountered by communicators.

- *Relationship between communicator and audience:* Are you writing to a superior, a subordinate, or an equal? (Try not to appear dictatorial to subordinates or not to shield superiors from bad news.) How well do you and your audience know each other? Can you joke around or should you be dead serious? Do you get along or have a history of conflict? Do you trust one another? What you say and how you say it—and how it is interpreted—will be influenced by the relationship.

- *Audience's personality:* Researchers claim that "some people are easier to persuade than others, regardless of the topic or situation" (Littlejohn 136). Any reader's ability to be persuaded might depend on such personality traits as confidence, optimism, self-esteem, willingness to be different, desire to conform, open- or closed-mindedness, or regard for power (Stonecipher 188–89). The less persuasible your audience, the harder you have to work. If you sense that your audience is totally resistant, you may want to back off—or to abandon your effort.

- *Audience's sense of identity and affiliation as a group:* How close-knit is the group? Does it have a strong sense of identity (as, say, union members or conservationists or engineering majors)? Will group loyalty or pressure to conform prevent certain appeals from working? Address the group's collective concerns.

- *Perceived size and urgency of the problem or issue:* In the audience's view, how big is this issue or problem? Has it been understated or overstated? Big problems are more likely to cause people to exaggerate their fears, anxieties, loyalties, and resistance to change—or to desperately seek some quick and easy solution. Appraise the problem realistically: You don't want to downplay it, but you don't want to cause panic, either.

Writers who can assess a situation's constraints avoid serious blunders and can develop their message for greatest effectiveness.

## SUPPORT YOUR CLAIMS CONVINCINGLY

A persuasive argument is one that makes the best case—in the audience's view. And the strength of your case can be measured by the quality of the *reasons* you offer in support of your claims.

> When we seek a project extension, argue for a raise, interview for a job, justify our actions, advise a friend, speak out on issues of the day . . . we are involved in acts that require good reasons. Good reasons allow our audience and ourselves to find a shared basis for cooperating. . . . In speaking and writing, you can use marvelous language, tell great stories, provide exciting metaphors, speak in enthralling tones, and even use your reputation to advantage, but what it comes down to is that you must speak to your audience with reasons they understand (Hauser 71).

To see how reasons serve in persuasion, imagine yourself in the following situation: As documentation manager for Bemis Software, a rapidly growing company, you supervise preparation and production of all user manuals. The present system of manual preparation is inefficient because three respective departments are involved in (1) assembling the required material, (2) word processing and designing, and (3) publishing the manuals. As a result, much time and energy are wasted as a manual goes back and forth among software specialists, communication specialists, and the art and printing department. After studying the problem and calling in a consultant, you decide that a more integrated approach could be achieved if desktop publishing software were installed in all computer terminals. This way, all employees involved could contribute to all three phases of the process. And so you set out to persuade bosses and coworkers with good reasons, in the form of *evidence* and *appeals to readers' needs and values* (Rottenberg 83).

### Offer Convincing Evidence

Evidence is any information that supports your claim. Common types of evidence are statistics, examples, and expert testimony.

**Statistics.**   Numbers can be convincing. Before reading about other details, many workplace readers are most interested in the "bottom line" (Goodall and Waagen 57).

> After a cost/benefit analysis, our accounting office estimates that an integrated desktop publishing network will save Bemis 30 percent in manual production costs and 25 percent in production time—savings that will enable the system to pay for itself within one year.

Give the numbers

Any statistics you present have to be accurate and easy for the audience to understand and verify. Always cite your source.

**Examples.**   By showing specific instances of your point, examples help audiences *visualize* the idea or the concept. The best way to explain what you mean by "inefficiency" in your company is to show one or more instances of it occurring:

> The figure illustrates the inefficiency of Bemis's present system for producing manuals:

Show what you mean

> A manual typically goes back and forth through this cycle three or four times, wasting both time and effort in all three departments.

Good examples have persuasive force; they give readers something to hold onto, a way of understanding even the most surprising or unlikely claim. Always present examples that the audience can identify with, and that fit the point they are designed to illustrate.

**Expert testimony.**   Any position gains authority to the extent that it can be supported by the views of recognized experts.

> Ron Catabia, nationally recognized networking consultant, has studied our needs and strongly recommends we move ahead with the integrated network.

Cite the experts

To be credible, however, an expert has to be unbiased and considered reliable by the audience.

While solid evidence can be persuasive, evidence alone isn't always enough. At Bemis, for example, the "bottom-line" might be very persuasive for company executives, but will mean little to some managers and employees who will be asking: Does this threaten my authority? Will I have to work harder? Will I fall behind? Is my job in danger? These readers will have to perceive some benefit beyond company profit.

## Appeal to Common Goals and Values

Audiences expect a writer to share their goals and values. If you hope to connect, you have to identify some goals you and the audience have in common: "What do we all want most?"

Bemis employees, like most people, share these goals: job security, being appreciated, a sense of belonging, control over their jobs and destinies, a growing and fulfilling career. Any recommendation you make will have to take these goals into account. For example,

Appeal to shared goals

> I'd like to show how desktop publishing skills, instead of threatening anyone's job, would only increase career mobility for all of us.

Our goals are shaped by certain values: friendship, loyalty, ambition, honesty, self-discipline, equality, fairness, achievement, a commitment to quality—among others (Rokeach 57–58). Beyond appealing to common goals, you can appeal to shared values:

At Bemis, for example, you might appeal to the commitment to quality and achievement shared by the company and by individual employees:

Appeal to shared values

> None of us needs reminding of the fierce competition in the software industry. The improved collaboration among networking departments will result in better manuals, keeping us on the front line of quality and achievement.

For your persuasive effort to succeed, you have to give the audience reasons that have real meaning for them.

Figure 3.5 shows a persuasive letter from Ewing Power Systems, a firm that distributes systems for generating electrical power from a plant's recycled steam (cogeneration). President Tom Ewing writes a persuasive answer to a potential client's question: "Why should I invest in the cogeneration system you are proposing for my plant?" As you read the letter, notice the kinds of evidence and appeals that support the author's opening claim. Notice also how the writer focuses repeatedly on reasons that are important to his reader.

Persuasion, at best, is risky business because no single collection of strategies always works. Your approach will depend on the people involved, the setting, the relationships, and the full range of variables discussed in this chapter. And with a truly resistant audience, even the best arguments may fail.

But even though you have no way of knowing whether your efforts will succeed, careful planning can greatly improve your chances.

**FIGURE 3.5** Supporting a Claim with Good Reasons

## EWING
### POWER SYSTEMS

July 20, 1989

Mr. Richard White, President
Southern Wood Products
Box 84
Memphis, TN 37162

Dear Dick,

**The writer states his claim**

In our meeting last week, you asked me to explain why we have such confidence in the project we are proposing. Let me outline what I think are excellent reasons.

**Offers first reason**

**Gives example**

First, you and Don Smith have given us a clear idea of your needs, and our recent discussions confirm that we fully understand these needs. For instance, our proposal specifies an air-cooled condenser rather than a water-cooled condenser for your project because water in Memphis is expensive. And besides saving money, an air-cooled condenser will be easier to operate and maintain.

**Offers second reason**

**Appeals to shared value (quality)**

**Gives example**

**Gives statistic**

**Further examples**

**Appeals to reader's goal (security)**

Second, we have confidence in our component suppliers and they have confidence in this project. We don't manufacture the equipment; instead, we integrate and package cogeneration systems by selecting for each application the best components from leading manufacturers. For example, Alias Engineering, the turbine manufacturer, is the world's leading producer of single-stage turbines, having built more than 40,000 turbines in 70 years. Likewise, each component manufacturer leads the field and has a proven track record. We have reviewed your project with each major component supplier, and each guarantees the equipment. This guarantee is of course transferable to you and is supplemented by our own performance guarantee.

**Offers third reason**

**Cites experts**

Third, we have confidence in the system design. We developed the CX Series specifically for applications like yours, in which there is a need for both a condensing and a backpressure turbine. In our last meeting, I pointed out the cost, maintenance, and performance benefits of the CX Series. And although the CX Series is an innovative design, all components are fully proven in many other applications, and our suppliers fully endorse this design.

**FIGURE 3.5** Supporting a Claim with Good Reasons *Continued*

Richard White, July 20, 1989, p. 2

**Closes with best reason**

**Appeals to shared values (trust) and shared goals (success)**

Finally, and perhaps most importantly, you should have confidence in this project because we will stand behind it. As you know, we are very anxious to establish ourselves in Memphis-area industries. If we hope to succeed, this project must succeed. We have a tremendous amount at stake in keeping you happy.

If I can answer any questions, please phone me. We look forward to working with you.

Sincerely,

EWING POWER SYSTEMS, INC.

*Tom Ewing*

Thomas S. Ewing
President

## OBSERVE PERSUASION GUIDELINES

Later chapters offer specific guidelines for various persuasive documents. But beyond attending to the particular requirements of this or that document, you must always keep in mind this one principle:

**IF IT DOESN'T WORK IN THE EYES OF ITS AUDIENCE,
IT DOESN'T WORK AT ALL.**

Some or all of the following guidelines apply to any persuasive situation; they can help you see things the reader's way.

1. Size up the political realities. What is your organization's climate (receptive, repressive, cooperative)? How outspoken can you afford to be without alienating anyone? Who will be most affected by your document? Might you be hurting anyone unintentionally? How are readers likely to react? Do you risk alienating anyone? Are you shirking any responsibility? Are readers likely to regard you as self-serving?

   The more precisely you assess your readers' political feelings, the less likely your document is to backfire. Once you have analyzed the political problems, do what you can to win each reader's confidence and goodwill:

   - Be diplomatic; avoid making anyone look bad.
   - Don't overstep your bounds as a member of the organization.
   - Don't expect anyone to be perfect—including yourself.
   - Ask your intended reader(s) to review an early draft.

   When reporting about company negligence, dishonesty, stupidity, or incompetence, you might risk political fallout by standing up for your principles. In such cases, you might face the hard choice of keeping your job or your dignity—about which more in Chapter 4.

2. Identify the unspoken rules and other constraints governing what you say, to whom you say it, and how and when you say it.

3. Never make a claim or ask for a response that you know readers will reject outright. Ask for something your audience can live with.

4. Spend a lot of time anticipating your audience's reaction and any possible objections, and tailor your argument to address those objections. Find ways to lessen the resistance and to create contact.

5. Decide on a connection (or combination of connections): power connection (merely telling what you want), relationship connection (reaching out), or rational connection (appealing to common sense and reason).

6. Avoid extreme personas. *Persona* is the image or impression of a writer that readers get from the tone. Be sure your tone comes across as reasonable, not belligerent or superior. When we're feeling strongly

*for* something, and our audience feels just as strongly *against* it, we tend to become impatient or frustrated; the urge to "sound off" can seem overpowering. But audiences dislike aggressive people and tune them out—no matter how sensible the argument. Try to be likable. Admit the imperfections in your case. A little humility never hurts.

7. Find points of agreement with your audience. If you can begin at a point at which you both share some value or goal or experience, you improve chances for reducing conflict and for winning agreement on later points.

8. Never misrepresent an opposing viewpoint. A sure way to alienate readers is to make the opponent seem more of a "bad guy" or simpleton than the facts warrant. Be sure your interpretation of the opponent's position is in no way distorted.

9. Try to concede *something* to the opponent. Surely your readers have at least one good reason for their position. Acknowledge the merits of their case *before* arguing for the superiority of your own. Try not to seem like a know-it-all. Show some empathy.

10. Try to anticipate the audience's biggest objection to your position, and address it in detail.

11. Use only your best material. Not all your reasons or appeals will be equally strong or significant to your audience. Decide which material— from your *audience's* view—best supports your case.

12. Make no claim unless you can support it with good reasons.

Figure 3.6 shows a good example of persuasive writing that observes the previous guidelines. Our writer, Brook Biddle, applied earlier to a state university for mid-year acceptance as a full-time transfer student. But, facing severe budget cuts, the university decided not to review any applications for mid-year transfer. And so Brook appeals his case directly to the university president.

As you read Brook's letter, think about the ways he establishes a relationship and appeals to reason. Befitting the relationship, his tone is forceful, but respectful and *reasonable*. Brook is careful to ask for something that doesn't exceed the reader's probable range of acceptance: instead of special consideration for himself only ("Make an exception for me") he asks that *all* transfer applications be reviewed ("Give us all a fair chance"). And by citing the views of other officials at the university, Brook shows he has done his homework on this issue.

**FIGURE 3.6** A Persuasive Request

Dear President Mason:

**Introduces himself and his problem**

As fall semester ends, I face bleak prospects: After three semesters of excellent work as a special student, I had planned to begin the new year fully enrolled at State U. Enter the budget cuts.

Having reached the credit limit for special students, I'm now looking at the end of my academic career for at least a semester, if not forever.

**States his claim (as a question)**

I realize that the budget cuts mandated drastic measures on your part. However, the decision not to review mid-year transfer applications may have already served its purpose. If so, might there be time to review those applications and admit the very best applicants on a space-available basis?

**Establishes early agreement**

**Acknowledges the reader's dilemma**

Of all the tough decisions you've had to make in this financial crisis, the decision not to review mid-year applications must have been one of the toughest. On one hand, you have a responsibility not to "water down" the education of those students already admitted. On the other hand, you have a responsibility toward students who have been working hard at junior colleges or as special students, all in hopes of being admitted this winter. It's not fair for the presently enrolled students to have their classes overcrowded by midyear transfers; but then again it's not fair for transfer students to be denied a chance.

**Appeals to the reader's sense of fairness**

**Concedes a danger in what he's asking**

The legislature understandably needs to receive the message loud and clear that State U is struggling. And I see the danger of legislators interpreting transfer admissions as a sign of "business as usual"—a sign that State U has weathered these cuts and thus might be able to handle further cuts.

**Tries to represent the reader's position fairly**

I don't envy your role in the decision process, and I'm in no position to fault your handling of the issue. Maybe this is just the kind of "blood" our legislature had to see.

**Concedes the merit of the reader's position**

Through state house demonstrations (of which I was a part), countless protest letters and phone calls, and political SOS's from higher education officials, we have implored the legislators to ease up. And they have been forced to listen. Almost as soon as they had announced the latest round of cuts, the legislature felt compelled to soften the blow by almost one-third. Maybe the university's political hardball helped.

**Restates his claim**

But the point has now been made. State U has profited from its decision not to review mid-year applications, and there still may be time to counteract the losses.

**FIGURE 3.6** A Persuasive Request *Continued*

Supports his claim with appeals to shared goals and values

Besides benefiting the students themselves, mid-year transfers would benefit the university. A transfer student occupying an otherwise empty seat costs State U no money; tuition dollars in fact bring money in. And assuming that only the best students are admitted, they enhance the entire student body.

Anticipates and addresses a probable objection

Cites expert opinion

Of course, some majors already are overcrowded, and so transfers would have to be admitted on a space-available basis. The Registrar assures me that he has worked with the admissions office in earlier years to fill last-minute spaces with transfers as late as mid-January. And the Admissions Director claims that his office's role in this process is "the easy part."

Concedes difficulties and restates his claim

I realize that even the "easy part" might not be so easy at this late stage, but I would love to see the university give it a try.

Supports his claim with examples

Picture State U after the budget cuts and what do you see? The football, basketball, and hockey teams are still playing. The radio station is still broadcasting.

What's missing from this picture? Students. A couple of dozen highly motivated transfer students. So what's the big deal? The university can get by without these students.

Appeals to shared values

Closes with his best reason

The big deal is that the university exists to educate. Without a football team you can still have a university. Without a track team, without a swimming pool, even without a radio station or newspaper you can still have a university. As long as teachers are teaching students, there is still a university. But take away the students and what's the point? Students should come first, and they should go last.

Respectfully,

*Christopher B. Biddle*

Christopher B. Biddle

## EXERCISES

**1.** Find an example of a persuasive letter you consider effective. (Check around campus, at your part-time job, or look through sales letters you or your family has received.) In a memo to your instructor, explain why and how the message succeeds. Base your evaluation on the persuasion guidelines, pages 49–50. Attach a copy of the document to your evaluation memo, and submit all materials to your instructor. Be prepared to discuss your evaluation in class.

Now, evaluate a poorly written document, explaining how and why it fails.

**2.** Think about some particular change you would like to see on your campus or at your part-time job. Perhaps you would like to make something happen such as a campus-wide policy on plagiarism, changes in course offerings or requirements, more access to computers, a policy on sexist language, or a day-care center. Or perhaps you would like to improve something such as the grading system, campus lighting, or the system for student evaluation of teachers, or the promotion system at work. Or perhaps you would like to stop something from happening such as noise in the library or sexual harassment at work. (Brook Biddle's letter on pages 51–52 offers a useful model.)

Of course having a viewpoint and making a claim mean nothing until you connect with your audience. Decide who, exactly, you want to persuade, and write a memo to that audience. Anticipate carefully your audience's implied questions, questions like these:

- Do we really have a problem or need?
- If so, should we care enough about it to do anything?
- Can the problem be solved?
- What are some possible solutions?
- What benefits can we anticipate? What liabilities?

You probably will envision additional audience questions, but these should get you started. Do an audience-and-use analysis based on the profile sheet, page 57.

Don't think of this memo as the final word on the subject; think of it instead as an introduction to the issue, a kind of consciousness-raising that gets the reader to acknowledge that the issue deserves attention. At this early stage, highly specific recommendations would be premature and inappropriate.

**3.** Assume you were setting out to challenge an attitude or viewpoint that is widely held by your particular audience. Maybe you want to persuade your classmates that the time required to earn a Bachelor's degree should be extended to five years or that grade inflation is watering down your school's education. Maybe you want to claim that the campus police should (or should not) wear guns. Or maybe you want to ask students to support a 10-percent tuition increase in order to make more computers and software available.

Do an audience-and-use analysis based on the profile sheet, page 57. And write specific answers to the following questions: What are the political realities? What kind of resistance could you anticipate? How would you connect with readers? What about their range of acceptance? Other constraints? What kinds of reasons could you offer to support your claim?

In a memo to your instructor, submit your plan for presenting your case. Be prepared to discuss your plan in class.

## TEAM PROJECT

Much of workplace writing calls for more than merely sharing information. Often, readers need to be *persuaded* to accept recommendations that are controversial or unpopular. This project offers practice in dealing with the persuasion problems of communicating within organizations.

Divide into teams of four to six. Assume that your team agrees strongly about one of these recommendations, and is seeking support from classmates and instructor (and administrators, as potential readers) for implementing the recommendations.

### Choose One Goal

*a.* Your campus Writing Center always needs qualified tutors to help freshman composition students with problems in their writing. On the other hand, students of professional writing need to sharpen their own skill in editing, writing, motivation, and diplomacy. All students in your class, therefore, should be assigned to the Writing Center during the semester's final half, to serve as tutors for twenty hours (beyond normal course time).

*b.* To prepare students for communicating in an automated work environment, at least one course assignment (preferably the long report) should be composed, critiqued, and revised on disk. Students not yet skilled on a word processor will be required to develop the skill by midsemester.

*c.* This course should help individuals improve at their own level, instead of forcing them to compete with stronger or weaker writers. All grades, therefore, should be Pass/Fail.

*d.* To prepare for the world of work, students need practice in peer evaluation as well as self-evaluation. Because this textbook provides definite criteria and checklists for evaluating various documents, students should be allowed to grade each other and to grade themselves. These grades should count as heavily as the instructor's grades.

*e.* To help prepare students of writing for meeting expectations in the workplace, no one should be allowed to limp along, just getting by with minimal performance. This course, therefore, should carry only three possible grades: A, B, or F. Those whose work would otherwise merit a C or D would instead receive an Incomplete, and be allowed to repeat the course as often as needed to achieve a B grade.

*f.* To ensure that all graduates have adequate communication skills for survival in a world where information is the ultimate product, each student in the college should pass a writing proficiency examination before graduation.

### Analyze Your Audience

Your audience here consists of classmates and instructor (and possibly administrators). From your recent observation of this audience, what reader characteristics can you deduce? Use these questions as a guide for developing your audience-and-use profile. (One possible set of responses is given for Goal *e*.)

- *Who is my audience?* Classmates and instructor (and possibly some administrators).

- *How will readers use my information?* Readers will decide whether to support our recommendation for limiting the range of grades in this course to three: A, B, or F.
- *How much is the audience likely to know already about this topic?* Everyone here is already a grade expert, and will need no explanation of the present grading system.
- *What else does the audience need to know?* The instructor should need no persuading; he or she knows all about the quality of writing expected in the workplace. But some of our classmates probably will have questions like these: Why should we have to meet such high expectations? How can this grading be fair to the marginal writers? How will I benefit from these tougher requirements? Don't we already have enough work here?

  We will have to answer such questions by explaining how the issue boils down to "suffering now" or "suffering later," and that one's skill in communication will determine one's career advancement.
- *What attitude or misconceptions about the topic is our audience likely to have?*

  If they are indifferent ("Who cares?"), we have to encourage them to care by helping them understand the eventual career benefits of our recommended grading system.

  If they are interested ("Tell me more!"), we have to hold their interest and gain their support.

  If they are skeptical ("How could this plan ever work?"), we have to show concretely how the plan could succeed.

  If they are biased ("I hate writing"), we have to emphasize how greatly writing matters in the workplace.

  If they are misinformed ("I'll have secretaries or word processors to fix up my writing"), we have to provide the correct information, backed by solid evidence.

  If they are defensive ("Why pick on me?"), we have to persuade them that our idea is constructive, that we are genuinely supportive, and that we have their best interests in mind.

  If they have realistic objections ("Weaker writers will be unfairly penalized," or "Some students can't afford more than one semester for this course"), we need to offer a plan addressing these objections, and to show how the benefits can outweigh the objections.

  Whichever combination of attitudes our audience holds, we have to do our best to satisfy each reader's objections—as well as we can identify them. And if some readers are dead set against the idea, we can't expect to convert them, but we can encourage them to at least consider our position.
- *What is the audience's attitude toward the writer(s), before anyone has read the document? (Is the writer seen as trustworthy, sincere, threatening, arrogant, meek, underhanded, or what?)* We need to gain our readers' confidence, to make sure they interpret our motives not as totally self-serving or elitist, but as sincere and caring about the welfare of the whole group.
- *What is the organizational climate? (Is it competitive, repressive, cooperative, creative, resistant to change, or what?)* In this class, we can all feel comfortable about speaking out.
- *Who will be most affected by this document?* The primary audience, our classmates.

- *In this situation, how can we characterize the audience's temperament and probable reaction?* Many classmates are likely to feel threatened, and probably will react initially with resentment and resistance.
- *Do we risk alienating anyone?* Yes, especially students whose writing never has earned a grade of B or better.
- *How did this document originate, and how long should it be?* Because we writers initiated the document, we can't expect readers to tolerate a long, involved presentation. To be persuasive, however, we do have to make our case concrete.
- *What material will be most important to this audience?* They will want a clear picture of the benefits in such an apparently radical plan.
- *What arrangement would be most effective for this audience and purpose?* We should propose the new grading system, offer our reasons, point out the benefits, show how the system could operate, and close with a request for support.
- *What tone would this audience expect?* We are all at least acquainted; a friendly, conversational tone seems best.
- *What is this document's intended effect on its audience?* If it manages to connect with its audience, this document will win their support for our recommendation.

Follow the model on page 57 for designing a profile sheet to record your audience-and-use analysis, and to duplicate for use throughout the semester. (Feel free to improve on the design and content of our model.)

### Devise a Plan for Achieving Your Goal

From the audience traits you have identified, develop a plan for justifying your recommendation. Express your goal and plan in a statement of purpose.

We need to convince classmates that our recommendation for an A/B/F grading system deserves their support. We will explain how skill in workplace writing affects career advancement, how higher standards for grading would help motivate students, and how our recommendation could be implemented realistically and fairly.

### Plan, Draft, and Revise Your Document

Brainstorm for worthwhile content, do any research that may be needed, write a workable draft, and revise as often as needed to produce a document that stands the best chance of connecting with your audience.

Appoint a member of your team to present the finished document (along with a complete audience-and-use analysis) for class evaluation and response.

# AUDIENCE-AND-USE PROFILE SHEET

## Audience Identity and Needs

Primary audience: _____ *(name, title)*

Other potential reader(s): _____

Relationship: _____ *(client, employer, other)*

Intended use of document: _____ *(take action, solve a problem, other)*

Prior knowledge about this topic: _____ *(knows nothing, a few details, other)*

Additional information needed: _____ *(background, only bare facts, other)*

Probable questions: _____?

_____?

_____?

_____?

_____?

## Audience's Probable Attitude and Personality

Attitude toward topic: _____ *(indifferent, skeptical, other)*

Probable objections: _____ *(cost, time, none, other)*

Probable attitude toward this writer: _____ *(intimidated, hostile, receptive, other)*

Organizational climate: _____ *(receptive, repressive, creative, other)*

Persons most affected by document: _____

Temperament: _____ *(cautious, impatient, other)*

Probable reaction to document: _____ *(resistance, approval, anger, guilt, other)*

Risk of alienating anyone: _____

## Audience Expectations About the Document

Reason document originated: _____ *(audience request, my idea, other)*

Acceptable length: _____ *(comprehensive, concise, other)*

Material important to this audience: _____ *(interpretations, costs,*

_____ *conclusions, other)*

Most useful arrangement: _____ *(problem-causes-solutions, other)*

Tone: _____ *(businesslike, apologetic, enthusiastic, other)*

Intended effect on this audience: _____ *(win support, change behavior, other)*

# Writing for Readers: The Ethics Problem

■ Recognize Unethical Communication

Expect Pressure for Unethical Communication

Never Confuse Team Play with *Groupthink*

Try to Balance Your Loyalties

Anticipate Some Hard Choices

Never Depend Only on Legal Guidelines

Know Your Communication Guidelines

Know Your Rights as an Ethical Employee

Know How to Take a Stand and Survive

Chapters 2 and 3 show how audience analysis helps us tailor messages to inform and influence readers (so we can complete the project, win the contract, or the like). But an *effective* message (one that achieves its purpose) isn't necessarily an *ethical* message. Think of examples from advertising: "Our artificial sweetener is composed of proteins that occur naturally in the human body (amino acids)" or "our potato chips contain no cholesterol." Such claims are technically accurate but misleading: amino acids in certain sweeteners can alter body chemistry to cause headaches, seizures, and possibly brain tumors; potato chips are loaded with saturated fat—which produces cholesterol. While the advertisers' facts may be accurate, they often are incomplete and they imply misleading conclusions.

Whether the miscommunication occurs deliberately or through neglect, a message is unethical when it prevents readers from making their best decision. Therefore, writers ultimately face the threefold problem outlined in Figure 4.1. Ethical communication is measured by standards of honesty, fairness, and concern for everyone involved (Johannesen 1).

## RECOGNIZE UNETHICAL COMMUNICATION

Thousands of people are injured or killed yearly in avoidable accidents—the result of communication that prevented intelligent decision making. On the next page are some tragedies caused ultimately by unethical communication.

**FIGURE 4.1** Three Problems Confronted by Writers

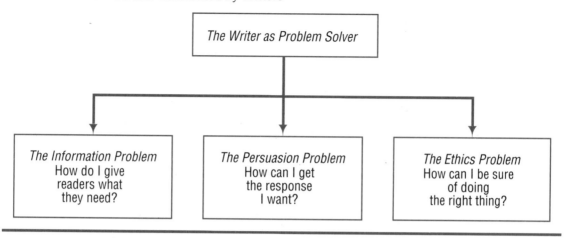

- On March 3, 1974, as a Turkish Airlines DC-10 leaving Paris reached 12,000 feet, a cargo door burst open, causing a crash that killed all 346 people. Immediate cause: a poorly designed latching mechanism gave way under high pressure. Ultimate cause: the faulty design had been recognized and documented since 1969, and cargo doors had burst open during a 1970 test and a 1972 flight (in which the pilot managed to land safely). But, unwilling to admit design errors and under pressure from competitors to get the DC-10 flying as early as possible, manufacturers virtually suppressed these data in their reports (Unger 8–12).

- In the early 1980s, thousands of workers (or their survivors) filed personal-injury suits against a leading manufacturer of asbestos products. The plaintiffs held the company responsible for respiratory afflictions ranging from lung cancer to emphysema. Immediate cause: exposure to asbestos fibers. Ultimate cause: for roughly 50 years the company hid from its workers the deadly facts about asbestos exposure — and even denied workers access to their own medical records. In 1963, the company's medical director offered this reason to justify the suppression of information: "As long as [the employee] is not disabled, it is felt that he should not be told of his condition so that he can live and work in peace, and the company can benefit from his many years of experience." When the coverup finally was revealed, the company tried to evade lawsuits by filing for bankruptcy (Mokhiber 15).

These catastrophes make for dramatic headlines, as did the revelation that a government-operated nuclear plant in Hannaford, Washington, had knowingly leaked radiation for years without local residents being informed. But more "routine" examples of deliberate miscommunication rarely are publicized. Messages like the following succeed by *saying whatever works.*

- A person lands a great job by exaggerating his credentials, experience, or expertise.

- A marketing specialist for a chemical company negotiates a huge bulk sale of its powerful new pesticide by downplaying its carcinogenic hazards.
- To meet the production deadline on a new auto model, the test engineer suppresses, in her final report, data indicating that the fuel tank could explode upon impact.
- A manager writes a strong recommendation to get a friend promoted, while overlooking someone more deserving.

Can you think of examples of miscommunication, perhaps from places you have worked?

To save face or escape blame or get ahead, anyone might be tempted to say what people want to hear, or to suppress bad news or make it seem "rosier." Some of these decisions are not simply black and white. Do you emphasize to a customer the need for extra careful maintenance of this highly sensitive computer—and risk losing the sale? Or do you downplay maintenance requirements, focusing instead on the computer's positive features? Do you tell a white lie so as not to hurt a colleague's feelings or do you "tell it like it is" because you're convinced that lying is wrong in any circumstance? The decisions we make in these ethical gray areas are often colored by the pressures we feel.

## EXPECT PRESSURE FOR UNETHICAL COMMUNICATION

Pressure to get the job done can cause normally honest people to break the rules. At some point in your career you might have to choose between doing what your employer wants ("just follow orders" or "look the other way") and doing what you know is right. Maybe you will be pressured to take questionable—and unreported—shortcuts to meet a project deadline:

> I know that if we continue our development program, we can design a blood pressure monitor that won't have that shock hazard and that will cost less. But if we don't go into production now with what we have, ABC Electronics will grab the market with their piece of junk (Unger 102).

Well over 50 percent of managers studied nationwide feel "pressure to compromise personal ethics for company goals" (Golen, et al. 75). And these pressures take varied form (Lewis and Reinsch 31):

- the drive for profit
- the need to beat the competition (other organizations or coworkers)
- the need to succeed at any cost, as when superiors demand more productivity or savings without questioning the methods
- an appeal to loyalty—to the organization and to its way of doing things

Figure 4.2 depicts how such pressures can add up.

**FIGURE 4.2** How Workplace Pressures Can Add Up

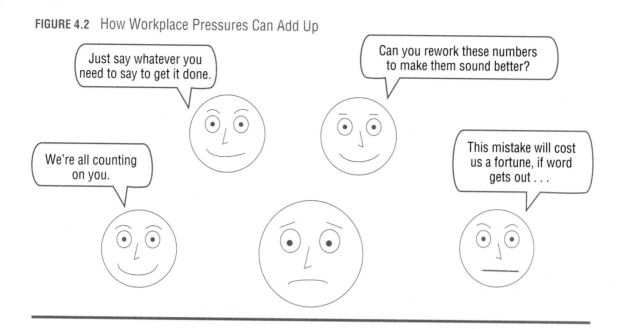

Here is a vivid reminder of how organizational pressure can result in disastrous communication: On January 28, 1986, the space shuttle *Challenger* exploded 43 seconds after launch, killing all seven crew members. Immediate cause: two rubber O-ring seals in a booster rocket permitted hot exhaust gases to escape, igniting the adjacent fuel tank. (See Figure 4.3) Ultimate cause: the O-ring hazard had been recognized since 1977 and documented by engineers, but largely ignored by management. (Managers had claimed that the O-ring system was safe because it was "redundant": each primary O-ring was backed up by a secondary O-ring.)

Moreover, in the final hours, engineers strongly advised against launching because that day's low temperature would drastically increase the danger of both primary and secondary O-rings failing. But, under pressure to meet schedules and deadlines, managers chose to relay only a highly downplayed version of these warnings to the NASA decision makers who were to make *Challenger's* fatal launch decision.[1]

The following analysis of key events and documents underscores the role of miscommunication in a tragic but avoidable accident.

1. More than six months before the explosion, officials at Morton Thiokol, Inc. (manufacturer of the booster rockets) received a memo from engineer R. M. Boisjoly (Presidential Commission 49). Boisjoly clearly described how exhaust gas leakage on an earlier flight had eroded O-

---

[1]For detailed analysis of miscommunication in the *Challenger* disaster, see D. A. Winsor, Roger C. Pace, Robert C. Rowland, and Dennis S. Gouran et al. Full citations appear in Works Cited, pages 560–564.

**FIGURE 4.3** Views of the Challenger

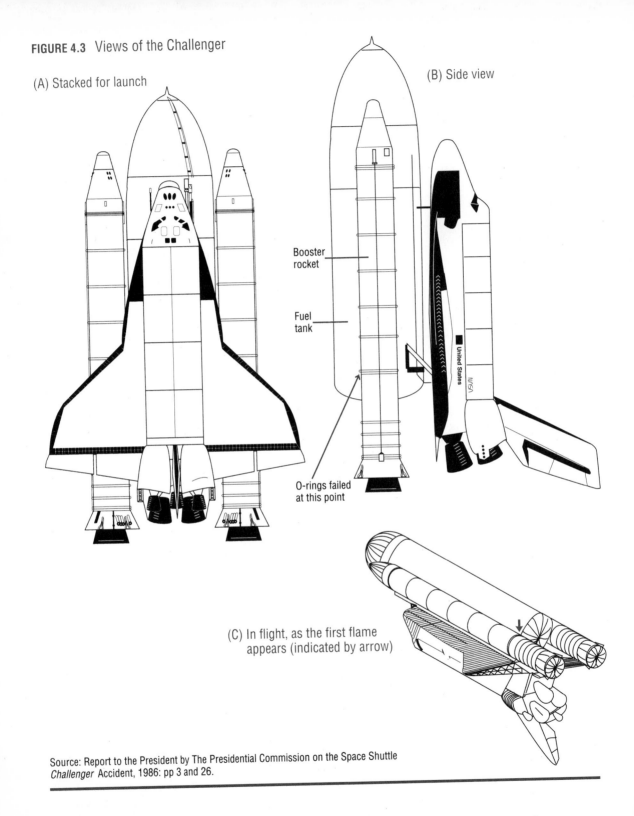

(A) Stacked for launch

(B) Side view

Booster rocket

Fuel tank

O-rings failed at this point

(C) In flight, as the first flame appears (indicated by arrow)

Source: Report to the President by The Presidential Commission on the Space Shuttle *Challenger* Accident, 1986: pp 3 and 26.

rings in certain noncritical joints. The memo emphatically warned of possible "catastrophe" in some future shuttle flight if O-rings should fail to seal a critical joint. Boisjoly called for renewed attention to the O-ring problem. Marked COMPANY PRIVATE, Boisjoly's urgent message was never passed on to top-level decision makers at NASA.

2. During the months preceding *Challenger's* fatal launch, Boisjoly and other Morton Thiokol engineers complained, in writing, about the lack of management attention or support on the O-ring issue. But these complaints were largely ignored within the company—and they never reached the customer, NASA decision makers.

3. On the evening of January 27, managers and engineers at Morton Thiokol debated whether to recommend the January 28 launch. In addition to the yet-unsolved problem of O-ring erosion from exhaust gases, engineers were especially concerned that predicted low temperatures could harden the rubber O-rings, preventing them from sealing the joint at all. No prior flight had launched below 53 degrees Fahrenheit. At this temperature—and even up to 75 degrees—some blow-by (exhaust leakage) had occurred. O-ring temperature on January 28 would be barely above 30 degrees.

   Arguing against the launch, Roger Boisjoly presented a chart to his Thiokol colleagues (Presidential Commission 89). The chart showed that the O-rings would take longer to seal because lower temperatures would make the rubber hard and less pliable, and that the worst exhaust leakage had occurred on a January 1985 flight, when the O-ring temperature was 53 degrees. At 30 degrees, the O-rings might not seal at all, Boisjoly warned.

   Boisjoly and his supporters made their point, and Thiokol management decided to recommend no launch until the temperature reached at least 53 degrees.

4. In a teleconference with NASA's Marshall Space Center, Thiokol managers relayed their no-launch recommendation to the next level of decision makers. But the recommendation was rejected. One Thiokol manager's testimony:

   > . . . Mr. Mulloy [a NASA official] said he did not accept that recommendation, and Mr. Hardy said he was appalled that we would make such a recommendation (Presidential Commission 94).

   NASA then asked Thiokol to reconsider its recommendation because NASA had been under the impression that the range of safe temperatures for the O-rings was 40 to 90 degrees.

5. The Thiokol staff met once again, with engineers and one manager continuing to oppose the launch. The managers then met *without* the engineers, and the reluctant manager was told to "take off your engineering hat and put on your management hat" (Presidential Commission 93). Despite the engineering evidence to the contrary, Thiokol managers finally concluded that the O-ring system had a substantial margin of safety. (Presidential Commission 108).

6. And so, despite continued engineering objections, Thiokol management reversed its recommendation (Presidential Commission 97) to Marshall and Kennedy Space Centers. Top NASA decision makers never were told of Thiokol engineers' objections to the low-temperature launch or the level of concern about O-ring erosion in prior shuttle flights. And the rest is history.

Here are some conclusions of the Presidential Commission investigating the fatal launch decision (104):

- The Commission was troubled by what appears to be a propensity of management at Marshall to contain potentially serious problems and to attempt to resolve them internally rather than communicate them forward.

- The Commission concluded that the Thiokol management reversed its position and recommended the launch . . . at the urging of Marshall and contrary to the views of its engineers in order to accommodate a major customer.

Unethical communication seems to have had a key role in the *Challenger* disaster.

## NEVER CONFUSE TEAM PLAY WITH *GROUPTHINK*

Any successful organization relies on team spirit, everyone cooperating to get the job done. But team spirit is not the same as blindly following orders. Team players don't use loyalty and sense of duty to the company as excuses for avoiding the issues or denying responsibility.

While team play is productive, *groupthink* is destructive. Psychologist Irving Janis defines groupthink as "a mode of thinking that people engage in when they are deeply involved in a cohesive in-group, when members' strivings for unanimity override their motivation to realistically appraise alternative courses of action" (9).

Groupthink occurs when pressure from the group prevents individuals questioning, criticizing, or "making a wave." In Figure 4.4, group members feel a greater need for acceptance and a sense of belonging than for critically examining the issues. In a conformist climate, critical thinking is impossible. Anyone who has lived through adolescent peer pressure has already experienced a version of groupthink.

Giving in to pressure can be especially tempting in a large company or on a big project, where individual responsibility is easy to camouflage in the crowd:

**FIGURE 4.4** Groupthink Obscures Critical Thinking

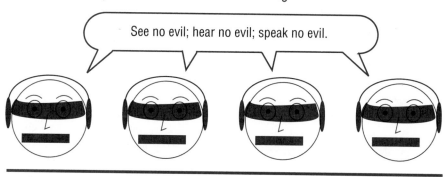

Lack of accountability is deeply embedded in the concept of the corporation. Share-holders' liability is limited to the amount of money they invest. Managers' liability is limited to what they choose to know about the operation of the company. And the corporation's liability is limited by Congress (the Price-Anderson Act, for example, caps the liability of nuclear power companies in the aftermath of a nuclear disaster), by insurance, and by laws allowing corporations to duck liability by altering their corporate structure (Mokhiber 16).

All kinds of people work at all levels on a major project (for instance, the production of a new passenger plane). As one ethics expert points out, countless decisions at any level have far-reaching effects on the whole project (as in the decision to ignore the DC-10's faulty latching mechanism). But with so many people involved, identifying those responsible for an error is often impossible—especially when the error is one of omission, that is, of *not* doing something that should have been done.

> Evasion of responsibility has become an important skill among managers and politicians, as is epitomized in such current terms as "deniability" and "the need not to know." Recent history has furnished us with too many instances where the responsibility for disastrous consequences has been diffused among large numbers of not particularly villainous people (Unger 137).

After completing their assigned task, people too often assume their job is done. Figure 4.5 depicts the kind of thinking that enables people to deny personal responsibility for the consequences of their communication.[2]

No matter how large the group or how strong the pressures, we are no less responsible for our individual actions—even if the way the group achieves results is by cutting corners, overlooking hazards, or suppressing bad news.

[2]My thanks to Judith Kaufman for this idea.

FIGURE 4.5 Groupthink Can be a Handy Hiding Place

Groupthink was exemplified by the decision to relay to top NASA officials only downplayed warnings about *Challenger's* launch. In such cases, the group suffers an "illusion of invulnerability," which creates "excessive optimism and encourages taking extreme risks" (Janis 197), as shown in Figure 4.6.

FIGURE 4.6 Groupthink in Action

Related to groupthink is the distortion of messages from subordinates to superiors. Research shows that subordinates tend to downplay or suppress bad news in their reports to superiors, instead "stressing what they think the superior wants to hear" (Littlejohn and Jabusch 159). Morton Thiokol's final launch recommendation to NASA serves as a memorable example.

## TRY TO BALANCE YOUR LOYALTIES

Our ethics decisions never occur in a void. Any decision can have many consequences for different parties to whom we owe loyalty (Christians 17–18):

- *Loyalty to ourselves* requires us to act in our own self-interest and according to our own good conscience.

- *Loyalty to clients and customers* requires us to stand by the people to whom we have contractual obligations—and who pay the bills.

- *Loyalty to our company or organization* requires us to further the company's goals and abide by its policies insofar as these goals and policies are fair and productive, to protect confidential information, and to expose misconduct that would harm the company.

- *Loyalty to coworkers* requires us to act in the interest of their safety and well-being.

- *Loyalty to society* requires us to consider the public good in all our professional actions.

How can we be honest and fair to everyone involved? There is no easy answer. Figure 4.7 depicts the difficulty.

**FIGURE 4.7**  The Difficulty of Balancing Your Loyalties

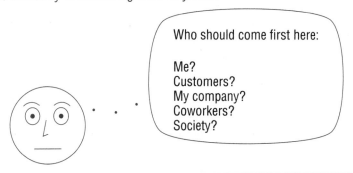

When the interests of these parties conflict—as they almost inevitably do—we have to decide very carefully about where to place our loyalties. In the *Challenger* disaster, for example, loyalty to the customer and to the organization caused people to ignore their responsibility to all other parties.

Worse than being *confused* about where our loyalties lie, we can sometimes be downright mistaken:

Someone observes, for example, that waste from the local mill is seeping into the water table and polluting the water supply. This is a serious situation and requires a remedy. But before one can be found, extremists condemn the mill for lack of conscience and for exploiting the community. People get upset and clamor for the mill to be shut down and its management tried on criminal charges. The next thing you know, the plant does close, 500 workers are without jobs, and no solution has been found for the pollution problem (Hauser 96).

The previous example is a case of misplaced loyalty on the part of the extremists: loyalty only to their sense of what is right—to the exclusion of everyone else involved.

## ANTICIPATE SOME HARD CHOICES

Communicators' ethical choices basically are concerned with honesty in choosing to reveal or conceal information:

- What, exactly, do I report, and to whom?
- To whom do I owe loyalty?
- How much do I reveal or conceal?
- How do I say it?
- Could misplaced loyalty to one party be causing me to lie to others?

For illustration of a hard choice that working professionals might face when they communicate, consider the following scenario:

### A Hard Choice

You are an assistant structural engineer working on the construction of a nuclear power plant near a far northern city. After years of construction delays and cost overruns, the plant has finally received its limited operating license from the Nuclear Regulatory Commission.

During your final inspection of the nuclear core containment unit on February 15, you discover a 10-foot-long hairline crack in a section of the reinforced concrete floor, within 20 feet of the area where the cooling pipes enter the containment unit. (The especially cold and snowless winter likely has caused a frost heave under a small part of the foundation.) The crack has either just appeared or was overlooked by NRC inspectors on February 10.

The crack could be perfectly harmless, caused by normal settling of the structure; and this is, after all, a "redundant" containment system (a shell within a shell.) But then again, the crack could signal some kind of serious stress on the entire containment unit, which furthermore could damage the entry and exit cooling pipes or other vital structures.

You phone your boss, who is just about to leave on a ski vacation, and who tells you, "Forget it; no problem," and hangs up.

You know that if the crack is reported, the whole start-up process scheduled for February 16 will be delayed indefinitely. More money will be lost; excavation, reinforcement, and further testing will be required—and many people with a stake in this project (from company executives to construction officials to shareholders) will be furious—especially if your report turns out to be a false alarm. All segments of plant management are geared up for the final big moment. Media coverage will be widespread. As the bearer of bad news—and bad publicity—you suspect that, even if you turn out to be right, your own career could be hurt by what some people will see as your overreaction that has made them look bad.

On the other hand, ignoring the crack could compromise the system's safety, with unforeseeable consequences. Of course, no one would ever be able to implicate you: the NRC has already inspected and approved the containment unit, leaving you and your boss and your company in the clear. You have very little time to decide: start-up is scheduled for tomorrow, at which time the containment system will become intensely radioactive. What will you do?

---

Working professionals commonly face similar choices, the product of conflicting goals and expectations, the pressure to meet deadlines and get results, to be a "loyal" employee, a "team player," to consider "the bottom line." And, as often as not, these choices have to be made alone or on the spur of the moment, without the luxury of meditation or consultation.

## NEVER DEPEND ONLY ON LEGAL GUIDELINES

Can the law tell you how to communicate ethically? Sometimes. If you stay within the law, are you being ethical? Not always. Legal standards "sometimes do no more than delineate minimally acceptable behavior." In contrast, ethical standards "often attempt to describe ideal behavior, to define the best possible practices for corporations" (Porter 183). Some perfectly legal actions:

- The investigative TV show "20/20" exposed the common and legal practice for trucks to haul garbage or toxic substances (such as formaldehyde — an embalming chemical) one way and then haul food products (such as juice concentrates) on the return trip — without ever informing the customer.

- It is perfectly legal to advertise a cereal made with oat bran (which allegedly lowers cholesterol) without mentioning that another ingredient in the cereal is coconut oil (loaded with cholesterol).

Lying is rarely illegal, except in cases of lying under oath or breaking a contractual promise (Wicclair and Farkas 16). But putting aside these and other illegal lies, such as defamation of character or lying about a product so as to cause injury, we see plenty of room for the kinds of "legal" lies in Figure 4.8. Later chapters cover other kinds of legal lying, such as page design that distorts the real emphasis or words that are deliberately unclear or misleading or ambiguous.

What then are a communicator's legal guidelines? Besides obscenity laws (not especially relevant here), workplace writing is regulated by these four types of laws:

- *Laws against libel* prohibit any false written statement that maliciously attacks or ridicules anyone. A statement is considered libelous when it damages someone's reputation, career, or livelihood or when it causes humiliation or mental suffering. Material that is damaging but *truthful*

**FIGURE 4.8** Some Legal Lies in the Workplace

| "Count on us to meet your deadline!" | "This product will last you for years!" | "Trust our experts to solve your problem!" | "You're our #1 priority!" |
|---|---|---|---|
|  |  |  |  |
| **Promises you know you can't keep** | **Assurances you haven't verified** | **Credentials you don't have** | **Inflated claims about your commitment** |

would not be considered libelous. In the event of a libel suit, a writer's ignorance is no defense: even when the damaging material has been obtained from a source presumed reliable, the writer (and publisher) are legally accountable.[3]

- *Copyright laws* protect the ownership rights of authors—or of their employers, in cases where the writing was done as part of one's employment (Girill 48).

- *Laws against deceptive or fraudulent advertising* make it illegal, for example, to falsely claim or imply that a product or treatment will cure cancer, or to represent and sell a used product as new.

- *Liability laws* define the responsibilities of authors, editors, and publishers for damages resulting from the use of incomplete, unclear, misleading, or otherwise defective information. The misinformation might be about a product, such as a failure to warn about the toxic fumes from a spray-on oven cleaner or a procedure, such as the wrong instructions in an owner's manual for using a tire jack. And even if misinformation is given out of ignorance, the writer is liable (Walter and Marsteller 164–165). Later chapters deal with specific liabilities for particular documents, especially product and mechanism descriptions and instructions.

Laws regulating communication practices are few because such laws traditionally have been seen as threats to our freedom of speech (Johannesen 86).

Many companies have legal departments you can turn to if you have doubts about the legality of a document.

Recognizing that ethical behavior is a matter of commitment, and not of legislation, most professions have developed their own ethics guidelines. If your field has its own formal code, be sure to obtain a copy before entering the workplace.

---

[3]Thanks to my colleague Peter Owens for the material on libel.

## KNOW YOUR COMMUNICATION GUIDELINES

Ethical communicators are concerned with more than merely "transporting information effectively," without considering "those who might be affected by the consequences of the communication" (Clark 191).

How do we balance self-interest with the interests of others—the organization, the public, our customers? How can we be "practical" and "responsible" at the same time? Here are two essential guidelines for ethical communication (Clark 194):

1. *Give the audience everything it needs to know.* To be able to see things your way, people need more than just a partial view; don't bury them in needless details but make sure they get the facts straight.

2. *Try to make sure the audience understands the information accurately.* Even complete information can be misinterpreted. Do all you can to see that the audience understands your information as well as you do. Try to anticipate their needs as you write.

We have seen how the *Challenger* tragedy resulted from various decision makers knowing too little, or misunderstanding what they did know (Clark 194): "At low temperatures the O-rings may not be capable of sealing correctly." What did this information mean?

- To the engineers—the real experts, but not the decision makers—it meant that the problem was serious enough to warrant a launch delay.

- To lower-level decision makers, it meant that the problem carried an acceptable risk because of the benefit of launching "on schedule."

- To the ultimate decision makers at NASA, it meant very little, because lower-level decision makers provided information that was not sufficient or clear enough for the problem to be understood—or even recognized.

The communicators responsible presumably had good *intentions,* but—deliberately or by neglect—they allowed vital information to be overlooked or misinterpreted.

The Ethics Checklist on page 74 brings together additional guidelines from various chapters. Use the checklist for any document you prepare or for which you are responsible.

## KNOW YOUR RIGHTS AS AN ETHICAL EMPLOYEE

How do you respond if your employer asks you to do something wrong, say, altering a report to cover up a violation of federal pollution standards? If you choose not to cooperate, your choices seem limited: resign or go public (i.e., blow the whistle).

These alternatives, however, aren't realistic (Rubens 330). Walking away from a lucrative job is not all that easy in a world of mortgages, car payments, and tuition bills. And whistle-blowing, although an act of supreme courage, can seriously damage (at least temporarily) a career; many organizations simply refuse to hire anyone branded as a whistle-blower (Wicclair and Farkas 19). And even if you don't end up being fired, your work relations are likely to be unpleasant.

Moreover, the law offers little recourse for whistle-blowers (Unger 94). In general, employers have immunity from lawsuits by employees who have been dismissed unfairly and who have no contract or labor union agreement specifying length of employment. The *employment-at-will* rule stipulates that employees can be dismissed at any time and for any cause, or even no cause, at the employer's discretion.

Certain protections, however, do supersede the employment-at-will rule:

- Anyone punished for reporting employer violations to various regulatory agencies (Federal Aviation Administration, Nuclear Regulatory Commission, Environmental Protection Agency, and so on) can request a labor department investigation. Employees whose claims are ruled valid can win reinstatement and reimbursement for back pay and legal expenses.

- A 1980 Michigan law that protects whistle-blowers from employer retaliation is now being considered for adoption by other states (Unger 94–95).

- The Federal False Claims Act allows an employee to initiate a suit, in the government's name, against a contractor for defrauding the government (as in overcharging for military parts). The employee can receive up to 25 percent of money recovered by the government as a result of the suit. Also, this law allows employees of government contractors to sue when they are punished for whistle-blowing (Stevenson 7–10).

In at least a few cases, an ethics dilemma need not boil down to an either/or proposition: "either keep your mouth shut or kiss your job goodbye."

## KNOW HOW TO TAKE A STAND AND SURVIVE

Ethics expert Stephen Unger argues that it is sometimes possible to take a stand without damaging your career. The higher your professional standing, the more seriously your complaint or objection is likely to be taken. But even junior employees stand a better chance of speaking out and surviving if they follow these suggestions[4]:

- *Make sure your view is accurate.* Have you invented a problem where none existed? Check all facts and figures, and verify the risks. Study opposing arguments, and keep an open mind. Seek the advice of an attorney, but do so discreetly.

[4]Adapted excerpts from CONTROLLING TECHNOLOGY by Stephen Unger, copyright © 1982 by Holt, Rinehart and Winston, Inc., reprinted by permission of the publisher.

- *Don't overreact.* Stay calm; be reasonable; don't attack or blame or threaten anyone—and don't sermonize about morality.

- *Don't overstate the problem.* You might in fact begin by understating the dangers, etc., thus allowing later evidence to strengthen your case.

- *Define the problem in terms of the company's best interests.* Instead of a pious and judgmental emphasis ("This is a racist and sexist policy, and you'd better get your act together"), focus on what the company stands to gain or lose ("Promoting too few women and minorities makes us vulnerable to legal action").

- *Don't procrastinate.* Because of ego and expenses and deadlines, everyone's stake in a project grows—no matter how faulty the course of action. Don't wait for positions to become so hardened that your objections merely infuriate people.

- *Seek advice and support.* Ask colleagues for suggestions and help. Objections or complaints are harder to ignore when made by a group instead of one person. But don't expect an army to line up on your side: taking on the company requires real courage.

- *Keep a clear "paper trail."* Make sure all your suggestions, objections, appeals, or recommendations are on record, and that your position is spelled out. Documentation is vital in the event of legal proceedings.

- *Aim your appeal at the right person.* If you must take your complaint over the heads of immediate supervisors, and if you have any choice about where to take it, choose someone who has enough technical knowledge to understand the problem, who is likely to sympathize with your position, and who has enough clout to make something happen.

- *Know when to back off.* Maybe you won't get all you wanted. But if threat of harm to the people involved has been eliminated, and if your appeal elicits a response you can live with, you might want to back off. Crusaders make too many enemies.

- *Get professional advice before taking your complaint outside.* Contact your professional society and seek the advice of your attorney.

Before taking a job, inquire about the company's ethical support system. Some companies have ombudsmen, who can help employees lodge complaints. Other companies have hotlines for advice on ethics problems or for reporting violations. And more and more companies are developing ethics codes for personal and organizational behavior. But lacking such supports, don't expect to last long as an ethical employee in an unethical organization.

**A Final Note:** No one has any sure way of always knowing what to do. This chapter is only an introduction to the inevitable hard choices that, throughout your career, will be yours—to make and to live with.

## AN ETHICS CHECKLIST FOR COMMUNICATORS[5]

Use this checklist to verify that your documents reflect the best possible ethical judgment. (Page numbers in parentheses refer to the first page of discussion).

☐ Do I avoid exaggeration, understatement, sugarcoating, or any distortion that leaves readers at a disadvantage? (58)

☐ Am I sufficiently informed to know what I'm talking about – instead of merely "faking it"? (64)

☐ Do I actually believe in what I'm saying, instead of being a mouthpiece for groupthink? (64)

☐ Would I accept personal responsibility for this document? (65)

☐ Am I being honest and fair to everyone involved? (67)

☐ Can I be reasonably sure this document harms or endangers no one, now or later? (67)

☐ Do I respect everyone's right to privacy and confidentiality? (42)

☐ Do I give readers all the information and understanding they need for making an informed decision? (61)

☐ Do I make readers aware of the consequences (as I am able to forecast them) of what I'm advocating? (66)

☐ Could I publicly defend the position I'm advocating here? (65)

☐ Do I state the case as clearly as I can, instead of hiding behind the language? (221)

☐ Do I encourage readers to question or challenge my position? (49)

☐ Do I make readers aware of alternatives, if there are any? (39)

☐ Am I providing copies of this document to everyone who has the right to know about it? (359)

☐ Do I give credit to all contributors and sources of ideas and information? (127)

[5]This list was largely adapted from Stephen H. Unger 39–46; Richard L. Johannesen 21–22; George Yoos 50-55. Full citations appear in Works Cited, page 562.

## EXERCISES

**1.** Prepare a memo (one or two pages) for distribution to incoming freshman in which you introduce the ethical dilemmas they will face in college. For instance,

- "If you receive a final grade of 'A' by mistake, would you inform your professor?"
- "If the library loses the record of books you've signed out, would you return them anyway?"
- "Would you plagiarize—and would that change in your professional life?"
- "Would you support the lowering of standards for underprivileged students if you felt your school's status would be compromised?"
- "Would you allow a friend to submit a paper you've written for some other course?"

What other ethical dilemmas can you think of? Tell your audience what to expect, and give them some *realistic* advice for coping. No sermons, please.

**2.** Only about two dozen companies cause one-third of toxic-waste pollution in the U.S.—and it's all perfectly legal. Who are the biggest polluters, and why do they get away with it?

**3.** In your workplace communications, you may end up facing hard choices concerning what to say, how much, how to say it, and to whom. And whatever your choice, it will have definite consequences. Be prepared to discuss the following cases in class. Can you think of similar choices you or someone you know has already faced? What happened?

- While traveling on assignment that is being paid for by your employer, you visit an area in which you would really like to live and work, an area in which you have lots of contacts but never can find time to visit on your own. You have five days to complete your assignment, and then you must report on your activities. You complete the assignment in three days. Should you spend the remaining two days checking out other job possibilities, without reporting this activity?
- As a marketing specialist, you are offered a lucrative account from a cigarette manufacturer; you are expected to promote the product. Should you accept the account? Suppose instead the account was for beer, junk food, suntanning parlors, or ice cream. Would your choice be different? Why?
- You have been authorized to hire a technical assistant, and so you are about to prepare an advertisement. This is a time of threatened cutbacks for your company. People hired as "temporary," however, have never seemed to work out well. Should your ad include the warning that this position could be only temporary?
- You are one of three employees being considered for a yearly production bonus, which will be awarded in six weeks. You've just accepted a better job, at which you can start anytime in the next two months. Should you wait until the bonus decision is made before announcing your plans to leave?
- You are marketing director for a major importer of coffee beans. Your testing labs report that certain African beans contain roughly twice the caffeine of South American varieties. Many of these African varieties are big sellers, from countries whose coffee bean production helps prop otherwise desperate economies. Should your advertising of these varieties inform the public about the high caffeine content? If so, how much emphasis should this fact be given?

**4.** Study the following scenario and complete the assignment: You belong to the Forestry Management Division in a state whose year-round economy depends almost totally on forest products (lumber, paper, etc.) but whose summer economy is greatly enriched by tourism, especially from fishing, canoeing, and other outdoor activities. The state's poorest area is also its most scenic, largely because of the virgin stands of hardwoods. Your division has been facing growing political pressure from this area to allow logging companies to harvest the trees. Logging here would have good and bad consequences: for the foreseeable future, the area's economy would benefit greatly from the jobs created; but traditional logging practices would erode the soil, pollute waterways, and decimate wildlife, including several endangered species—besides posing a serious threat to the area's tourist industry. Logging, in short, would give a desperately needed boost to the area's standard of living, but would put an end to many tourist-oriented businesses and would change the landscape forever.

You've been assigned to weigh the economic and environmental impacts of logging, and prepare recommendations (to log or not to log) for your bosses, who will use your report in making their final decision. To whom do you owe the most loyalty here: the unemployed or underemployed residents, the tourist businesses (mostly owned by residents), the wildlife, the land, future generations? The choices are by no means simple. In cases like this, it isn't enough to say that we should "do the right thing," because we are sometimes unable to predict the consequences of a particular action— even when it seems the best thing to do. In a memo to your instructor, tell what action you would recommend and explain why. Be prepared to defend your ethical choice in class.

## TEAM PROJECT

In the workplace, decisions with ethical implications often are made *collaboratively* by members of a project or management team. This project offers practice in collaborative decision making on a growing ethical problem spawned by new technology.

Divide into groups. Assume you are a junior manager at Killawatt, a major producer of consumer electronics, from digital watches to cordless phones to computer games to the newest gadgetry. Your team is composed of managers from various departments such as research and development, computer services, human resources, accounting, production, and corporate relations. You have been brought together to tackle the following problem.[6]

### Analyze the Problem

Killawatt is facing a threefold management problem:

**1.** A two-million dollar loss in the recent quarter, caused by an economic downturn, necessitates severe austerity measures. Upper management wants to eliminate internal losses from such sources as

- excessive and unauthorized use of photocopy and fax machines, digitizers, laser printers, and other costly equipment

---

[6]Adapted from "Monitoring on the Job," by Gary T. Marx and Sanford Sherizen, (Nov./Dec. 1986). Reprinted with permission from *Technology Review,* copyright 1986.

- personal phone calls on the WATS line
- personal or excessive use of database retrieval services such as Compuserv and Dialog (costing as much as $5.00 per minute), for stock market quotations, investment information, and the like
- extra-long lunch hours and coffee breaks
- personal use of company cars and trucks
- unlimited purchases of company products at a 40 percent employee discount

In rosier economic periods all these "perks" were tolerated by Killawatt. But today's hard times make such benefits no longer affordable.

**2.** A more crucial and long-term problem for the company is to find ways of identifying its most and least productive employees. Presumably, this step could not only increase productivity, but might also provide a more selective basis for salary increases, bonuses, promotions, and even any layoffs that might later be unavoidable.

**3.** A third and somewhat related problem is for Killawatt to plug all security leaks. In the past two years, major company breakthroughs in product technology have been leaked to competitors. And details of the company's recent fiscal problems have been leaked to the press, hurting stock value and scaring investors.

In light of these problems with perks, productivity, and security, Killawatt is exploring ways to monitor its employees. (Monitoring equipment, although expensive, could be capitalized, depreciated, or written off as a business expense; resulting tax benefits would largely offset equipment costs.) At this preliminary stage, you have been asked to assess the feasibility of such a plan in terms of its impact on employee relations.

Like many coworkers, you have heard your share of horror stories about perfectly "legal" monitoring in some other companies. For example, security officers for one government contractor rummage through employees' desk drawers after work hours. Abuses at other companies include

- Random checking of employees' computers for unauthorized files (such as personal correspondence, coursework, home designs, and the like).
- Devices that monitor computer workstations by keeping track of the number of keystrokes, errors, corrections, words typed per minute, customers processed per hour — even the number and length of visits to the bathroom.
- Companywide polygraph (lie detector) testing to identify dishonest employees.
- Cameras or microphones to keep track of productivity, behavior, and even conversations.
- Telephone devices that keep track of each call: the extension from which it was made, the number dialed, time and length of call, and even a recording of the conversation. Such devices presumably help expose security leaks by keeping track of who is talking to whom about what.
- Programs that tell workers how their performance is measuring up against coworkers', or some with "relaxation" sounds or other subliminal messages designed to increase concentration or speed or morale.

While advocates of monitoring at Killawatt acknowledge the potential for abuse, they argue that a conscientious program could benefit the employees as well as the company. Besides keeping everyone "honest," monitoring could provide an objective measure of productivity, without the interpersonal and sometimes discriminating clashes between supervisor and employee. The technology, in effect, would ensure equal treatment across the board. Annual performance appraisals could then be based on reliable criteria that would enhance a supervisor's "subjective" impressions of the employee's contribution.

But these arguments haven't persuaded you or your team members. In your view the plan raises troubling ethical questions:

- What are the employees' rights in this issue?
- Is the plan fair?
- Could the monitoring plan backfire? Why?
- How would the plan affect employee's perceptions of the company?
- How would the plan affect employee morale and loyalty and productivity?
- What should be done to avoid alienating employees?
- Are there alternatives to monitoring?

Your team struggles with these and other questions you identify when you meet to plan your response.

How will your team respond? What position will you advocate? What are the human costs and benefits of the proposed plan? What plan can your team offer instead? How can you be persuasive here?

## Plan Your Response

As you plan your response to management, think of key points you want to emphasize.

## Analyze Your Audience(s)

Your first audience consists of executives and managers, many who see nothing wrong with monitoring. If your team's memo (followed by a meeting) manages to persuade your bosses, then you will be expected to win employee acceptance of your plan—and so your second audience will consist of employees who mostly are strongly opposed to any type of monitoring. Prepare audience-and-use profiles (pages 165, 166) of each audience.

## Prepare and Present Your Documents

In a memo to your bosses, argue against their plan and make a *persuasive* case for your alternative. Be explicit about your objections and about the *ethical* problems you envision. Brainstorm for worthwhile content, do any research that may be needed, write a workable draft, and revise using the Ethics Checklist (page 74).

Assume that your argument succeeds with your bosses, and prepare a memo to employees that is reassuring and persuasive in eliciting their cooperation.

Appoint a team member to present the finished documents (along with complete audience profiles) for class evaluation and response.

# The Writing Process Illustrated

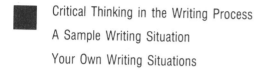

Critical Thinking in the Writing Process

A Sample Writing Situation

Your Own Writing Situations

**W**e have seen that an efficient and useful document never is randomly prepared—never just happens. If writers merely recorded information, their documents would do little communicating or persuading—and a good deal of alienating. To connect with their audience, writers select and shape their material, designing their document to meet the readers' specific need.

Every writing situation requires deliberate decisions for *working with the information* and for *planning, drafting,* and *revising* the document. Some of these decisions are illustrated in Figure 5.1. During this process you transform the material you discover (by experience, observation, research, inspiration, trial and error, or other method) into a message your audience will find useful.

Every writer struggles with the same essential decisions, but rarely in a neat, predictable sequence. Instead, each writer approaches the process through a sequence of decisions that works best for *that* person. And no group of decisions is complete until *all* groups are complete. This looping structure of the writing process is illustrated in Figure 5.2.

## CRITICAL THINKING IN THE WRITING PROCESS

The writing process is a *critical thinking* process: the writer makes a series of deliberate decisions in response to a situation. And the actual writing (putting words on the page) is only a small part of the overall process—probably the least significant part.

In this chapter, we will follow one working writer through an important writing situation; we will see how he solves an array of communications problems and how he designs a useful and efficient document.

**FIGURE 5.1** Typical Decisions During the Writing Process

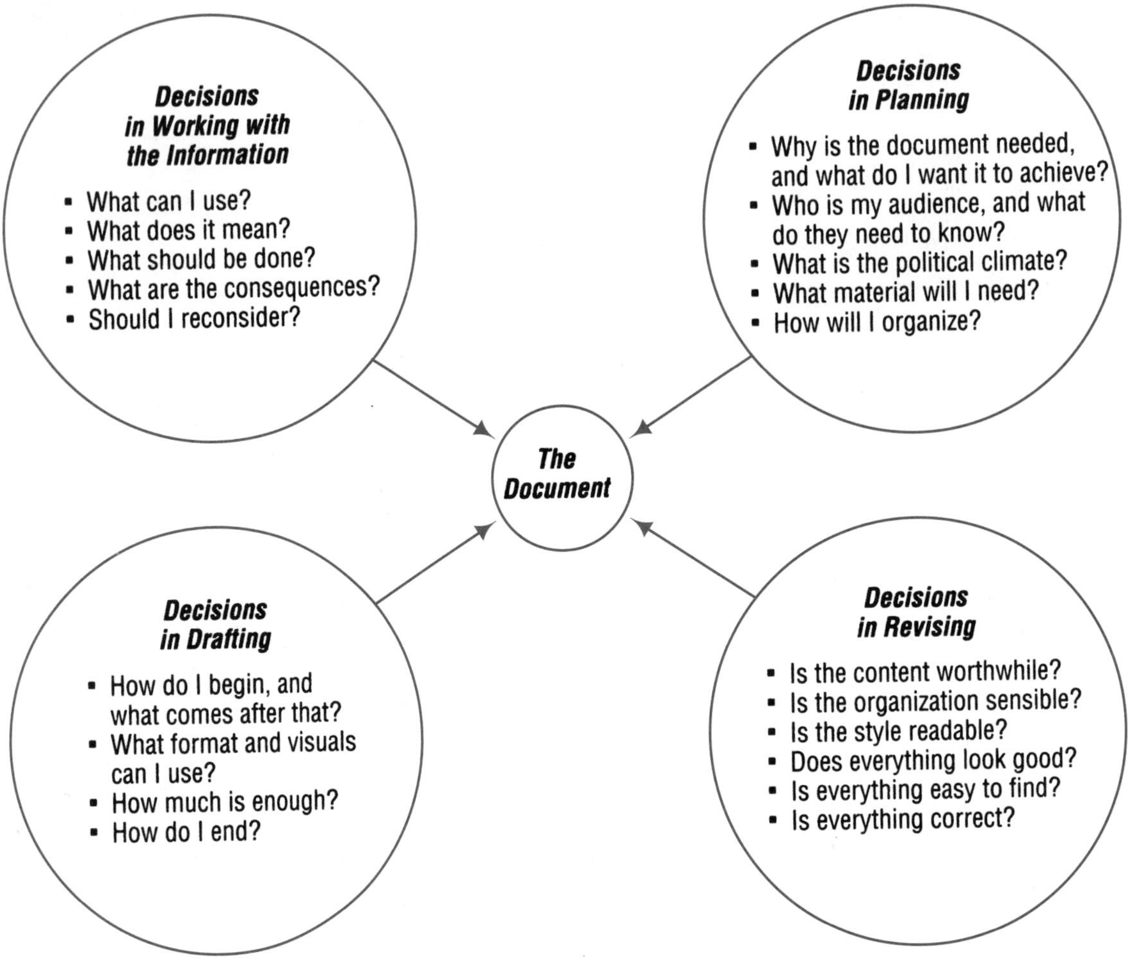

**Decisions in Working with the Information**

- What can I use?
- What does it mean?
- What should be done?
- What are the consequences?
- Should I reconsider?

**Decisions in Planning**

- Why is the document needed, and what do I want it to achieve?
- Who is my audience, and what do they need to know?
- What is the political climate?
- What material will I need?
- How will I organize?

**The Document**

**Decisions in Drafting**

- How do I begin, and what comes after that?
- What format and visuals can I use?
- How much is enough?
- How do I end?

**Decisions in Revising**

- Is the content worthwhile?
- Is the organization sensible?
- Is the style readable?
- Does everything look good?
- Is everything easy to find?
- Is everything correct?

## A SAMPLE WRITING SITUATION

The company is Microbyte, maker of dedicated word processors and portable microcomputers. The writer is Glenn Tarullo (B.S., Management; Minor: Computer Science). Glenn has been on the job two months as Assistant Training Manager for Microbyte's Marketing and Customer Service Division.

For three years, Glenn's boss, Marvin Long, periodically has offered a training program for new managers. Long's program combines an introduction to the company with instruction in management skills (time management, motivation, communication). Long seems satisfied with his two-week program, but has asked Glenn to evaluate it and write a report as part of a company move to upgrade personnel procedures.

**FIGURE 5.2** A Flowchart for the Writing Process

The planning, drafting, and revising decisions in the writing process are recursive: No one stage of decisions is complete until all stages are complete.

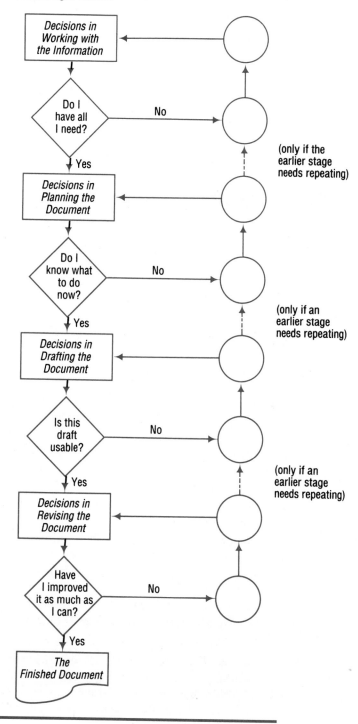

Glenn knows his report will be read by Long's boss, George Hopkins (Assistant Vice President, Personnel), and Charlotte Black (Vice President, Marketing, the person who devised the upgrading plan). Copies will go to other division heads, and to the division's chief executive.

Glenn spends two weeks (Monday, October 3, to Friday, October 14) sitting in and taking notes on Long's classes. On October 14, the trainees evaluate the program. After reading these responses and reviewing his notes, Glenn concludes that the program was successful, but could stand improvement. But he will have to recommend improvements—without offending anyone (instructors, his boss, or guest speakers). Figure 5.3 depicts Glenn's threefold problem.

**FIGURE 5.3** Glenn Has to Solve a Threefold Problem

**His persuasion problem**
"How can I recommend improvements everyone can live with?"

**His information problem**
"How do I make readers understand the positive features of the program, as well as those features that need improving?"

**His ethics problem**
"How can I tell the truth without harming anyone?"

Glenn is scheduled to present his report in conference with Long, Black, and Hopkins on Wednesday, October 19. Right after the final class (1 P.M., Friday, the 14th), Glenn begins work on his report.

Glenn spends half of Friday afternoon fretting over the many details of his situation, the many readers and other people involved, and the political realities and constraints. (He knows that no love is lost between Long and Black, and he wants to steer clear of their ongoing conflict.) By 3 P.M., Glenn hasn't written a word. Desperate, he decides to write whatever comes to mind:

> Although the October Management Training Session was deemed quite successful, several problems have emerged which require our immediate attention.
>
> • Too many of the instructors had poor presentation skills. A few never arrived on time. Mr. Thomas didn't stick to the topic but rambled incessantly. Mr. Jones and Ms. Wells seemed poorly prepared. Instructors in general seemed to lack any clear objectives. Also, because too few visual aids were used, many presentations seemed colorless and apparently bored the trainees.

- The trainees (all new people) were not at all cognizant of how the company was organized or functioned. So the majority of them often couldn't relate to what the speakers were talking about.

- It is my impression that this was a weak session due to the fact that there were insufficient members (only five trainees). Such a small class makes the session a waste of time and money. For instance, Lester Beck, Senior Vice President of Personnel, came down to spend over one hour addressing only a handful of trainees. Another factor is that with fewer trainees in a class, less dialogue occurs, with people tending to just sit and get "talked at."

- Last but not least, executive speakers generally skirted the real issues, saying nothing about what it was really like to work here. They never really explained how to survive politically (e.g., never criticize your superior; never complain about the hard work or long hours; never tell anyone what you *really* think; never observe how few women are in executive or managerial positions, or how disorganized things seem to be). New employees shouldn't have to learn these things the hard way.

In the final analysis, if these problems can be addressed immediately, it is my opinion we can look forward in the future to effectuating management training sessions of even higher quality than those we now have.

Glenn completes this draft at 5:10 P.M. Displeased with the results, but not sure how to improve the piece, he asks an experienced colleague for a critique. Blair Cordasco, a senior project manager, has already helped Glenn several times. Cordasco agrees to read Glenn's draft over the weekend.

At 8:05 Monday morning, Cordasco reviews the document with Glenn. First, she points out obvious style problems: wordiness ("due to the fact that"), jargon ("effectuating"), triteness ("in the final analysis"), implied bias ("weak presentation," "skirted"), among others. Can you identify other style problems in Glenn's draft?

Cordasco points out other problems as well: the piece is disorganized; the emphasis is too critical (offensive to Glenn's boss, making Long look bad to his bosses); the views are too subjective (no one is interested in what Glenn thinks the company's political problems are); the report lacks persuasive force because it contains little useful advice; material about course follow-up and *positive* features of the program is missing. In this form, the report will only alienate several people, and could harm Glenn's career.

The overall tone is too bossy and judgmental. Glenn is in no position to make this kind of *power connection* (p. 38); he needs to be more diplomatic and rational.

Rethinking his situation, Glenn decides to spell out a deliberate plan for this document.

## Planning the Document

Glenn realizes he needs to begin by *focusing* on his writing situation. His audience-and-use analysis goes like this:[1]

> To begin, I'd better figure out *exactly* what my primary reader wants to know.
>
> Long requested the report, but only because Black developed the scheme for division-wide improvements. And so I really have two primary readers here. My boss and the big boss.
>
> My major question here is, am I including enough details for all the bosses? And the answer to this question will require answers to more specific questions:

Anticipated readers' questions

What are we doing right, and how can we do it better?

What are we doing wrong, and does it cost us money?

Have we left anything out, and does it matter?

How, specifically, can we improve the program, and how will those improvements help the company?

> Because all readers have participated in these sessions (as trainees, instructors, or guest speakers), they don't need background explanations.
>
> To answer the readers' questions, I should begin with the *positive* features of the last session. Then I can discuss the problems and make recommendations. Maybe I can eliminate the bossy and judgmental tone by *suggesting improvements* instead of *criticizing weaknesses*. Also, I could be more persuasive by describing the *benefits* of following my suggestions.

Glenn realizes that if he wants successful future programs, he can't afford to alienate anyone. After all, he hopes to work here a long time.

> Now I have a clear enough sense of what to do.

Statement of purpose

My purpose is to provide my supervisor and interested executives with an evaluation of the October training workshop by describing its strengths, suggesting improvements, and explaining the benefits of these changes.

> Working from this plan, I should be able to revise my first draft into a useful report. But . . . that first draft lacks important details. I should brainstorm to get *all* the details (including the *positive* ones) I want to include.

**Glenn's Brainstorming List.** Glenn's first draft touched on several topics; incorporating them into his brainstorming list, he came up with

1. better-prepared instructors and more visuals

---

[1]Throughout this chapter, Glenn's analysis will address *all* the areas illustrated on the audience-and-use profile sheet (page 33).

2. on-the-job orientation *before* the training session

3. more members in training sessions

4. executive speakers should spell out qualities needed for success

5. beneficial emphasis on interpersonal communication

6. need follow-up evaluation (in six months?)

7. four types of training evaluations:

    a. trainees' reactions

    b. testing of classroom learning

    c. transference of skills to the job

    d. impact of training on the organization (high sales, more promotions, better-written reports)

8. videotaping and critiquing of trainee speeches worked well

9. acknowledge the positive features of the session

10. ongoing improvement ensures quality training

11. division of class topics into two areas was a good idea

12. additional trainees would increase classroom dialogue

13. the more trainees in a session, the less time and money wasted

14. instructors shouldn't drift from the topic

15. on-the-job training to give a broad view of the division

16. clear course objectives, to increase audience interest and to measure the program's success

17. Marvin Long has done a great job with these sessions over the years

By 9:05 A.M., the office is hectic. Glenn puts his list aside to spend the day on work piling up. Not until 4 P.M. does he return to his report.

Now what? I should delete whatever my audience already knows or doesn't need, or whatever seems unfair or insincere: 7 can go (this audience needs no lecture in training theory); 14 is too negative and critical—besides, the same idea is stated more positively in 4; 17 is obvious brown-nosing, and I'm in no position to make such grand judgments.

Maybe I can unscramble this list by arranging items within categories (strengths, suggested changes, and benefits) from my statement of purpose.

**Glenn's Brainstorming List Rearranged.** Notice here how Glenn discovers additional *content* (in italic type) while he's deciding about *organization*.

**Strengths of the Workshop**

- division of class topics into two areas was useful
- emphasis on interpersonal communication
- videotaping of trainees' oral reports, followed by critiques

Well, that's one category done. Maybe I should combine *suggested changes* with *benefits,* since I'll want to cover them together in the report anyway.

**Suggested Changes/Benefits**

- more members per session would increase dialogue and use resources more efficiently

- varied on-the-job experiences before the training sessions would give each member a broad view of the marketing division

- executive speakers should spell out qualities required for success *and future sessions should cover professional behavior, to provide trainees with a clear guide*

- follow-up evaluation in six months *by both supervisors and trainees would reveal the effectiveness of this training and suggest future improvements*

- clear course objectives and more visual aids to increase *instructor efficiency* and audience interest

Now that I have a fairly sensible arrangement, I can get this list into report form, although I'm sure to think of more material to add as I work. Since this is *internal* correspondence, I'll use a memo format.

## Drafting the Document

After a series of rough drafts, Glenn produces a usable draft—one that has just about everything he wants to cover. (Sentences are numbered for our later reference.)

[1]In my opinion, the Management Training Session for the month of October was somewhat successful. [2]This success was evidenced when most participants rated their training as "very good." [3]But improvements still are needed.

[4]First and foremost, a number of innovative aspects in this October session proved especially useful. [5]Class topics were divided into two distinct areas. [6]These topics created a general-to-specific focus. [7]An emphasis on interpersonal communication skills was the most dramatic innovation. [8]This innovation helped class members develop a better attitude toward things in general. [9]Videotaping of trainees' oral reports, followed by critiques, helped clarify strengths and weaknesses.

[10]There is a detailed summary of the trainees' evaluations attached. [11]Based on these and on my past observations, I have several suggestions.

- [12]All management training sessions should have a minimum of ten to fifteen members. [13]This would better utilize the larger number of managers involved and the time expended in the implementation of the training. [14]The quality of class interaction with the speakers would also be improved with a larger group.

- [15]There should be several brief on-the-job training experiences in different sales and service areas. [16]These should be developed *prior* to the training session. [17]This would provide each member with a broad view of the duties and responsibilities in all areas of the marketing division.

- [18]Executive speakers should take a few minutes to spell out the personal and professional qualities essential for success with our company. [19]This would provide trainees a concrete guide to both general company and individual supervisors' expectations. [20]Additionally, by the next training session we should develop a presentation dealing with the attitudes, manners, and behavior appropriate in the business environment.

- [21]Do a six-month follow-up. [22]Get feedback from supervisors as well as trainees. [23]Ask for any new recommendations. [24]This would provide a clear assessment of the long-range impact of this training on an individual's job performance.

- [25]We need to demand clearer course objectives. [26]Instructors should be required to use more visual aids and improve their course structure based on these objectives. [27]This would increase instructor quality and audience interest.

[28]These changes are bound to help. [29]Please contact me if you have further questions.

Although now developed and organized, this version still is some way from the finished document on pages 89–90. Glenn has to make further decisions about his style, content, arrangement, audience, and purpose.

## Revising the Document

At 8:15 Tuesday morning, Glenn decides on a sentence-by-sentence revision for worthwhile content, sensible organization, and readable style.

Sentence 1 begins with a needless qualifier, has a redundant phrase, and sounds insulting ("somewhat successful"). Sentence 2 should be in the passive voice, to emphasize the training—not the participants. Also, 1 and 2 are choppy and repetitious, and so I'll combine them.

> Original    In my opinion, the Management Training Session for the month of October was somewhat successful. This success was evidenced when most participants rated their training as "very good."(28 words)
>
> Revised    The October Management Training Session was successful, with training rated "very good" by most participants. (15 words)[2]

Sentence 3 is too blunt. I need an orienting sentence here that forecasts content diplomatically. I can be honest without being so negative.

> Original    But improvements still are needed.
>
> Revised    A few changes—beyond the recent innovations—should result in even greater training efficiency.

In sentence 4, "First and foremost" is trite, "aspects" is a clutter word, and word order needs changing to improve the emphasis (on innovations) and to lead into the examples.

> Original    First and foremost, a number of innovative aspects in this October session proved especially useful.
>
> Revised    Especially useful in this session were several program innovations.

---

[2]Notice throughout how careful revision sharpens the writer's meaning while cutting needless words.

Sentences 5 and 6 need combining, and content in 5 needs beefing up: *name* the two areas. If 7 is labeled the "most dramatic innovation," it ought to come last, for emphasis. In 8, "things in general" is too indefinite. And 7 and 8 need combining. Sentence 9 seems okay. All three examples would be more readable in a *list,* and with parallel phrasing (maybe starting each with an "ing" phrase to signal the "action" verb, as in 9).

Original    Class topics were divided into two distinct areas. This created a general-to-specific focus. An emphasis on interpersonal communication was the most dramatic innovation. This helped class members develop an enthusiastic and relaxed attitude toward things in general. Videotaping of trainees' oral reports, followed by critiques, helped clarify strengths and weaknesses.

Revised  • Dividing class topics into two areas created a general-to-specific focus: *the first week's coverage of company structure and functions created a context for the second week's coverage of management skills.*[3]
         • Videotaping and critiquing trainees' oral reports clarified strengths and weaknesses.
         • Emphasizing interpersonal communication skills *(listening, showing empathy, and reading nonverbal feedback) generated enthusiasm and a sense of ease about the group, their training, and the company.*

And while I'm thinking about format, a couple of clear headings ("Workshop Strengths," "Suggested Changes/Benefits") would segment the text, improving readability and appearance.

Glenn follows this same editing and revising process throughout (Question 3 in the Exercises asks you to identify these remaining changes.) Wednesday morning, after much revising and proofreading of his hard copy, Glenn hits the PRINT command on his word processor for the *final* draft shown in Figure 5.4.

Glenn's final effort produces a report that is informative and persuasive. But this document did not leap magically from the writer's head onto the page. Glenn made deliberate decisions about purpose, audience, content, organization, and style. Most important, he *revised* more than once.[4]

This report can have an important effect on Glenn's career. His bosses will evaluate Glenn by his ability to produce clear, helpful recommendations.

---

[3]Notice that this revision has more words, but also much more concrete and specific detail (in italic type). Completeness of information *always* takes priority over word count.

[4]A special thanks to Glenn Tarullo for his perseverance. I made his task doubly difficult by having him explain each of his decisions during this writing process.

**FIGURE 5.4** Glenn's Final Draft

# :::::.MICRO**BYTE**

Memorandum

October 19,1987

To:       Marvin Long
From:     Glenn Tarullo
Subject:  October Management Training Program: Evaluation
          and Recommendations

**Begins on a positive note, and cites evidence**

**States his claim**

The October Management Training Session was successful, with training rated as "very good" by most participants. A few changes, beyond the recent innovations, should result in even greater training efficiency and profits.

## Workshop Strengths
Especially useful in this session were several program innovations:

**Gives clear examples of "innovations"**

—Dividing class topics into two areas created a general-to-specific focus: The first week's coverage of company structure and functions created a context for the second week's coverage of management skills.

—Videotaping and critiquing trainees' oral reports clarified their speaking strengths and weaknesses.

—Emphasizing interpersonal communication skills (listening, showing empathy, and reading nonverbal feedback) created a sense of ease about the group, the training, and the company.

Innovations like these ensure high-quality training. And future sessions could provide other innovative ideas.

## Suggested Changes/Benefits
From the trainees' evaluation of the October session (summary attached) and my observations, I recommend these additional changes:

**Cites the bases for his recommendations**

—We should develop several brief (one-day) on-the-job rotations in different sales and service areas before the training session. These rotations would give each member a real-life view of duties and responsibilities throughout the company.

—All training sessions should have at least 10 to 15 members. Larger classes would make more efficient use of resources and improve class-speaker interaction.

**FIGURE 5.4**  Glenn's Final Draft   *Continued*

Long, Oct. 19, 1987, page 2

**Supports each recommendation with good reasons**

—We should ask instructors to follow a standard format (based on definite course objectives) for their presentation, and to use visuals liberally. These enhancements would ensure the greatest possible instructor efficiency and audience interest.

—Executive speakers should spell out personal and professional traits essential in our company. Such advice would give trainees a concrete guide to both general company and individual supervisor expectations. Also, by the next training session, we should assemble a presentation dealing with the attitudes, manners, and behavior appropriate in business.

—We should do a six-month follow-up of trainees (with feedback from supervisors as well as ex-trainees) to gain long-term insights, to measure the influence of this training on job performance, and to help design advanced training.

**Closes by appealing to shared goals (efficiency and profit)**

Inexpensive and easy to implement, these changes should produce more efficient training and higher profit.

Copies:     B. Hull, C. Black, G. Hopkins, J. Capilona, P. Maxwell,
            R. Sanders, L. Hunter

## YOUR OWN WRITING SITUATIONS

Although we can recognize a planning, drafting, and revising sequence in the writing process, we should avoid enslaving ourselves to any formula. For any writing, we generally begin with a purpose and a sense of our audience. And we follow some sort of progression from generating content, to organizing, to improving style. But at various times we might discover new ideas; we might redefine our purpose; we decide that sentences need shuffling and reorganizing; and we decide on clearer and more precise phrasing. In the work of discovering what we mean, and struggling to express that meaning, we find ourselves revising. Although decisions about *content, organization,* and *style* are treated separately in other chapters, during the process (the planning, drafting, and revising), these decisions often are indistinguishable.

The procedure you follow in designing any document might be quite different from Glenn's. Writers work in different ways. Some begin by brainstorming. Some come up with a statement of purpose immediately. Some begin with an outline. Others hate outlining and simply write and rewrite, using scissors and tape to organize. Some even write a quick draft before thinking through their writing situation. Introductions and titles more often are written last. No one "step" in the process is ever complete until *the whole* is complete. (Notice, for instance, how Glenn sharpens his content *and* his style while he organizes.) Every document you write will require *all* these decisions but rarely will you make them in the same sequence from day to day.

No matter what a writer's sequence, *revision* is a fact of life. It is the one *constant* in the writing process. When you've finished a draft of your document, you have, in a sense, only begun. Sometimes you will have more time to compose your report than Glenn did, sometimes much less. Whenever your deadline allows, leave time to revise. Revision checklists at the end of subsequent chapters will help sharpen your skill in editing and revising.

## EXERCISES

**1.** Think of a course you've taken that had both definite strengths and weaknesses. Assume that the instructor has asked you to evaluate the course and recommend improvements. Your secondary audience will be your instructor's chairperson and the Dean of Faculty. The instructor is up for tenure and is someone you like very much. You need to be candid in your report, but you don't want to make the instructor look bad. Write the report, being as specific as possible. (*Note:* Use a fictional name for the instructor and the course.)

**2.** Assume that you are a training manager for XYZ Corporation. After completing this first section of the text and the course, what advice about the writing process would you have for a beginning writer who will frequently need to write reports on the job? In a one- or two-page (single-spaced) memo to new employees, explain the writing process briefly, and give a list of general suggestions these beginning writers can follow.

**3.** Compare Glenn's second draft (page 86) with his final draft (pages 89–90). Identify all improvements in content, arrangement, and style besides those discussed on pages 87–88.

**4.** Identify your own writing process compared with ideas presented in this chapter. What are your specific strengths and weaknesses? In a memo to your instructor, describe your writing process, its strong and weak areas, and where you think you need most improvement. Include definite suggestions about ways in which this course might help you achieve these needed improvements.

## TEAM PROJECT

### Revising a Document

Having spent several weeks in your Technical Writing class, you have a good sense of *what* material is covered and of *how* the material is taught. Now, imagine that at a recent meeting your school's Advanced Writing Committee passed this motion:

> Because of the growing popularity of Technical Writing, and the 200 percent enrollment increase within the past two years, many new sections of the course have been added to the schedule. To ensure a unified approach in *each* section, we suggest that all sections follow a standard syllabus, and that general teaching approaches be as similar as possible.

For help in evaluating proposed course content and approaches, the committee has decided to survey all students about to complete the Technical Writing course. Each student has been asked to submit, in memo form, an evaluation of his or her section, along with any concrete suggestions for improvement. The collected responses will form a databank for the committee's decisions about a course model that best fulfills students' needs.

As a guide for evaluation, the committee has provided all students with a copy of this question:

> How well do the content and teaching approach in your Technical Writing course fulfill your needs and expectations? How can this course best be taught, and what material should be covered? Please be specific in your evaluation of the course's strengths and weaknesses. Along with suggestions for improvement, be sure to explain how a specific change or improvement would benefit you.

Assume that Frank White, a student in some other section of the course, has responded to the assignment with this memo:

Date: April 25, 19XX
To: The Advanced Writing Committee
From: Frank White, Technical Writing Student
Subject: Section 1499: Evaluation and Recommendation

This course is providing me with a great deal of useful information. Also, the teacher usually manages to hold the attention of the class very well. Learning about writing can be a pretty boring experience, but this class hardly ever is boring because the teacher does such a good job of making the material interesting. Also, the teacher is a very nice person. I'm happy to say that I've learned a number of approaches that have helped me improve my writing.

Writing skills are important in just about anyone's career, and so I'm glad we have the chance to take a Technical Writing course before graduation. Having the course as a requirement is a good idea, because many students (myself included) probably would avoid any course that requires this much writing. It isn't until we get here that we realize how worthwhile (and difficult!) this course is for everyone. I'd like to see every section taught the way this one is: by a teacher who knows how to get a tough job done.

The only complaint I have against this teacher is the fact that he talks too much about computers. I know that computers are important, but I'd like to learn to use one for my writing instead of just being exposed to computers in general. And too many writing assignments have been saved for the latter part of the course. Everything is just stacking up, leaving me buried and confused.

Also, the class is much too large, and the layout is awful. With so many students, the teacher has no way of providing individual attention, and so too many students "get lost in the crowd." We need a classroom layout that would make group editing easier.

I really enjoyed the audiovisual presentations in class, and would like to see more of them. All in all, the concrete stuff was always the most useful.

In general, the material has been covered well, except for the material on oral communication.

With a few minor changes, this course would be excellent.

Before Frank can submit this memo, it will need heavy revision for content that is informative and persuasive and ethical, organization that is easy to follow, and style that is readable. Working in teams of four to six students, your task is to revise Frank's memo.

a. As a first step, complete an audience-and-use profile sheet (page 33) based on the following data, as well as on the details given earlier.

*Primary audience:* The writing committee and the department chairperson. Several committee members also are on the tenure committee, and they are likely to use this information for an additional purpose: to evaluate Frank's instructor for tenure. These are the decision makers. Above all, Frank wants to convey a *positive* impression of his instructor.

*Secondary audience:* Frank's instructor (whom Frank likes, but who deserves an honest and detailed evaluation.) Frank wants to be fair to his instructor, but he also wants to make some realistic suggestions for improving a course that is so important to everyone's career.

b. Assume that you have helped Frank prepare the brainstorming list he should have prepared *before* writing the previous draft. To visualize what it was that specifically pleased or displeased Frank, imagine that these items would have appeared on his list—if Frank had taken the time to brainstorm:

- instructor is always willing to help students individually
- always takes time in class to answer all questions thoroughly
- emphasis on planning, drafting, and revising for a specific audience is very helpful
- instructor spends a lot of time encouraging us
- spends too much time talking about computers and automated offices—what I need is to develop strong writing skills by *using* the computer

- now and then we should have a guest lecturer from business and industry
- we should spend more time discussing the documents *we* ourselves have prepared
- too much time spent emphasizing mistakes—not enough on the positive
- should spend more time on writing, and less than the present four weeks on oral communication
- when they are used (which is not often enough), the opaque and overhead and slide projectors make things more vivid and interesting
- instructor gives a lot of feedback on our papers
- the class should have an IBM PC and a Macintosh computer (since these are the kinds we have in our campus micro labs), to illustrate how automation can affect the writing and revising process
- should have a class with several round tables that seat 4–6 students for editing groups
- class size should be reduced from 30 to no more than 20 students
- tutoring should be available on a regularly scheduled basis for students with any type of writing problem
- we should spend at least one full week on job applications
- too few writing assignments early in the course; thus, too many at the end
- should begin discussing the long report (term project) as early as the first or second week

From Frank's original draft and from his brainstorming list, your team should select only what is useful and appropriate for *this* audience and purpose. Compose a final draft of Frank's memo. Appoint one team member to present the finished document in class.

# Information Retrieval and Synthesis

# 6

# Gathering Information

- Planning Your Research
- Searching the Literature
- Using Electronic Information Services
- Interviewing
- Using Questionnaires and Surveys
- Using Other Primary Sources
- Recording Your Information
- Documenting Your Sources

Research is the effort to discover any fact or set of facts: the cost of building the first computer or the price range of waterfront lots on Boca Grande Island in March 1990. Any kind of significant research is done for some purpose—to answer a question, to make an evaluation, to establish a principle. We set out to discover whether diesel engines are efficient and dependable because we have a reason: we're thinking about buying a diesel-powered car or we're trying to decide whether to produce diesel-powered cars, or the like. The purpose behind all research is to check our opinions against the facts, to reach a conclusion that has the best chance of being valid. In the workplace, almost no major decision is made without research, with the findings recorded in a written report.

Depending on its sources, research is classified as *primary* or *secondary*. Primary research is a first-hand study of the subject or problem; primary sources include observation, interviews, questionnaires, letters of inquiry, personal experiments, analysis of samples, fieldwork, or company records. Secondary research is constructed from information that other researchers—by their primary research—have published. Most workplace research draws from both primary and secondary sources, and requires much more than merely finding books in your library.

Research strategies and resources differ widely among disciplines. In this chapter therefore we focus on research activities for preparing a formal report (Chapter 20). This process is summarized in Figure 6.1.

Each stage in this process is explained in the sections that follow.

**FIGURE 6.1** The Research Process for a Formal Report

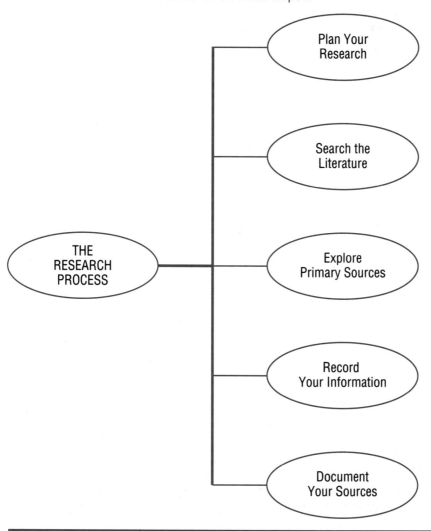

PLANNING YOUR RESEARCH

Perhaps the most crucial stage in your research is the planning. Figure 6.2 shows some of the key planning questions you will need to answer. These planning guidelines will save you time and energy.

**FIGURE 6.2** Questions as You Plan Your Research

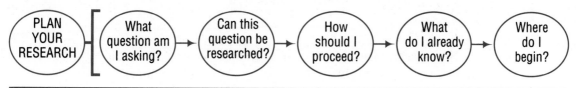

## Isolate the Topic

Choose a practical topic in which your knowledge can make a difference. Or ask professional acquaintances if they need research done. If you are interested in workplace communication practices, for instance, you might settle on this topic: How Our School's Communication Program Relates to Local Business and Industrial Needs.

## Define the Question

To find a direction for your research, you need to define a specific question, which you can then set out to answer. Be sure you can answer your proposed question within the time allowed for completing your report. After some hard thinking about the topic above, you might settle on this question: How can our university's communication program better serve students and local businesses? The audience for your report will be faculty, administrators, and executives.

## Verify That Your Question Can Be Researched

Make sure beforehand that you can find the sources of information you need. (Some topics might be too recent to generate any published information.) Read the next section on gathering information, and check for resources in your library, community, or electronic database service. You might in fact choose your topic or question only after a preliminary search for primary and secondary sources.

List your probable sources (including interviews) in a tentative bibliography, a bibliography that will grow as your research continues.

## Write a Statement of Purpose

Remember that you are not merely collecting views, but screening and evaluating facts for a definite purpose. Identify your goal and your plan for achieving it. Then, your research will have a direction.

> My purpose is to learn how our communication program can better serve students and area companies. I will survey local workplace needs, and evaluate our offerings in relation to these needs. To enhance my study, I will include national data on other professional communication programs.

## Make a Background List

List facts you already know about your topic. Brainstorming should produce some good ideas for investigation.

## Compose a Working Outline

Now that you have identified your definite purpose, you need a road map. By partitioning your topic into subtopics, you compose a working outline:

*I.* Profile of Communication in Local Companies

*II.* Internal Communication in Those Companies

*III.* Frequency and Types of Writing

*IV.* Perception of Writing Effectiveness

*V.* Improving Service for the Companies and Students

Each of these topics can be further divided (Chapter 9). Your working outline is only tentative. Depending on your findings, the outline may change radically before your final draft. But with a working outline on paper, you can begin gathering information.

## SEARCHING THE LITERATURE

No matter how much you already know, you can invariably learn something useful from visits to your library. Inexperienced researchers, however, are sometimes confused about where to begin searching the literature. The various options are shown in Figure 6.3.

**FIGURE 6.3** Ways You Can Search the Literature

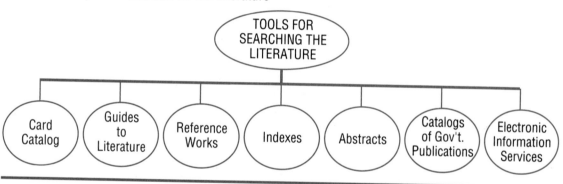

Where you begin your search depends on whether you seek background and basic facts or the very latest information. If you are an expert in the field, you might simply do a computerized database search or browse through the specialized journals. But if you have only limited knowledge or if you need to focus your topic, you will probably want to begin with books and other general reference works. These sources can be located through the card catalog.

## The Card Catalog

All books, reference works, indexes, and periodicals owned by a library usually are listed in its card catalog under three headings: author, title, and subject. You thus have three ways in which to locate an item, as shown in Figures 6.4, 6.5, and 6.6.

**FIGURE 6.4** A Catalog Card Classified by Author

| | |
|---|---|
| Call number ———————— | HM131<br>G56 |
| Author ———————————— | **Goldhaber, Gerald M.** |
| Title —————————————— | Organizational communication.<br>Dubuque, Iowa: |
| Publisher ——————————— | W. C. Brown [1974] |
| No. of chapters, pages, etc. —— | xi, 391 p. illus. 24 cm. |
| Supplementary material ———— | Includes bibliographies. |
| Other headings for this work —— | 1. Organization.  2. Communication |
| Lib. of Congress call no. ———— | HM131.G56 |

**FIGURE 6.5** Part of a Catalog Card Classified by Title

| | |
|---|---|
| HM131<br>G56 | **Organizational communication.**<br>Goldhaber, Gerald M. |

**FIGURE 6.6** Part of a Catalog Card Classified by Subject

| | |
|---|---|
| HM131<br>G56 | **Communication**<br>Goldhaber, Gerald M.<br>Organizational communication |

In your search, first decide whether you are seeking a specific title, an author, or material about a subject. Then check the arrangement of the card catalog to determine where to look. (In place of index cards, more and more libraries are computerizing their card catalogs. Ask your librarian.)

If you know neither authors nor titles of works on your subject, use the subject listing. To identify other subject headings under which you might find useful material on your topic, consult large volumes titled the *Library of Congress Subject Headings* (usually kept in the card catalog area). In your search for material on organizational communication, for example, you might scan the *Subject Headings* volume that has listings for *Communication;* among other entries, you would see this one:

**Communication in Management**
    Automatic data collection systems
    Business report writing
    Communication in library administration
    Communication in personnel management
    Management – Communication systems
    Proposal writing in business
    Television in management
    Administrative communication
    Business communication
    Communication, Administrative
    Communication, Business
    Communication, Industrial
    Communication in industry
    Industrial communication
    Communication in organizations
    Industrial management
    Personnel management

You now have an array of subjects under which to search for useful material in the card catalog.

## Guides to Literature

If you simply don't know which books, journals, indexes, and reference works are available for your topic, consult a guide to literature. For a general list of books in various disciplines, see Walford's *Guide to Reference Material* or Sheehy's *Guide to Reference Books.*

To see listings for sources in scientific and technical literature, consult Malinowsky and Richardson's *Science and Engineering Literature: A Guide to Reference Sources* or White's *Sources of Information in the Social Sciences.* For sources in specific disciplines, consult specialized guides such as *Using the Chemical Literature: A Practical Guide* or the *Encyclopedia of Business Information Sources.* Ask your librarian about literature guides for your discipline.

## Reference Works

Reference works include bibliographies, encyclopedias, dictionaries, handbooks, almanacs and directories, as shown in Figure 6.7. These can be a good starting point because they provide background and can lead to more specific information. The one major drawback to reference books is that some may be out of date.

Reference works usually are found in a special reference section of the library. All reference works will be indexed in the *Subject* card catalog, with a "Ref." designation just above the call number. On the following page is a partial list of reference works.

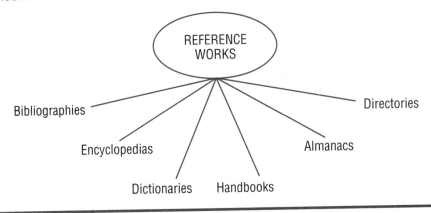

**FIGURE 6.7** Common Reference Works for Technical Disciplines

**Bibliographies.** Bibliographies are lists of publications about a subject, within specified dates. Although they provide a comprehensive view of the major sources for a subject, bibliographies quickly become dated. Ask a librarian for the most current bibliography on your subject.

Annotated bibliographies (which include an abstract for each entry) are most helpful because they can help you identify the sources most useful to you. A sample listing of bibliographies (shown with annotations):

*Bibliographic Index.* A bibliography of bibliographies; to see which bibliographies are published in your field, begin here.

*A Guide to U.S. Government Scientific and Technical Resources.* A list of everything published in these broad fields by the government.

*Bibliographic Guide to Business and Economics.* A list of all major business and economic publications.

*Bibliography of Plant Viruses.* One of many bibliographies focused on a highly specific subject.

Shorter, more specific bibliographies are found at the back of books and journal articles. To locate bibliographies that are whole books in themselves, look in the card catalog under "Bibliography" as a subject or title heading. In your research on organizational communication, for example, you might consult the *Bibliographic Guide to Business and Economics.*

**Encyclopedias.** Use encyclopedias to quickly find background on general or specific subjects. Sample listings:

*Encyclopaedia Britannica*
*Encyclopedia of Building and Construction Terms*
*Encyclopedia of Banking and Finance*
*Encyclopedia of Food Technology*

Journals, newsletters, and other publications from professional organizations (such as the American Medical Association or the Institute of Electrical and Electronics Engineers) are a valuable source of specialized information. The *Encyclopedia of Associations* offers a yearly listing of societies and organizations worldwide that range from agricultural to scientific and technical.

**Dictionaries.** Besides carrying general definitions, dictionaries can be focused on specific disciplines or they can give biographical information. Sample listings:

> *Webster's Third New International Dictionary of the English Language.* Considered the best general dictionary.
> *Dictionary of Engineering and Technology*
> *Dictionary of Telecommunications*
> *Dictionary of Scientific Biography*

**Handbooks.** Handbooks amass significant facts (including formulas, tables, advice, and examples) about a field in condensed form. Often aimed at users experienced in the field, some handbooks may not be useful to newcomers. Sample listings:

> *Business Writer's Handbook*
> *Civil Engineering Handbook*
> *The McGraw-Hill Computer Handbook*

**Almanacs.** Ranging from general to specific, almanacs have factual and statistical data. Sample listings:

> *World Almanac and Book of Facts*
> *Almanac for Computers*
> *Almanac of Business and Industrial Financial Ratios*

**Directories.** Directories offer information about organizations, companies, people, products, services, or statistics, often including addresses and phone numbers. This material usually is updated annually. Sample listings:

> *Directory of Computer Software*
> *Standard & Poor's Register of Corporations, Directors, and Executives*
> *Directory of New England Manufacturers*
> *Directory of American Firms Operating in Foreign Countries*
> *Directory of International Statistics*

A growing number of directories, such as the *Construction Directory,* are accessible by computer.

Reference books can be found for every discipline. In researching organizational communication practices, you might start with titles such as the *Encyclopedia of Management* or the *Business Writer's Handbook.* Look for your subject in the card catalog, then check for any subheadings such as "Handbooks," "Manuals," or "Dictionaries." These often will be the books that get you started.

**FIGURE 6.8** Useful Indexes for Technical Disciplines

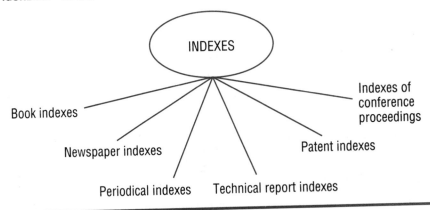

## Indexes

Indexes are lists of books, newspaper articles, or journal articles on general or particular subjects, as shown in Figure 6.8. They are excellent sources for the most current information. Because different indexes list sources in different ways, always read the introductory pages for advice on locating specific items. Or ask a librarian for help.

**Book Indexes.** All books currently being published (up to a set date) are listed in book indexes by author, title, or subject. Sample indexes (shown with annotations):

> *Books in Print.* An annual listing of all books published in the United States.
> *Cumulative Book Index.* A monthly worldwide listing of books written in English.
> *Forthcoming Books.* A listing every two months of U.S. books about to be published.
> *Scientific and Technical Books and Serials in Print.* An annual listing of literature in science and technology.
> *New Technical Books: A Selective List with Descriptive Annotations.* Issued 10 times yearly.
> *Technical Book Review Index.* A monthly listing (with excerpts) of book reviews.

In your research on organizational communication, you might check the current issue of *Business Books in Print* or *Scientific and Technical Books and Serials in Print.* But keep in mind that most books are not current enough to have the latest information.

**Newspaper Indexes.** Most newspaper indexes list articles by subject. The *New York Times Index* is best known, but other major newspapers have their own indexes. Sample titles:

*Boston Globe Index*
*Christian Science Monitor Index*
*Wall Street Journal Index*

For your research in organizational communication, you might check recent editions of the *Wall Street Journal Index* for relevant articles.

**Periodical Indexes.** For the most recent information appearing in magazines and journals, consult the periodical indexes. A broad array of indexes is needed to list the countless articles published. To find the indexes most useful for your research, first decide whether you seek general or specialized information.

One of the most general indexes is the *Magazine Index,* a subject index (on microfilm) of 400 general periodicals. A popular index is the *Readers' Guide to Periodical Literature,* listing articles from 150 popular magazines and journals. Because the *Readers' Guide* is updated every few weeks, you can locate the most current material on many topics. In your research on organizational communication, you would find these entries in the January 1987 *Readers' Guide* (the checked entry might be useful for this topic):

**Communication**
     *See also*
    Communications satellites
    Interstellar communication
    Telecommunication
**Communication, Chemical** *See* Pheromones
**Communication, Fiber optic** *See* Fiber optics
**Communication (Theology)**
    *See also*
    Television broadcasting—Religious programs
**Communication in management**
    *See also*
    Telecommunication in business
✓  Communication specialists and corporate challenges [address, September 8, 1986] W. E. Jurma. *Vital Speeches Day* 53:104–6 D 1 '86
    Listen to your whistleblower [corporate ombudsmen] M. Brody. il *Fortune* 114:77–8 N 24 '86
    The power of questions. M. Grassell. il *Nations Bus* 74:56+ N '86

As in many indexes, subjects and their subheads in the *Readers' Guide* are listed alphabetically. Each article is listed in this order: article title, author (if known), periodical title, volume and page numbers, and date. Codes for abbreviated journal titles are in the front pages of each volume. General indexes such as the *Readers' Guide* list only nonspecialized sources of information about your topic.

For specialized information and research data, consult the works that list journal articles in specific disciplines, such as *Ulrich's International Periodicals Directory.* Another comprehensive source of specialized information, the

*Applied Science and Technology Index* carries a monthly listing, by subject, of articles in more than 200 journals from many scientific and technical fields. In research on organizational communication, you would find these entries in the August 1986 issue of the *AS&T Index* (checked entries seem to be titles useful for your topic):

**Communication in management**
    *See also*
    Factory communication
    CIM leading to changes for workers. *Mod Mater Handl* 41:13+ Je '86
✓    Communication barriers direct link to motivation. L. M. Hanifin. *Am City Cty* 101:38 F '86
    Computer conferencing; an alternative communications paradigm. W. H. Landgraf. diags *ASTM Stand News* 14:36–9 Ap '86
✓    Effective communications leads to smoother operations [fire departments] G. Carlson. *Fire Eng* 139:8 Je '86
✓    Productivity improvements begin with communication. W. Mutch. *Plast World* 44:55–7 Je '86

For business articles, consult the *Business Periodicals Index,* with its monthly subject listing of articles and book reviews from 270 business periodicals. In researching organizational communication, you would see these entries in the August 1985–July 1986 *BP Index* (checked entries seem useful):

**Business communication**
    *See also*
    Annual reports
    Boundary spanning activity
    Commercial correspondence
    Employee communication
✓    ABC = auditing business communications. M. E. Campbell and R. W. Hollmann. *Bus Horiz* 28:60–4 S-O '85
    Before you say a word. *Business* 35:54–5 O-N-D '85
    Communication pays off. J. W. Farrell. *Traffic Manage* 24:33+ N '85
✓    Communications is everybody's business [chemical industry] P. P. McCurdy. *Chem Week* 136:3 My 15 '85
    Communications: zooming in on decisions. L. Kuzela. *Ind Week* 226:46–8 S 2 '85
✓    Did you say what you said or did you say what I think you said? T. D. Lewis. *Manage Account* 67:40–1+ Ag '85
✓    Don't bother trying without communication skills [data processing professionals] H. M. Weiss. *Data Manage* 23:16–17 My '85

Other broad indexes that cover specialized fields in general include the *Business Index* (on microfilm), the *General Science Index*, the *Science Citation Index,* and its counterpart, the *Social Sciences Citation Index.* The *Statistical Reference Index* lists statistical works not published by the government.

Along with these broad indexes, some disciplines have their own specific indexes. Sample listings:

*Agricultural Index*
*Education Index*
*Energy Index*
*F&S Index of Corporations and Industries*
*Environment Index*
*Index to Legal Periodicals*
*International Nursing Index*

Ask your librarian about the best indexes for your topic, and about those (such as the *Science Citation Index*) which can be searched by computer.

**Technical Report Indexes.**  Countless government and private-sector reports written worldwide offer a continuing source of specialized and highly current information. Sample indexes for these reports:

*Scientific and Technical Aerospace Reports*
*Government Reports Announcements and Index*
*Monthly Catalog of United States Government Publications*

U.S. government indexes are discussed on page 109.

**Patent Indexes[1].**  Over 75,000 patents yearly are issued in the United States for new inventions, products, or processes. These patents are just a fraction of the roughly one-half million issued worldwide. As information specialists Schenk and Webster explain, patents are an excellent and often overlooked source of current information: "Since it is necessary that complete descriptions of the invention be included in patent applications, one can assume that almost everything that is new and original in technology can be found in patents." Sample indexes:

*Index of Patents Issued from the United States Patent and Trademark Office*
*NASA Patent Abstracts Bibliography*
*World Patents Index*

On-line information about patents in fiber optics, lasers, or other technologies can be obtained through databases such as Hi Tech Patents, Data Communications, and through WPI (World Patents Index).

**Indexes of Conference Proceedings.**  Schenk and Webster point out that many of the papers presented at the more than 10,000 yearly professional conferences are indexed in printed or computerized listings such as these:

[1]Adapted from Margaret T. Schenk and James K. Webster. *Engineering Information Resources.* New York: Decker, 1984. Consult their work for detailed treatment of patent information and its sources, and for invaluable discussions of information sources in general.

*Proceedings in Print*
*Index to Scientific and Technical Proceedings*
*Engineering Meetings* (an Engineering Index database)

The very latest advances in a field sometimes are unveiled during such proceedings.

## Abstracts

Beyond listing articles found in periodicals, abstracts briefly describe each article. A bound collection of abstracts can be harder to use than an index (the abstract's front pages contain instructions). But the abstract of an article, once located, can save you from going all the way to the journal; by describing the article, the abstract can help you decide whether to read the article or to skip it.

Abstracts usually are titled by discipline. A sample list:

*Biological Abstracts*
*Computer Abstracts*
*Engineering Index*
*Environment Abstracts*
*Excerpta Medica*
*Forestry Abstracts*
*International Aerospace Abstracts*
*Metals Abstracts*

In researching organizational communication, you would see this abstract in the 1985 *Engineering Index Annual,* under the subject heading "Information Dissemination":

**056296 (MIS)MANAGEMENT OF COMMUNICATIONS.** This paper discusses the role of communication systems in management. Decision making and information collection are cited as two categories of activities contributing to management efficiency. Common abuses of communications include misappropriation of information and imprecise information which can be misinterpreted.
Goshen, Charles E. (Vanderbilt Univ., Columbia, MD, USA). *Eng Manage Int* v 3 n 2 Feb 1985 p 139–145.

An increasing number of abstracts (such as *Energy Research Abstracts* and *Pollution Abstracts*) can be searched by computer. Check with your librarian.

For some current research, you might consult abstracts of doctoral dissertations in *Dissertation Abstracts International.*

## Locating the Indexed or Abstracted Article

To find the periodicals containing the articles you have selected, check in the card catalog for the journal titles. Or ask at the reference desk for the list of periodicals (also called *serials*) held by your library. If your library does not hold the article you need, use the OCLC terminal (pages 111–112) to identify a holding library, and request the article through interlibrary loan. Your librarian will be glad to help.

## The Reference Librarian

Your best bet for help with library research is the reference librarian, who specializes in *referring* people to sources of information. Don't hesitate to ask for assistance; even professional researchers sometimes need help. The reference librarian can save you hours of wasted time finding whatever you need, or even save you from choosing the wrong topic.

## Access Tools for United States Government Publications

The federal government publishes maps, periodicals, books, pamphlets, manuals, monographs, annual reports, research reports, and a bewildering array of other information. Kinds of information available to the public include presidential proclamations, congressional bills and reports, judiciary rulings, some reports from the Central Intelligence Agency, and publications from all other government agencies (Departments of Agriculture, Commerce, Transportation, and so on). A few of the countless titles available in this gold mine of information:

> *Effects of New York's Fiscal Crisis on Small Business*
> *Economic Report of the President*
> *Major Oil and Gas Fields of the Free World*
> *Decisions of the Federal Trade Commission*
> *Journal of Research of the National Bureau of Standards*
> *Siting Small Wind Turbines*

Much of this information can be searched on-line as well as in printed volumes. Your best bet for tapping this valuable but complex resource is to ask the librarian in charge of government documents for help. If your library does not own the publication you seek, it can be obtained through an interlibrary loan.

Here are the basic access tools for documents issued or published at government expense as well as for many privately sponsored documents.

- *The Monthly Catalog of the United States Government,* the major access to government publications and reports, is indexed by author, subject, and title. These indexes provide you with the catalog entry number that leads you, in turn, to a complete citation for a work.

- *Government Reports Announcements & Index* is a listing published every two weeks by the National Technical Information Service (NTIS),[2] a federal clearinghouse for scientific and tchnical information—all stored in a computer database. The collection has summaries of over 900,000 federally sponsored research reports published and patents issued since 1964.

[2] A branch of the U.S. Department of Commerce.

About 70,000 new summaries are added annually in 22 subject categories, from aeronautics to medicine and biology. Full copies of reports are available from NTIS.

- *The American Statistics Index,* a yearly guide to statistical publications by the U.S. government, is divided into two sections: *Index* and *Abstracts* (an index with summaries). The *Index* volume lists material by subject and provides geographic (U.S., state, and so on), economic (income, occupation, and so on), and demographic (sex, marital status, race, and so on) breakdowns. The *Index* volume refers you to a number and entry in the *Abstracts* volume.

In addition to these three basic access tools, the government issues *Selected Government Publications,* a monthly list of 150 titles (with descriptive abstracts). These titles range from highly general *(Questions About the Oceans)* to highly technical *(An Emission-Line Survey of the Milky Way).*

The government also publishes bibliographies on hundreds of subjects, from "Accidents and Accident Prevention" to "Home Gardening of Fruits and Vegetables." Ask your librarian for information about these subject bibliographies.

## USING ELECTRONIC INFORMATION SERVICES

The Information Age has created the need for modern cataloging and research methods. In the United States alone, 35,000 to 45,000 books are published *each year.* Add to this number foreign publications and countless special-interest magazines, journals, newspapers, films, recordings, and reports.

To help us quickly find what we need, more and more libraries offer computerized information services. If your library has an electronic card catalog, you can rapidly search the holdings to find specific works. Electronic retrieval services provide access to information far beyond everything held by your library. Much of today's information can be retrieved electronically from compact disks or from databases in mainframe computers as shown in Figure 6.9. Electronic data retrieval is revolutionizing our way of doing research.

**FIGURE 6.9** Ways of Searching the Literature Electronically

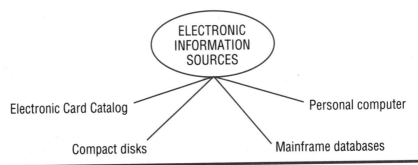

## Databases on Compact Disks

A single, laser-driven compact disk, smaller than a 45-rpm record, easily stores an entire encyclopedia. Using a computer, you can access even the smallest portion of that encyclopedia in just a few seconds. Instead of thumbing through printed reference books and indexes, you can now locate more and more basic resources on compact disks. The growing array of disk-based reference materials now includes electronic versions of *Books in Print* and *Ulrich's International Guide to Periodicals.*

One useful index stored on compact disk is InfoTrac™, an index to articles from 900 business, technical, and popular magazines and journals. InfoTrac is updated monthly. As in printed indexes, its entries are arranged by subject headings and subheadings. If, for instance, you type in the subject heading "organizational communication," the system will list dozens of entries. With a simple command, you can print out your own hard copy of the list.

Other databases on compact disk include the *Government Publications Index*™, a monthly listing of U.S. government publications, and LegalTrac™, a monthly listing of entries from 750 legal publications. Ask if your library subscribes to any of these disk-based indexing services.

## Mainframe Databases

Many college libraries now subscribe to retrieval services that offer public access to more than 3,000 individual databases stored on mainframe computers. At a terminal in a subscribing library, you can access information from countless indexes, journals, books, monographs, dissertations, and reports. Compared with databases on disks, mainframe databases usually include more specialized sources, and they can offer the most current entries—sometimes updated daily.

Mainframe retrieval services can offer either *bibliographic* or *full-text* databases. In a bibliographic database, entries are arranged according to different fields. Titles and abstracts help you identify the works that seem most useful. In a full-text database, the entire article or document can be displayed directly on your computer screen.

Three popular services with mainframe databases are discussed below. The first service (OCLC) helps you locate titles you have already identified as useful. The two subsequent services (Dialog and BRS) help you identify useful titles.

**The On-Line Computer Library Center (OCLC).** You can speed up your research tremendously if a nearby library belongs to the On-Line Computer Library Center. The OCLC database, in Columbus, Ohio, has eight million records with the same information found in a conventional card catalog. Using the library's computer terminal, you simply key in author, title, or subject. Within seconds, you get a listing of the publication you seek and information about where to find it—all at no charge. If your library doesn't have the publication, your librarian can activate the Interlibrary Loan System (ILS). The system forwards requests to libraries holding the material. Once a lender indicates (via its terminal) that it will supply the material, the system stops forwarding the request, and notifies your librarian that the request has been filled. Your order will arrive at the library by mail in a week or two.

**Dialog.**   Some libraries also belong to Dialog, a comprehensive technical database in Palo Alto, California. Through Dialog, you can get information from many fields by typing in key terms that enable the computer to scan bibliography lists for titles containing those terms. Say you need information on possible *health hazards from video display terminals* (i.e., monitors). You instruct the computer to search Medline, a medical database (one of 150 Dialog databases in science, technology, medicine, business, and so on), for titles including the words italicized above (or synonymous words, such as *risk, danger, cathode-ray tubes*). The system would provide you with full bibliographies and abstracts of the most recent medical articles on your topic.

Dialog gives bibliographic information on published works, dissertations, and conference papers, and it can also provide financial and product information about companies, names and addresses of company officers, statistical data and patent information. Here are just a few of Dialog's databases:

*Career Placement Registry*
*Claims/U.S. Patents*
*Conference Papers Index*
*Electronic Yellow Pages* (for Retailers, Services, Manufacturers)
*Environmental Bibliography*
*Foreign Traders Index*
*International Software Database*
*Oceanic Abstracts*
*U.S. Exports*
*Water Resources Abstracts*

Because of its expense ($150 an hour or more for many of its databases), Dialog is not yet widely used by college libraries. But many major companies who have their own management information departments with full-time database researchers also subscribe to Dialog.

**BRS.**   Bibliographic Retrieval Services (BRS) is another popular database providing bibliographies and abstracts from life sciences, physical sciences, business, or social sciences. These are a few from the more than fifty BRS databases:

*American Chemical Society Journals*
*Dissertation Abstracts International*
*Government Reports Announcements & Index*
*Harvard Business Review*
*International Pharmaceutical Abstracts*
*Military and Federal Specifications and Standards*

*Monthly Catalog of United States Government Publications*
*PATDATA* (U.S. Patents)
*Pollution Abstracts*
*Robotics Information Database*

Many college libraries now offer BRS service.

**A Sample Automated Search.** Assume that you are continuing research on organizational communication. You have done a brief search through the "manual" indexes, and for a comprehensive view you decide on an automated search, using your library's BRS service. You ask a librarian for help, and the two of you sit at the terminal and begin.

After logging into the BRS system, you instruct the computer to search all the databases under the group heading *Business,* using the key words *writing skills, improvement, results, methods.* The computer responds with a listing of each database and the number of articles in that database. The *Writing Skills* database lists 195 articles. You are most interested in the effects of good business writing, and you narrow the search to *Methods,* instructing the computer to list the titles. Here is the partial list:

```
TI¹ Secrets of Successful Writing, Speaking, Listening
TI² Written Communications and In-Person Contacts
TI³ Inflation Hurts Business Language, Too
```

The title that seems most relevant is number 3. You instruct the computer to print the full bibliographic information (including the abstract) on this article. The computer responds:

| | | |
|---|---|---|
| **Accession number** | AN | SPM82H008 |
| **Author** | AU | Bacon, M.S. |
| **Title** | TI | Inflation Hurts Business Language, Too |
| **Source** | SO | Supervisory Management, Vol. 27, No. 8, Aug. 1982, pp. 8-15. |
| **Descriptors** | DE | Report Writing, Business, Writing Skills; Improvements; Communication Methods |
| **Abstract** | AB | Writing is an excellent means of communicating ideas in business. Business writing has become unbearably long, a waste of time, money, and copier machines. Writing |

```
                    skills need to be improved to
                    help organizations save
                    money, become more efficient,
                    and improve rapport between
                    employees and customers.
                    Ways to improve business
                    report writing skills are
                    presented.
Publication    PT   Journal
      type
                    END OF DOCUMENT
```

After reading the abstract, you decide it would be worthwhile to read the entire article, and so you make a note to check your library's holdings or to order a reprint or photocopy through interlibrary loan. (Some reprints can be ordered right through the computer terminal.) You turn again to the list of titles to see if any others seem promising.

## Advantages and Disadvantages of Automated Searches

On-line searches have several advantages over manual searches (that is, flipping pages by hand).

- They are rapid: you can review ten or fifteen years of an index in minutes.
- They are detailed: beyond listing titles and sources, an automated search often provides abstracts, for a fuller view of the material.
- They are current: the index usually comes on line about six weeks before the printed copies. This feature is especially useful in science and technology, in which developments are so rapid.
- They are thorough: because the system can search not only for titles but for key words (or word combinations) found in the title *or* the abstract, your chance of discovering useful material is improved.
- They are efficient: you can extract from the database just the information you need (as shown in the sample search above), without wasted effort tracking material that turns out to be of no use.

But automated searches have limitations as well. Most computerized bibliographies include no entries before the mid-1960s; for information published earlier, a manual search is necessary. Also, a manual search provides you with the whole "database" (the bound index or abstracts). As you flip pages and browse, you often *randomly* discover something useful. This kind of randomness, of course, is impossible with an automated search, which can give you the illusion of having surveyed all that is known on your topic. Finally, automated searches can be expensive, depending on how many databases you search and how much time you spend on-line. (The average cost of a BRS search is about $30.) Some schools offer students one free search, but if your school doesn't, you will have to pay the cost.

For any automated search, keep in mind that a manual (random) search is almost always needed as well. And a good automated search calls for a preliminary conference with a trained librarian.

### Personal Information Retrieval Services

On a personal computer, you can do some research at home in source databases. For a small fee, you can join *Compuserve Information Service* or *The Source Telecommunications Service,* and gain access to energy news, stock-market quotes, corporate news releases, new-product news, business reports, medical news, and many other reference sources.

Computerized information retrieval will be indispensable in future research. Ask your librarian for information about commercial databases in your field.[3]

---

## INTERVIEWING

Although libraries are a good secondary research source, you may also have to consult firsthand, or primary, sources as shown in Figure 6.10. One excellent primary source can be the personal interview. The following suggestions should help you conduct an interview.

**FIGURE 6.10**  Sources for Firsthand Research

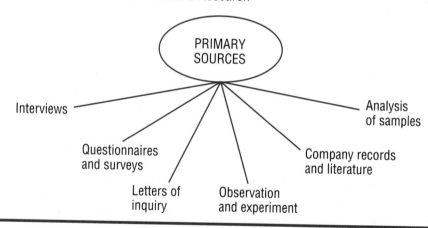

---

[3]For a major source of information on databases in various disciplines, see these works: Martha E. Williams et al., eds., *Computer-Readable Data Bases: A Directory and Data Sourcebook.* Urbana, Ill.: American Society for Information Science; *The Database Book,* a directory of the world's databases, published by On-line, Inc; The *Data Base Directory,* published by Knowledge Industry; *Directory of On-line Databases* Santa Monica, Calif.: Cuadra Associates, Inc., published quarterly since 1979.

### Identify Your Purpose

Know *exactly* what you are looking for. Suppose, for example, that you are thinking of opening a microcomputer retail store. Determine the information you hope to obtain:

> The purpose of my interview with Abner Jones, city clerk, is to learn how licensing requirements, fees, and zoning restrictions would affect the operation of my proposed microcomputer retail store. Also, I need to know how to meet the legal requirements for this business.

When you are sure enough about your purpose that you can write it clearly in one or two sentences, you can contact your respondent(s).

### Choose Your Respondent Carefully

Don't waste time interviewing the wrong people. For your microcomputer store feasibility study you might select these respondents: the head of your chamber of commerce, representatives for the brands of machine you expect to sell, a member of your local licensing board, the loan officer at a local bank. Any of these respondents might give helpful leads for interviews with other essential people.

One way to ensure that you interview the appropriate person is to state the purpose of your interview when you call for an appointment:

> I am studying the possibility of opening a microcomputer retail store in Winesburg, and I would appreciate the opportunity to meet with you, at your convenience, because I feel that you could answer some important questions about the future of such a business in our town.

If you communicate your purpose, your intended respondent can tell you whether he or she knows enough to make an interview worthwhile. If you are planning several interviews, begin with the least crucial so that you can sharpen your skills for the later ones.

### Prepare Well

After identifying your purpose and choosing a respondent, you have to plan questions that will elicit the information you need. To ask useful questions, you have to know your subject! Complete your background investigation *before* the interview, and when your homework is done, write out your questions, *on paper*. Too many beginners set out for an interview armed only with a few blank notecards, a pencil, and a confident feeling that "the questions are all in my head." This mistake leads to embarrassing silences. Don't expect your respondent to keep the ball rolling.

Write each question on a separate notecard. You then can summarize a response on each card.

Make your questions specific and direct. For a computer wholesale representative, you might phrase a question in one of two ways:

1. Does your company have a provision for buying back unsold stock from the retailer?
2. What provisions, if any, does your company have for buying back unsold stock from the retailer?

Version 1 can yield a simple yes or no answer, whereas version 2 requires a detailed explanation of the yes answer. Likewise, a question from a problem-solving interview can be worded in one of two ways:

1. Do you think this problem can be solved?
2. What steps would you recommend for solving this problem?

Version 2 again calls for specific information. Any answer you receive will be only as good as the question you ask.

Phrase your questions so as not to influence the answers:

1. Don't you think this is a good plan?
2. What do you think are the principal (faults, merits) of this plan and how should they affect my success?

Version 1 signals your respondent that he or she should agree with you, whereas version 2 encourages a candid response.

## Maintain Control

You—not your respondent—are responsible for conducting a competent and productive interview. These suggestions should help:

1. Dress neatly and arrive on time.
2. Begin by thanking your respondent, in advance.
3. Restate the purpose of your interview.
4. Tell your respondent why you feel he or she can be helpful.
5. Discuss your plans for using the information.
6. Ask (preferably before you arrive) if your respondent objects to being quoted or taped. (Although taping is the most accurate means for recording responses, it makes some people uncomfortable.)[4]

---

[4] In an interview about your prospective business, it would be inappropriate and unnecessary to tape the responses of the bank representative, sales representative, or licensing official. You might, however, wish to tape an interview with a member of the Small Business Advisory Service who is giving detailed advice about how to proceed with your plan.

7. Avoid small talk.

8. Ask your questions clearly and directly, following the order in which you prepared them.

9. Be assertive but courteous. Ask pointed questions, but remember that the respondent is doing you a favor.

10. Let your respondent do most of the talking. Keep your opinions to yourself.

11. Guide the interview. If your respondent begins to wander, politely bring the conversation back on track (unless the additional information is useful).

12. Be a good listener. Don't stare out the window, doodle, or ogle office staff while your respondent is speaking.

13. Be prepared to explore new areas of questioning, if needed. Sometimes a respondent's answers will uncover new directions for the interview.

14. Keep your note taking to a minimum. Record all numbers, statistics, dates, names, and other precise data, but don't transcribe responses word for word, which would annoy the respondent. Simply record significant words and phrases that later can refresh your memory.

15. If interviewing several people about the same issue, standardize your questions. Ask each respondent the same questions in the same order and in the same phrasing. Avoid random comments that could affect one person's responses.

16. Don't hesitate to ask for clarification or further explanation.

17. When your questions have all been answered, ask for any additional comments.

18. Offer to send a copy of the document in which this information will be used.

19. Finally, thank your respondent, and leave promptly.

## Record Responses

As soon as you leave the interview, find a quiet place to write out a summary of the responses. Don't count on your memory. Get material on paper while the responses are fresh in your mind.

If all or part of the interview is to be published, be sure to ask respondents to review the final draft before you quote them in print.

Here is the partial text of an interview on the survival problems of television service businesses. Notice how well-planned questions generate informative responses that in turn lead to more specific questions.

## An Informative Interview

(This interview was with Al Jones, the owner of Al's TV Shop.)

*Q.* I see you have an excellent location. Is location of a service shop important?

*A.* Sure. If I ever get back to selling, it will be very important. Doing just service minimizes the importance because the customer will search you out if he really wants you. I suppose the same holds true if I start selling.

*Q.* Well, you seem to be selling car radios and stereos. Isn't that "selling"?

*A.* Yes. But by selling I mean TVs. Floor plans, etc.

*Q.* Why don't you sell now? What's a floor plan?

*A.* The floor plan is why I won't sell now. I used to sell with a floor plan. A floor plan means a television distributor will sell me, say, $10,000 worth of sets for only 10 percent down payment. I pay the balance as I sell the sets. But the temptation to use the distributor's money is overwhelming. All this money in the checking account gives a false sense of security.

*Q.* In other words, you've had a bad experience?

*A.* Not only me. Most television businesses fail because they start using someone else's money. Where can a small business come up with six, eight, and even ten thousand dollars after all the stock has been sold?

*Q.* And so you just sell car radios and stereos and parts.

*A.* That's right. I pay for them as I use them and stay out of trouble. I realize it's better to have thirty-day charge accounts and use the distributor's money to capitalize my business, but this can also cause trouble.

*Q.* Okay. What about advertising?

*A.* I use the yellow pages. That's all I can afford. Sometimes the local radio stations offer attractive deals that cost me half the usual rate. I take advantage of these. Radio is good for the stereo radio sales.

*Q.* Wouldn't it be a good idea for a group of you technicians to get together and do some group advertising?

*A.* They have no time.

*Q.* Tell me something about your personnel situation.

*A.* Okay. Here's something. My TV man came up to me a few days ago and told me he was giving me thirty days' notice. He wants to go into business for himself. It's always like this. You hire and train a technician for years, and when he gets the itch, off he goes.

*Q.* Can you think of any mistakes you made with personnel?

*A.* I don't think I'll ever hire a beginner any more. They just aren't productive enough. From now on they must have at least formal training and the technician's license.

*Q.* What do you mean by productive?

*A.* I have to get so much production for every dollar I pay a technician. Unfortunately it has taken me many years to come to a good interpretation of what a reasonable return is. The minimum return I must have is $3.00 for every wage dollar.

---

## USING QUESTIONNAIRES AND SURVEYS

An interview has some advantages over a questionnaire: face-to-face, you can tell whether your respondent understands your questions, and you are more likely to get answers. Also, you may get unexpected information. In an interview, however, the respondent may feel threatened.

A questionnaire, on the other hand, saves time and is an inexpensive way of surveying a large cross-section. Respondents can answer privately and anonymously—thus more candidly. And they have plenty of time to think about their answers.

These advantages, however, can be offset by the problem of getting people to respond to a questionnaire. Whether you mail or otherwise distribute your questionnaire, expect less than a 30 percent response, perhaps far less—and allow plenty of time for responses to be returned.

You can increase your chances for a good response by following these suggestions:

1. Introduce your questionnaire, so that respondents will appreciate the significance and purpose of their answers. Also, thank the respondent.

   > Your answers to these questions about your views on proposed state handgun legislation will be appreciated. Your state representative will tabulate all responses so that she may continue to speak accurately for *your* views in legislative session. Thank you.

   Some writers include a cover letter with a questionnaire (as shown on pages 121–122).

2. Keep it short. Try to limit both questions and response space to two sides of a page. Also, phrase questions concisely.

3. Place easiest or more interesting questions first, to attract attention.

4. Make each question precise and unambiguous. *This is crucial.* If readers can interpret a question in more than one way, their responses are useless—or worse, misleading. Consider this ambiguous question: "Do you favor foreign aid?" *Foreign aid* can mean military, economic, or humanitarian aid, or all three. Some people might support one form or another, but not all three. Rephrase the question to allow for these differences:

Do you favor:
   a. Our current foreign aid program of both military and humanitarian assistance to other countries?
   b. Greater military assistance in our foreign aid program?
   c. Greater economic and humanitarian assistance in foreign aid programs?
   d. No foreign aid program at all?

5. Avoid influencing your reader's responses with leading questions:

   Do you think foreign aid is a waste of taxpayers' money?
   Do you think foreign aid is a noble gesture by a wealthy nation toward the less fortunate?

   Notice that each of these questions is phrased to elicit a definite response.

6. When possible, phrase questions so that you can easily classify and tabulate responses: yes-or-no; multiple-choice; true-false; fill-in-the-blank.

7. Include an "Additional Comments" section at the end.

8. Enclose a stamped, self-addressed envelope.

9. If possible, do the questioning by telephone to save time and ensure a response.

The questionnaire on pages 122–123, sent to presidents of 612 companies, is prefaced by the following cover letter explaining the questionnaire's purpose. The clearer the rationale for a questionnaire, the more inclined readers will be to respond.

_____

**Questionnaire Cover Letter**
As part of my course work, I am preparing a report on Southeastern Massachusetts University's professional communication program. This survey will help determine how SMU's communication program can better serve students and the community.

I plan to assess services and needs by studying

   1. the communication needs of local companies and industries
   2. the feasibility of on-campus and in-house seminars in communication
   3. the feasibility of offering a broader variety of professional communication courses at SMU

Please take a few minutes to respond to this survey. *Your response is important* because it provides the data needed for my study. Survey results, conclusions, and recommendations will be published in the fall issue of *The Business and Industry Newsletter*. (I will gladly send a copy.)

One last favor: Kindly mail the survey to
>Lynne Taylor
>House 10, SMU Dorms
>North Dartmouth, MA 02747

I had hoped to include an addressed, stamped envelope for your convenience, but my budget for this project has been depleted by printing and mailing costs.

Sincerely,

Lynne Taylor

———

The following questionnaire is designed to elicit specific responses that can then be easily classified and tabulated.

———

**Communication Questionnaire for Executives**

1. Describe your type of company (manufacturing, high tech, banking, and so on).

   _____

2. How many people do you employ? (Please check one.)
   _____ 5–25　　　_____ 100–150
   _____ 25–50　　　_____ 150–300
   _____ 50–100　　_____ 300–450

3. How do you usually communicate with employees?
   _____ by memo
   _____ in person
   _____ through supervisors and managers

4. Which types of writing are done in your company? (Label by frequency: never, rarely, sometimes, often, most often.)
   _____ office memos　　　_____ catalogs
   _____ manuals　　　　　 _____ advertisements
   _____ procedures　　　　_____ house organs
   _____ letters　　　　　　_____ other (specify)
   _____ reports　　　　　　_____

5. Who does most of the writing in your organization? (Give title.)

   _____

6. Please characterize the writing effectiveness in your organization.
   _____ good　　　_____ fair　　　_____ poor

7. Does your company have writing guidelines?
   _____ no　　　_____ yes (Please describe briefly below.)

   _____

8. Do you ever hire outside writers for specific projects?
   _____ no　　　_____ yes

9. Do you have an in-house program for teaching employees how to communicate?

_____ no      _____ yes

10. Rank the usefulness of the following topics in a writing program (with 1 as most important through 9 as least important).

_____ organizing information      _____ audience awareness
_____ summarizing information     _____ grammar
_____ precise phrasing            _____ sales writing
_____ formatting reports          _____ other (specify)
_____ editing for style           _____

11. Rate these skills in order of importance (1 being most important).

_____ reading      _____ speaking to groups
_____ writing      _____ speaking face to face
_____ listening    _____ dictating

12. Do you provide tuition reimbursement for employees?

_____ no      _____ yes

13. Would you participate in a contract learning program in which SMU writing students would work for you?

_____ no      _____ yes

14. Should SMU offer Saturday seminars in Professional Communication?

_____ no      _____ yes

15. Which courses, beyond introductory professional writing courses, should SMU offer? (Rank in numerical order of preference.)

_____ Report Writing      _____ Documentation
_____ Proposal Writing    _____ Sales Writing
_____ Procedure Writing   _____ other (specify)
_____ Public Speaking     _____

Additional comments:_____
_____

_____

*Caution:* Interpret statistical results carefully. With a political questionnaire, more people with extreme opinions will respond than will people with moderate views, and so your data may not truly reflect a cross-section. Supplement questionnaire responses with other sources whenever possible.

You may wish to include the full text of the interview or questionnaire (with questions, answers, and tabulations) in an appendix (Chapter 18). You can then refer to these data in your report discussion as necessary.

## USING OTHER PRIMARY SOURCES

Whenever possible, explore all primary sources by writing letters, checking records, or observing and analyzing directly.

## Letters of Inquiry

Letters of inquiry are handy for obtaining specific information. These are discussed in Chapter 16.

## Organizational Records and Publications

Company records (reports, memos, computer printouts, and so on) are a good primary source for data. Most organizations also publish pamphlets, brochures, annual reports, or prospectuses for consumers, employees, investors, or voters. But be alert for bias in company literature. If you were evaluating the safety measures at a local nuclear power plant, you would want the complete picture. Along with the power company's literature, you would want publications by federal, state, and local environmental groups. A pamphlet can be a good beginning for your research.

## Personal Observation and Experiment

If possible, amplify and verify your research findings with a firsthand look. Conclude your computer-store feasibility study by visiting possible shop sites. Measure consumer traffic by observing each site at identical hours on typical business days.

Observation should be your final step, because you now know what to look for. Without a solid background you could waste time. A visit to the nuclear power plant will be useful if you understand how meltdown can occur or how safety systems work. Have a plan. Know what to look for, and jot down observations immediately. You might even take photos or make drawings, but don't rely on your memory.

Informed observations can pinpoint real problems. Here is an excerpt from a report investigating poor management-employee relations at an electronics firm. This researcher's observations are crucial in defining the problem:

> The survey revealed that employees were unaware of any major barriers to communication. The 75 percent of employees with positive work attitudes said they felt free to talk to their managers, but the managers, in turn, estimated that only 50 percent of employees felt free to talk to them.
>
> The problem involves misinterpretation. Because managers don't ask for complaints, employees are afraid to make them, and because employees never ask for an evaluation, they never get one. Both sides have inaccurate perceptions of what the other wants, and because of ineffective communications, each side fails to realize that its perceptions are wrong.

Effective research requires informed observation.

An experiment is a highly controlled form of observation that is designed to *verify an assumption* (e.g., the effect of fish oil in the prevention of heart disease) or to *test something untried* (the relationship between piped-in music and productivity in the workplace). Each specialty, of course, has its own guidelines for experiment design.

## Analysis of Samples

Research in the workplace often involves collecting and analyzing samples: water or soil or air, for contamination and pollution; foods, for nutritional value; ore, for mineral value; or plants, for medicinal value. Investigators analyze material samples to find the cause of an airline accident. Engineers analyze samples of steel, concrete, or other building materials to determine their load-bearing capacity. Medical specialists analyze tissue samples for disease.

Virtually any professional in the workplace routinely employs a variety of primary sources for information gathering.

## RECORDING YOUR INFORMATION

For primary research, you can record your findings by using notebooks, photographs, drawings, tape recorder, videotape, or whichever medium is best employed for your subject and purpose. For example, you might use videotape for recording the hazards on a construction site; or audiotape for recording the decibel levels in neighborhoods surrounding a major airport.

For recording secondary research, notecards work well because they are easy to organize. Put one idea on each card and give each card a code number in the upper right corner. You can pretty well figure out the organization for your report before you write it by arranging the cards in logical order on a large table.

1. Begin by making bibliography cards, listing all works you plan to consult. Using a separate notecard for each source, write the complete bibliographic information (Figure 6.11). This entry is identical to the citation in the final draft of your report.

FIGURE 6.11 A Bibliography Card

Nickerson, Robert C.
*Fundamentals of Programming in Basic*. Boston: Little, 1981.

2. Skim the entire book, chapter, article, or pamphlet.

3. Go back and decide what to record. Use one card for each item.

4. Decide how to record the item: as a quotation or a paraphrase.

In quoting, copy the statement word for word (Figure 6.12). Place quotation marks around all directly quoted material, even a phrase or a word used in a special way. Otherwise, you could forget to credit the author, and face a charge of plagiarism (borrowing someone's words or ideas—even if paraphrased—without giving credit, intentionally *or* unintentionally). If you quote only sections of a sentence or paragraph, use an ellipsis—three dots (. . .) indicating that words have been left out of a sentence. If you leave out the last part of the sentence, the first part of the next sentence, or one or more whole sentences of paragraphs, use four dots (. . . .):

If you quote only sections . . . use an ellipsis. . . . If you leave . . . .

If you insert your own comments within the quotation, place brackets around them to distinguish your words from the author's:

"This job [aircraft ground controller] requires exhaustive attention."

Use direct quotations only when precision or clarity or emphasis requires the exact words from the original. (For placing long and short quotations, see pages 514–515.)

Introduce quotes, and integrate them with your own sentences by using phrases such as "Jones argues," "Smith agrees," or "Brown suggests." Integrate quotations so that your sentences are grammatical: "The agricultural crisis," Marx acknowledges, "resulted primarily from unchecked land speculation." Or

FIGURE 6.12  Sample Notecard for a Quotation

> II-A-1
>
> *Nickerson, Robert. Fundamentals*
> *p. 34*
> *"The first step in the programming*
> *process is understanding the problem*
> *to be solved. Understanding the*
> *problem involves determining the*
> *requirements of the problem and how*
> *these requirements can be met."*

**FIGURE 6.13**  Sample Notecard for a Paraphrase

*Nickerson, Robert. Fundamentals*
*p. 35*
*A programmer's documentation describes the program, and explains how it works. Later programmers use the documentation to understand the program if they need to correct or change it.*

*II-A-3*

"She has rejuvenated the industrial economy," Smith writes of Berry's term as economic adviser.

Figure 6.13 shows a paraphrased entry. The researcher condenses this original in his own words.

> Finally the programming process is completed by bringing together all the material that describes the program. This is called *documenting* the program, and the result of this activity is the program's *documentation*. Included in the documentation is the program listing and a description of the input and output data. Documentation enables other programmers to understand how the program functions. Often it is necessary to return to the program after a time to make corrections or changes. With adequate documentation, it is much easier to understand a program's operation.[5]

Most of your notes will be paraphrased, but some notecards may have both paraphrases and direct quotations. Paraphrased material needs no quotation marks, but it *must be documented,* to acknowledge your debt to a source. By paraphrasing, you demonstrate to readers that you have clearly understood the original passage. But be careful to preserve the original meaning.

## DOCUMENTING YOUR SOURCES

Most research draws on the information and ideas of others. Credit each source of direct quotations, paraphrases, and visuals. Proper documentation satisfies the professional requirements for ethics, efficiency, and authority.

[5]This passage and those in Figures 6.12 and 6.13 are adapted from *Fundamentals of Programming in BASIC,* by Robert C. Nickerson.

Documentation is a matter of *ethics* in that the originator of an idea always deserves credit. All published material is protected by the copyright law. Failure to credit your source could make you liable to a charge of plagiarism, even if your omission was unintentional.

Documentation is also a matter of *efficiency*. It provides a network for locating the world's printed knowledge. If you cite a particular article, your reference will enable readers to locate that source easily.

Finally, documentation is a matter of *authority*. In making any claim ("A Mercedes-Benz is a better car than a Ford Taurus") you invite challenge: "Says who?" Data on road tests, frequency of repairs, resale value, workmanship, and owner comments can help validate your claim by showing its basis in *fact*. Your credibility increases in relation to the expert references supporting your claims. For a controversial subject, you may need to cite several authorities, as in this next example, instead of forcing a simplistic conclusion on your material:

> Opinion is mixed as to whether a marketable quantity of oil rests under Georges Bank. Cape Cod geologist John Blocke feels that extensive reserves are improbable ("Geologist Dampens Hopes" 3). Oil geologist Donald Marshall is uncertain about the existence of any oil in quantity under Georges Bank ("Offshore Oil Drilling" 2). But the U.S. Interior Department reports that the Atlantic continental shelf may contain 5.5 billion barrels of oil (Kemprecos 8).

Document any source from which you have quoted words or borrowed facts and ideas that cannot be regarded as common knowledge (or general information) in that field. Common knowledge about a field can be found in any number of sources. In medicine it is common knowledge that foods high in fat can cause some types of cancer. Thus, in a research report on fatty diets and cancer, you probably would not need to document that fact. But you would document the results of tests on folic acid and vitamin E as possible cancer preventives.

If your information can be found in only one specific source, and not in various general reference sources, it should be documented. When in doubt, document the source.

## Choosing a System of Citation

Documentation practices vary widely. Many disciplines, institutions, and organizations publish their own documentation manuals. They are listed in the card catalog under "Technical Writing." Your readers might request that you use one of these style manuals or some other.

*Style Guide for Chemists*
*Geographical Research and Writing*
*Style Manual for Engineering Authors and Editors*
*IBM Style Manual*
*NASA Publications Manual*
*Guide for Preparation of Air Force Publications*

When a specific format is not stipulated, you can consult one of three general style manuals. *The MLA Handbook for Writers of Research Papers,* 3rd ed., by the Modern Language Association, outlines documentation in the humanities. *The Chicago Manual of Style,* 13th ed., by the University of Chicago Press, covers documentation in the humanities and related fields as well as natural sciences. *The Government Printing Office Style Manual* (1984), by the U.S. Government Printing Office, covers documentation for government writing.

For a more detailed list and discussion, see John A. Walter, "Style Manuals," *Handbook of Technical Writing Practices,* 2 vols., ed. Jordan, Kleinman, and Shimberg (New York: Wiley, 1971) 2: 1267–73.

In this chapter we discuss three systems of documentation: in-text citations, author-year designation, and numerical designation.

## In-Text Citations

The Modern Language Association (MLA) has replaced footnotes and endnotes with in-text citations (or "parenthetical references"). Instead of placing footnote numbers (e.g., [1]) in the text and listing the sources at the bottom of the page or as endnotes, list your abbreviated references within the text, then give full documentation on a "Works Cited" page at the end of the report.[6] (The "Works Cited" page usually replaces the traditional "Bibliography" page.)

An in-text or parenthetical citation usually includes the author's last name and the page cited, as in (Barrett 69). Here is how the citation would appear in the report:

The benefits of automation outweigh its costs (Barrett 69).

Readers needing the full citation for Barrett can turn to "Works Cited," listed alphabetically by author, to get complete publishing information.

Keep parenthetical references brief. If you mention the author's name in your discussion, don't repeat it in the citation; merely provide the page reference.

Barrett claims that the benefits of automation outweigh its costs (69).

If the work is by a corporate author or is unsigned (i.e., author unknown), use a shortened version of the title or corporate name in your citation, as in ("Information Systems" 18). But be sure that shortened titles correspond with the entries in "Works Cited" (e.g., "Information Systems for Tomorrow's Office," *Fortune* 18 Oct. 1982: 18–56).

Unless your readers request otherwise, use the following format for references on your "Works Cited" page.

**Works Cited Form—Books.**  A reference for a book should contain the following information (as applicable): author, title, editor or translator, edition, volume

---

[6]Footnotes are now used only to comment or elaborate on material in the text (as here). List them at page bottom or on a Works Cited page at document's end.

number, facts about publication (city, publisher, date). (This information is found on a book's title and copyright pages.) Abbreviate publishers' names in your list of works cited, as in "Little" for Little, Brown and Company; "Knopf" for Alfred A. Knopf, Inc.; "GPO" for Government Printing Office; or "Yale UP" for Yale University Press.

Type the first line of the entry flush with a 1-inch margin. Indent the second and all subsequent lines five spaces. Double-space within each entry as well as between entries. Skip two horizontal spaces after any period in an entry, and one space after any comma or colon. Here are examples:

### Single Author

```
Katzan, Harry.  Office Automation: A Manager's Guide.  New York:

     American Management Association, 1982.
```

### Two Authors

```
Adam, Everett E., Jr., and Ronald J. Ebert.  Production and

     Operations Management.  Englewood Cliffs: Prentice, 1978.
```

### Three or More Authors

```
Levin, Richard I., et al.  Quantitative Approaches to Management.

     5th ed. New York: McGraw, 1982.
```

### Author(s) Not Named

```
Computer Documentation.  Boston: Meredith, 1983.
```

### Two Books with the Same Author

```
Lamont, John W. Biophysics.  Boston: Little, 1989.

---. Diagnostic Techniques.  Boston: Little, 1990.
```

When citing more than one work by an author, do not repeat the author's name; simply type three hyphens followed by a period.

### An Editor

```
Meadows, A. J., ed.  The Random House Dictionary of New Information

     Technology.  New York: Vintage, 1983.
```

### A Quotation of a Quotation

```
Kline, Thomas.  Automated Office Systems.  New York: Random, 1983,

     p. 97, as cited in John Swenson, How to Increase White-Collar

     Productivity.  Boston: Little, 1988.
```

## *A Work in an Anthology (a collection of works by various authors)*

```
Anderson, Paul V.  ''What Survey Research Tells Us About Writing at
    Work.''  Writing in Nonacademic Settings.  Ed. Lee Odell and
    Dixie Goswami. New York: Guilford, 1985. 3-83.
```

The page numbers are for the one work cited from the anthology.

**Works Cited Form—Periodicals.** A reference for an article should give this information (as applicable): author, article title, periodical title, volume or number (or both) date (in day, month, year order), and page numbers for the entire article—not for pages cited. Here are examples:

### *A Magazine Article*

```
Main, Jeremy.  ''The Executive Yearning to Learn.''  Fortune 3 May
    1982: 234-48.
```

No punctuation separates the magazine title from the date. Notice also that the abbreviation *p.* or *pp.* is not used to designate pages.

If no author is given, list all other information:

```
''Information Systems for Tomorrow's Office.''  Fortune 18 Oct.
    1982: 18-56.
```

### *An Article in a Journal with New Pagination in Each Issue*

```
Thackman, John.  ''Computer-Assisted Research.''  American Librarian
    51.1 (1990): 3-9.
```

Because each issue for that year will have page numbers beginning with 1, readers have to know the number of this issue. The "51" is the volume number, and the "1" is the issue number.

Omit "The" or "A" or any other introductory article from a journal or magazine title.

### *An Article in a Journal with Continuous Pagination*

```
Barnstead, Marion H.  ''The Writing Crisis.''  Writing Theory 12
    (1989): 415-33.
```

When page numbers continue from one issue to another for the whole year, readers do not need to know the issue number because no other issue in that year will repeat the same page numbers.

### A Newspaper Article

Barrett, Marianne. ''The Decline in White-Collar Productivity.''
    <u>Boston News</u> 15 Jan. 1990, Evening ed., sec. 2: 3.

When a daily newspaper has more than one edition, cite the edition after the date. Omit any introductory article from the newspaper name (e.g., *New York Times*).

## Citing Miscellaneous Items

### An Encyclopedia, Dictionary, or Other Alphabetic Reference Work

''Communication.'' <u>The Business Reference Book</u>. 1977 ed.

If the entry is signed, begin with the author's name.

### A Personally Conducted Interview

Jones, Al. President, Al's TV Shop. Personal Interview. Swansea,
    Mass., 2 Apr. 1990.

### A Published Interview

Lescault, James. ''The Future of Graphics.'' <u>Executive Views of</u>
    <u>Automation</u>. Ed. Karl Prell. Boston: Haber, 1986. 216-31.

The name of the interviewee is placed in the entry's author slot.

### An Unpublished Letter

Rogers, Leonard. Letter to the author. 15 May 1990.

### A Questionnaire

Taylor, Lynne. Questionnaire sent to 612 southeastern
    Massachusetts business presidents. 14 Feb. 1990.

### A Pamphlet or Brochure

Waters, John L. <u>Investment Strategies for Tax Savings</u>. San
    Francisco: Blount Economics Division, 1989.

### A Lecture

Thompson, Edwin. ''Bureaucratic Follies and Blunders.'' Lecture
    at Southeastern Massachusetts University. North Dartmouth,
    17 Sept. 1988.

### A Database Source

Keyes, Langley Carlton. ''Profits in Prose.'' <u>Harvard Business</u>

    <u>Review</u> 39 (1961): 105-12; Latham, New York: Bibliographic

    Retrieval Service, 1984.

### Software

Levy, Michael C., et al. <u>Statmaster: Exploring and Computing</u>

    <u>Statistics</u>. Computer software. Boston: Little, 1982. IBM

    PC, 48KB, disk.

When documenting software, name the appropriate computer, the kilobytes ("48KB"), and the software format ("disk").

### Corporate Author or Government Publication

The Presidential Task Force on Acid Rain. <u>Acid Rain and Corporate</u>

    <u>Profits</u>. Washington: GPO, 1990.

### Other Items

Cite other items (reports, dissertations, and other unpublished works) thus:

Author (if known), title (in quotes), sponsoring

    organization, date, page number(s).

If the author is unknown, use the name of the organization or agency in place of the author's name.

In the list of works cited, arrange the entries alphabetically by the author's last name (as the first item of the entry). When the author's name is unknown, list the title alphabetically according to its first word (excluding *a, an,* and *the*). A title that begins with a number should be alphabetized as if the number were spelled out.

Here is the list of works cited in the report on pages 463–467.[7]

WORKS CITED

Bennett, James C. ''The Communication Needs of Business

    Executives.'' <u>Readings in Business Communication</u>. Ed. Robert

    D. Gieselman, Urbana, Ill.: Stipes, 1982.

---

[7]Normally, of course, this list would appear at the end of the report.

DiGaetani, John L.  ''The Business of Listening.''  In Norman

   Sigband, <u>Communication for Management and Business</u>.  3rd ed.

   Glenview, Ill.: Scott, 1982.

Halpern, Jeanne W.  ''What Should We Be Teaching Students in

   Business Writing?''  <u>Journal of Business Communication</u>

   18 (1981): 39-53.

Main, Jeremy.  ''The Executive Yearning to Learn.''  <u>Fortune</u> 3 May

   1982: 234-48.

Rochester, Jack B., and John L. DiGaetani.  ''Managerial

   Communication: Total Business Communication for the 1980s.''

   <u>ABCA Bulletin</u> 44 (1981): 9-10.

Sigband, Norman B.  <u>Communication for Management and Business</u>.

   3rd ed.  Glenview, Ill.: Scott, 1982.

## Author-Year Documentation System

One alternative to the works-cited system is the author-year system, in which the author's last name and the publication date appear in the textual citation:

Seventy-five percent of technicians interviewed expressed a desire for further training (Albey, 1986).

With this system, you can also include page numbers in the text:

Seventy-five percent of technicians interviewed expressed a desire for further training (Albey, 1986, p. 114).

When you mention the writer's name in the text of your discussion, do not repeat it within the parentheses:

Crashaw and co-workers (1987, pp. 93–94) claim that medical technologists "feel challenged by their work."

If you are citing two works published by the same author(s) in the same year, insert an "a" and a "b" after the dates, in both your reference list and your textual citation:

(J. Jones 1975a)

Because it emphasizes the date, author-year documentation is preferred in the sciences and social sciences, where information quickly becomes dated.

Every reference cited in the text appears alphabetically, by author's last name, in a list of references at the end of the document:

REFERENCES

Albey, J. (1990). Modern career choices. San Francisco: Hamilton.

Crashaw, H., et al. (1988). Technology and careers. Boston:
Little Brown.

---. (1989). Careers in the natural sciences. Dallas: Bovary.

Donne, M. (1989). Job prospects for college graduates. Education
Digest, 28(2), 86-89.

Marsh A., and C. Smith, eds. (1988). Advice for the job seeker.
Boston: Arngold.

Peters, Claire (1988). Wage scales for women in civil engineering.
Master's thesis, Brandon University.

The dates in the listed references are again prominently displayed. Notice that only the initial letter in the first word of book and article titles is capitalized, but that each important word in a journal title is capitalized.

For typical author-year entries, see the latest edition of *The Chicago Manual of Style*. Also, different disciplines have their own style manuals specifying formats for the author-year system. The format partially illustrated here is taken from the *Publication Manual of the American Psychological Association*, 3rd ed. (also called the APA Style Manual). In many fields in which writers wish to emphasize the publication dates of their references, the APA style is acceptable.

## Numerical Documentation System

Another alternative to the works-cited system is the numerical-reference system, in which each work is assigned a number upon first citation. And this same number is used for any subsequent reference to that work. Adding page numbers helps clarify the reference:

> Seventy-five percent of technicians interviewed expressed a desire for further training (2: 83).

In the list of references at report's end, works are numbered in alphabetical order or in the order in which first cited in the text. (Use one arrangement or the other, consistently.) Otherwise, the format resembles the author-year system. Here are entries for a reference list in order of first citation in the text. (These, of course, are not alphabetized.)

1. Donne, M. Job prospects for college graduates. Education Digest 28.2 (1989): 86–89.

2. Albey, J. Modern career choices. San Francisco: Hamilton, 1990.

3. Crashaw, H., et al. Careers in the natural sciences. Dallas: Bovary, 1989.

4. Marsh, A., and C. Smith, eds. Advice for the job seeker. Boston: Arngold, 1988.

5. Crashaw, H., et al. Technology and careers. Boston: Little Brown, 1988.

The numerical system generally is used in the physical sciences (astronomy, chemistry, geology, physics) and the applied sciences (mathematics, medicine, computer science). For specific formats in a discipline, consult one of these style guides or one your instructor recommends:

American Institute of Physics, *Style Manual*
American Medical Association, *Style Book*
*A Manual for Authors of Mathematical Papers*

Whichever documentation system you choose, be consistent throughout your report.

## EXERCISES

**1.** Begin researching for the analytical report (Chapter 20) due at semester's end. Complete these steps. (Your instructor might establish a timetable.)

*Phase One: Preliminary Steps*
  *a.* Choose a topic of *immediate practical importance*, something that affects you or your community directly. (See page 139 for a list of possible topics.)
  *b.* Identify a specific audience and its intended use of your information. Complete an audience-and-use profile (page 33).
  *c.* Narrow your topic, and check with your instructor for approval.
  *d.* Make a working bibliography to ensure sufficient primary and secondary resources. Don't delay this step!
  *e.* List things you already know about your topic.
  *f.* Write a clear statement of purpose and submit it in a proposal memo (pages 423–425) to your instructor.
  *g.* Make a working outline.

*Phase Two: Collecting Data*
  *a.* In your research, move from general to specific; begin with general reference works for an overview.

*b.* Skim your material, looking for high points.

*c.* Take selective notes. Don't write everything down! Use notecards.

*d.* Plan and administer (or distribute) questionnaires, interviews, and letters of inquiry.

*e.* Whenever possible, conclude your research with direct observation.

*Phase Three: Organizing Your Data and Writing Your Report*

*a.* Revise and adjust your working outline, as needed.

*b.* Compose an audience-and-use analysis like the sample on pages 467–469.

*c.* Fully document all sources of information.

*d.* Proofread carefully and add all needed supplements (title page, letter of transmittal, abstract, summary, appendix, glossary – Chapter 18).

*Due Dates: To Be Assigned by Your Instructor*

List of possible topics due:

Final topic due:

Proposal memo due:

Working bibliography and working outline due:

Notecards due:

Copies of questionnaires, interview questions, and inquiry letters due:

Revised outline due:

First draft of report due:

Final draft with supplements and documentation due:

**2.** Using the card catalog, locate and list the full bibliographic data (author's name, title of work, place and publisher, date) for five books in your field or on your semester report topic, all published within the past year.

**3.** Consult the *Library of Congress Subject Headings* for alternative headings under which you might find information in the card catalog for your semester report topic.

**4.** List five major reference works in your field or on your topic by consulting Sheehy, Walford, or a more specific guide to literature.

**5.** List the titles of each of these specialized reference works in your field or on your topic: a bibliography, an encyclopedia, a dictionary, a handbook, an almanac (if available), and a directory.

**6.** Identify the major periodical index in your field or on your topic. Locate a recent article on a specific topic (e.g., use of artificial intelligence in medical diagnosis). Photocopy the article and write an informative abstract.

**7.** Consult the appropriate librarian and identify two databases you would search for information on the topic in exercise 3.

**8.** Identify the major abstract collection in your field or on your topic. Using the abstracts, locate a recent article. Photocopy the abstract and the article.

**9.** Using technical report indexes, locate and summarize three recent reports on *one* specific topic in your field. Provide complete bibliographic information.

**10.** Using patent indexes, locate and describe three recently patented inventions in your field, and provide complete bibliographic information.

**11.** Using indexes of conference proceedings, locate and summarize three recent conference papers on *one* specific topic in your field. Provide complete bibliographic information.

**12.** Using the *Monthly Catalog* or *Government Reports Announcements and Index,* locate and photocopy a recent government publication in your field or on your topic.

**13.** Determine whether your library offers OCLC or InfoTrac services. Use whichever service is available to locate the titles of two current books or articles in your field or on your topic.

**14.** If your library offers students a free search of mainframe databases, ask your librarian for help in preparing an electronic search for your semester report.

**15.** Revise these questions to make them appropriate for inclusion in a questionnaire:
   - *a.* Would a female president do the job as well as a male?
   - *b.* Don't you think that euthanasia is a crime?
   - *c.* Do you oppose increased government spending?
   - *d.* Do you feel that welfare recipients are too lazy to support themselves?
   - *e.* Are teachers responsible for the decline in literacy among students?
   - *f.* Aren't humanities studies a waste of time?
   - *g.* Do you prefer Rocket Cola to other leading brands?

**16.** Arrange an interview with a successful person in your field. List general areas for questioning: job opportunities, chances for promotion, salary range, requirements, outlook for the next decade, working conditions, job satisfaction, and so on. Compose interview questions; conduct the interview; and summarize your findings in a memo to your instructor.

**17.** Using the library and other sources, locate and record the required information for *ten* of these items, and fully document each source:
   - *a.* population of your home town
   - *b.* *New York Times* headline on the day you were born
   - *c.* world pole-vaulting record
   - *d.* operating principle of a diesel engine
   - *e.* name of the first woman elected to the United States Congress
   - *f.* exact value of today's American dollar in French francs
   - *g.* comparative interest rates on an auto loan charged by three banks
   - *h.* major cause of small-business failures
   - *i.* top-rated compact car in the world today
   - *j.* origin of the word *capitalism*
   - *k.* distance from Boston to Bombay
   - *l.* definition of *pheochromocytoma*
   - *m.* effective plant for keeping insects out of a home garden
   - *n.* originator of the statement "There's a sucker born every minute"
   - *o.* names and addresses of five major United States paper companies
   - *p.* Australian job opportunities in your field
   - *q.* titles and specific locations of five recent articles on robotics
   - *r.* half-life of plutonium
   - *s.* brief description of your intended occupation
   - *t.* side effects of certain tranquilizers (e.g., Valium or Librium)
   - *u.* difference between an analog and a digital computer
   - *v.* topic of special interest that might form the subject for a longer report

# TEAM PROJECT

## Collaborative Writing

Divide into small groups. As a group, decide on a campus or community issue or some other topic worthy of research. Elect a group manager to assign and coordinate tasks. At project's end, the manager will provide a performance appraisal by summarizing, in writing, the contribution of each team member. Assigned tasks will include planning, information gathering from primary and secondary sources, document preparation (including visuals) and revision, and classroom presentation.

Do the research, write the report, and present your findings to the class. (Your instructor may assign Chapter 20 in conjunction with this project.)

Here are some possible research topics:

a. Survey student, faculty, and administration about some proposed curriculum change or about some other controversial campus issue. Compare your findings with nationwide or statewide statistics about attitudes on this issue.

b. As much as 30 percent of groundwater in some states is contaminated. Find out how the quality of local groundwater measures up to national averages. Has the quality increased or decreased over the last ten years? What are the major elements affecting local water quality? What is the outlook for the next decade? Can local residents feel safe drinking tap water?

c. Identify the main qualities employers seek in job applicants. Have employers' expectations changed over the last ten years? If so, why?

d. Find out which geographic area of the United States is enjoying the greatest prosperity and population growth (or which area is suffering the greatest hardship and population decrease). What are the major reasons? Trace the recent history of this change.

e. Older homes can present a frightening array of toxic hazards: termite spray, wood preservatives, urea formaldehyde insulation, radon gas, chemical contamination of well water, lead paint, asbestos, and so on. Research the major effects of these hazards for someone who is thinking of buying an older home.

f. How safe is your school (or your dorm)? Are there any dangers from insulation, asbestos, art supplies, cleaning fluids and solvents, water pipes, or the like? Find out, and write a report for your classmates.

g. Which area of your state has the cleanest air and groundwater, and which has the most polluted? Write for someone looking for the safest place to raise a family.

h. Has acid rain caused any damage in your area? Write for classmates.

i. Can peanut butter, black pepper, potatoes, or toasted bread cause cancer? Which of the most common "pure" foods can be carcinogenic? Find out, and write a report for the school dietitian.

j. Are there any recent inventions that could help decrease our reliance on fossil fuels—in ways that are economically feasible and practical? Find out, and prepare a report for your U.S. senator.

k. What is the very latest that scientists are saying about the implications of global warming from rain-forest destruction and ozone depletion? Find out, and prepare a report to be published in a national magazine.

# 7

# Summarizing Information

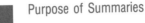

Purpose of Summaries

Elements of a Summary

The Summary Process

Applying the Principles

Summaries as Informative or Descriptive Abstracts

Placing Abstracts in Reports

A summary is a short version of a longer message. An economical way to communicate, a summary saves time, space, and energy.

## PURPOSE OF SUMMARIES

On the job, you have to write concisely about your work. You might record the minutes of a meeting, summarize a news article or report, or describe your progress on a project. A routine assignment for many new employees is to provide superiors (decision makers) with summaries of the latest developments in their field.

The Information Age generates more and more documents daily. Some reports and proposals can be hundreds of pages long. Those who must act on this information need to rapidly identify what is most important in a document. From a good summary, busy readers can get enough information to decide whether they should read the entire document, parts of it, or none of it.

Whether you summarize someone else's information or your own, your job is to communicate the *essential message*—to represent the original document accurately and in the fewest words. The principle is simple: include what your readers need and omit what they don't.

The essential message in any well-written document is easy enough to identify, as in the following passage:

The lack of technical knowledge among owners of television sets leads to their suspicion about the honesty of television repair technicians. Although television owners might be fairly knowledgeable about most repairs made to their automobiles, they rarely understand the nature and extent of specialized electronic repairs. The function and importance of the automatic transmission in an automobile are generally well known; however, the average television owner knows nothing about the flyback transformer in a television set. The repair charge for a flyback transformer failure is roughly $150—a large amount to a consumer who lacks even a simple understanding of what the repairs accomplished. In contrast, a $450 repair charge for the transmission on the family car, though distressing, is more readily understood and accepted.

Three significant ideas comprise the essential message: (1) television owners lack technical knowledge and are suspicious of repair technicians; (2) an owner usually understands even the most expensive automobile repairs; and (3) owners do not understand or accept expenses for television repairs. A summary of the paragraph might read like this:

> Because television owners lack technical knowledge about their sets, they often are suspicious of repair technicians. Although consumers may understand expensive automobile repairs, they rarely understand or accept repair and parts expenses for their television sets.

This summary is almost 30 percent of the original length because the original itself is short. With a longer original, a summary might be as short as 5 percent or less. But a summary's length is far less important than its informative value: does it give readers all they need?

Summaries are vital whenever people have no time to read in detail everything that crosses their desks.[1] For letters, memos, or other short documents that can be read quickly, the only summary needed is usually an opening thesis or topic sentence that gives a preview of the contents.

## ELEMENTS OF A SUMMARY

All effective summaries display the elements discussed below.

### Essential Message

A good summary answers the reader's implied question: "What does the original say?" The essential message is the minimum needed for the reader to understand the shorter version. It is the sum of the significant material—and *only* the significant material—in the original. Significant material includes controlling ideas (thesis and topic sentences); major findings and interpretations; im-

---

[1]A recent U.S. president reportedly required all significant world news for the last twenty-four hours to be compressed into one typed page and placed on his desk, first thing each morning. Another president had a writer who summarized articles from more than two dozen major magazines.

portant names, dates, statistics, and measurements; and conclusions or recommendations. Significant material does not include background; the author's personal comments, digressions, or conjectures; introductions; long explanations, examples, or definitions; visuals; or data of questionable accuracy. (These distinctions are illustrated on pages 144–147.)

## Nontechnical Style

When your audience is large, expect more people to read the summary than any other part of your report. Write at the lowest level of technicality. Translate technical data into plain English. "For twenty-four hours, the patient's serum glucose measured a consistent 240 mg%" can be translated: "For twenty-four hours, the patient's blood-sugar level remained critically high." When you do know your audience specifically, keep these people in mind. If they are expert or informed, you won't need to simplify as much. You are safer, however, to risk oversimplifying than to risk confusing your reader.

## Independent Meaning

In meaning as well as style, your summary should stand alone as a self-contained message. Readers should have to read the original only for a closer view—not to make sense of your message.

## No Added Material

Your job is to represent the original faithfully. Insert no personal comments or judgments ("This interesting report. . . ," or "The author is correct in assuming . . ."). Add nothing to the original.

## Introduction-Body-Conclusion Structure

Most good writing has an introduction, a body, and a conclusion; your summary should, too.

1. Begin with a clear statement of the controlling idea.
2. Present the supporting details in logical order.
3. Close with the original's conclusions and recommendations.

To improve coherence, use transitional expressions ("however," "in addition," "therefore," "although," "in contrast").

## Conciseness

Above all, a summary is concise. Because all messages are different, however, we cannot set a rule for length. Know your readers and their exact needs. Unless length is specified as part of the job, make the summary short enough to be economical and long enough to be clear and comprehensive. A long, clear summary is always better than a short, foggy one.

## THE SUMMARY PROCESS

Summarize your own work only after completing the original. Follow these instructions for paring down any piece—your own or another's.

1. *Read the entire original.* When summarizing another's work, read the whole thing before writing a word. Get a complete picture. You have to understand the original before you can summarize it.

2. *Reread and underline.* Reread the original, underlining significant points. Focus on the thesis and topic sentences.

3. *Edit the underlined data.* Reread the underlined material and cross out all but key phrases you can later rewrite as sentences.

4. *Rewrite in your own words.* Include all important data in the first draft, even if it's too long; you can trim later.

5. *Edit your own version.* When you have everything readers need, edit for conciseness.

    a. Cross out all needless words without harming clarity or grammar. Use complete sentences.

    The summer internship in journalism gives the ~~journalism~~ student ~~first-hand~~ experience ~~on a newspaper. at what goes~~ on ~~within~~ a ~~real~~ newspaper.

    b. Cross out needless prefaces such as "The writer argues . . ." or "Also discussed is . . . ."

    c. Use numerals for numbers, except to begin a sentence.

    d. Combine related ideas in order to emphasize relationships (pages 217–219).

6. *Check your version against the original.* Verify that you have preserved the essential message and added no comments.

7. *Rewrite your edited version.* Add transitional expressions to reinforce the connection between related ideas.

8. *Document your source.* If summarizing another's work, cite the source immediately below the summary, and place directly quoted statements within quotation marks. (See Chapter 6 for documentation formats.)

Although the summary is written last by the writer, it is read *first* by the reader. Though you may be tired, take the time to do a good job.

## APPLYING THE PRINCIPLES

Imagine that you work in the information office of your state's Department of Environmental Management (DEM). In the coming state election, citizens will vote on a referendum proposal for constructing municipal trash incinerators. Referendum supporters argue that incinerators would help solve the growing problem of waste disposal in highly populated parts of the state. Opponents argue that incinerators cause highly toxic air pollution.

To clarify the issues for voters, the DEM is preparing a newsletter to be mailed to each registered voter. You have been assigned the task of researching the recent data, and of summarizing them for newsletter readers. Here is one of the articles you decided to summarize (shown with steps 1, 2, and 3 from The Summary Process completed).

## INCINERATING TRASH: A HOT ISSUE GETTING HOTTER

**The Failure of Landfills; the Hope of Resource Recovery**

**Combine as orienting sentence (controlling idea)**

Alarmed by the tendency of landfills to contaminate the environment, both public officials and citizens are vocally seeking alternatives. The most commonly discussed alternative is something called a *resource recovery facility*. Nearly 100 U.S. cities have built such in the last 15 years, and another 150 or so are in various stages of planning.

**Include definition**

These recovery facilities are a new form of an old technology. Basically, they're incinerators. But, unlike the incinerators of old, they don't just burn waste. They also recover energy. The energy is sold as steam to an industrial customer, or it is converted to electricity and sold to the local utility. (A few facilities, not many, also recover metals or other materials before using the waste as fuel.)

**Include major fact**

**Include major statistic**

A ton of trash possesses the energy content of a barrel and a half of oil. This is not a trivial amount. The United States discards 150 million tons of municipal refuse a year. If all of it were converted to energy, we could replace the equivalent of 12 percent of our oil imports.

**Include major fact**

At the local level, selling energy or materials not only replaces nonrenewable resources; it also provides a source of income that partly offsets the cost of operating the facility.

**Include major fact**

**Delete explanation**

The new facilities are, on average, much cleaner than the municipal incinerators of old. Many have two-stage combustion units, in which the second-stage burns exhaust gases at high temperature, converting many potential organic pollutants to less harmful emissions such as carbon dioxide. Some, especially the larger and newer facilities, also come equipped with the latest in pollution control devices.

**Delete questionable point**

The environmental community is uneasy with this new technology. Environmentalists have argued for many years that the best method of handling municipal trash is to recycle it—i.e., to separate the glass, metal, paper, and other materials and use them again, either without reprocessing or as raw materials in producing new products. The thought of the potential resources in municipal solid waste simply being burned, even with energy recovery, has made many environmentalists opponents of resource recovery.

**Include key finding and explanation**

More recently, opponents have found a stronger reason to oppose burning waste: *dioxin* in the plants' emissions. The amounts present are extremely small, measured in trillionths of a gram per cubic meter of air. But dioxin can be deadly, at least to animals, at very low levels.

**What is Dioxin?**

Dioxin is a generic term for any of 75 chemical compounds, the technical name for which is poly-chlorinated dibenzo-p-dioxins (PCDDs). A related group of 135 chemicals, the PCDFs or furans, are often found in association with PCDDs. — Include definition / Delete technical details

The most infamous of these substances, 2,3,7,8-TCDD, is often referred to as the "most toxic chemical known." This judgment is based on animal test data. In laboratory tests, 2,3,7,8-TCDD is lethal to guinea pigs at a concentration of *500 parts per trillion.* A part per trillion is roughly equivalent to the thickness of a human hair compared to the distance across the United States. — Include major fact

The effects on humans are less certain, for many reasons: it is difficult to measure the amounts to which humans have been exposed; difficult to isolate the effects of dioxin from the effects of other toxic substances on the same population; and the latency period for many potential effects, such as cancer, may be as long as 20 to 30 years. — Include major point / Delete long explanation

Nevertheless, because of the extreme effects of this substance on animals, known releases of dioxin have generated considerable public alarm. One of the most publicized releases occurred at Seveso, Italy, in July 1976, where a pharmaceutical plant explosion resulted in the contamination of at least 700 acres of fields and affected more than 5,000 people. Dioxin was found in the soil in concentrations of 20 to 55 parts per billion. — Include continuation of major point / Delete long example

The immediate effects on humans were nausea, headaches, dizziness, diarrhea, and an acute skin condition called chloracne, which causes burn-like sores. The effects on animals were more severe: birds, rabbits, mice, chickens, and cats died by the hundreds, within days of the explosion. In response to the explosion, the Italian provincial authorities evacuated 730 people from the zone nearest the plant, and sealed off an area containing another 5,000 people from contact with nonresidents.

In this country, perhaps the best known dioxin contamination incident occurred at Times Beach, Missouri, where used oil, contaminated with dioxin, was sprayed on roads as a dust suppressant. Soil samples showed dioxin at levels exceeding 100 parts per billion. While no human health effects were documented at Times Beach, a flood, in December, 1982, led to widespread dispersal of the contamination, as a result of which the entire town was condemned, the population evacuated, and over $30 million of Superfund money used to purchase the condemned property. — Delete long example

Dioxin was among the substances of concern at Love Canal. And it was the major contaminant in the chemical defoliant, Agent Orange, the subject of a lawsuit by 15,000 Vietnam veterans and dependents and an out-of-court settlement of those complaints valued at $180 million. — Include the most striking and familiar example

As early as 1978, trace amounts of dioxin were found in the routine emissions of a municipal incinerator. Virtually every incinerator tested since that date has shown traces of dioxin. — Include key findings

### The Meaning of It All

Include major fact

At the request of Congress, the Agency began in 1984 a major research effort on dioxin, the National Dioxin Study. The study is intended to provide a context in which to place mounting concerns about dioxin. Research for the study was organized into seven "tiers," each tier including a group of sites at which dioxin contamination may be present.

Condense list

- Tier 1, production sites, includes the 10 sites at which 2,4,5-TCP, a pesticide known to have been contaminated by dioxin, was produced, and additional sites where waste materials from its production were disposed.
- Tier 2, precursor sites, includes 9 sites where 2,4,5-TCP was used as a precursor to make other chemical products, and related waste disposal sites. The chemical products included the herbicides 2,4,5-T and silvex, and hexachlorophene, a disinfectant that was widely used in soaps and deodorants, but was banned from nonprescription uses by the Food and Drug Administration in 1981.
- Tier 3 includes 60 to 70 sites at which 2,4,5-TCP and its derivatives were formulated into herbicide products, and associated waste disposal sites.
- Tier 4 includes a wide range of combustion sources, including internal combustion engines, wood stoves, fireplaces, forest fires, oil burners and other sources burning waste oil, and many others.
- Tier 5 includes 20 to 30 of the thousands of sites at which dioxin-contaminated pesticides have been used, for example, power line rights-of-way, forests, and rice and sugar cane fields.
- Tier 6 includes about 20 of the chemical and pesticide production facilities where improper quality control may have led to the accidental production of dioxin.
- And Tier 7 includes samples from sites where the Agency least expected to find dioxin, to determine whether there are background levels of dioxin in the environment. Soil samples have been taken at 500 randomly selected locations across the country—200 in rural areas, 300 in urban areas—and fish have been sampled from over 400 locations.

Include key finding

While the study is not yet complete, data from a variety of sources have already produced disturbing—yet, perhaps, in an odd way, reassuring—results. Dioxin in trace amounts appears to be widely present in the environment, even in remote locations where industrial activity and waste combustion are unlikely to be the source.

Delete long example

Environment Canada, the Canadian EPA, also has an extensive dioxin testing program under way. One of the more startling findings of their research, conducted at a resource recovery facility on picturesque Prince Edward Island, is that the garbage delivered to the plant contained more dioxin than the plant's emissions. The source of the dioxin in this case is not known,

Delete speculation

though it could include pesticide residues or other products contaminated with dioxin during manufacturing processes.

In short, dioxin is not just a problem created by burning municipal waste. It is not clear at this time whether municipal waste combustion is even the major source of dioxin in the environment.

Include key conclusion

Ultimately, the dioxin problem is like other toxic substance issues. We know less than we need to know to thoroughly evaluate the risk. The more we find out, the more complex the issues tend to become. There is no risk-free solution, since all the potential disposal methods may result in some release of toxic substances to the environment. Yet those who counsel delay, to allow the collection of more data, are met with the suspicion that their real agenda is to prevent action entirely.

Include conclusion

Delete personal comment

What we do know at present does not seem to suggest that we should stop planning to build resource recovery facilities. What it does suggest is that we proceed cautiously, inform the public of both what is known and what is unknown, install pollution controls if the plants' uncontrolled emissions are significant or if the exposed population wants added protection, and hope that continued examination of all the sources and effects of dioxin will eventually produce a consensus.

Include recommendations

*Source:* James E. McCarthy, *Congressional Research Service Review* Apr. 1986:19–21.

Assume that in two early drafts of your summary, you rewrote and edited; for coherence and emphasis, you inserted transitions and combined related ideas. Here is your final draft, with all steps from The Summary Process completed.

### INCINERATING TRASH: A HOT ISSUE GETTING HOTTER (A SUMMARY)

Because landfills often contaminate the environment, trash incinerators (resource recovery facilities) are becoming a popular alternative. Nearly 100 are operating in U.S. cities, and 150 more are planned. Besides their relatively clean burning of waste, these incinerators recover energy, which can be sold to offset operating costs. One ton of trash has roughly the energy content of 1.5 barrels of oil. Converting all U.S. refuse to energy could reduce oil imports by 12 percent.

Unfortunately, incinerator emissions contain very small amounts of dioxin (a generic name for any of 75 related chemicals). Even low dioxin levels can be deadly to animals. In fact, animal tests have helped label one dioxin substance "the most toxic chemical known." Although effects on human beings are less certain, news of dioxin in the environment creates public alarm, as evidenced at Love Canal and by the successful Agent Orange lawsuit by 15,000 Vietnam veterans. Almost every municipal incinerator tested since 1978 has shown traces of dioxin.

At Congress's request, the Environmental Protection Agency began in 1984 the National Dioxin Study. Sites included herbicide and pesticide production and waste-disposal facilities, combustion sources such as wood-stoves and forest fires, areas of dioxin-contaminated pesticide use such as rice and sugarcane fields, and random soil and fish samples nationwide. Findings indicate that trace amounts of dioxin are widely present in the environment, even in areas remote from industry and waste combustion. Waste incineration is by no means the only source of environmental dioxin—and may not even be the major source.

At this stage, we know too little to evaluate the risk. And no risk-free waste-disposal solution in fact exists. But no evidence so far suggests that we stop planning incinerators. We should, however, move cautiously, fully informing the public, installing pollution controls as needed, and searching for a better solution.

*Source:* James E. McCarthy, *Congressional Research Service Review* Apr. 1986: 19–21.

The version above is trimmed, tightened, and edited: word count is reduced to roughly 20 percent of the original. A summary this long serves well in many situations, but other audiences might want a briefer and more compressed summary—say, 10 to 12 percent of the original:

> Because landfills often contaminate the environment, trash incinerators (resource recovery facilities) are becoming a popular alternative across the United States. Besides their relatively clean burning of waste, these incinerators recover energy, which can be sold to offset operating costs.
>
> Unfortunately, incinerator emissions contain very small amounts of dioxin, a chemical proven so deadly to animals, even at low levels, that it has been labeled the most toxic chemical known. Although its effects on human beings are less certain, news of dioxin in the environment creates public alarm. Almost every municipal incinerator tested since 1978 has shown traces of dioxin.
>
> Findings of an EPA study begun in 1984 suggest that trace amounts of dioxin are widely present in the environment, even in areas remote from industry and waste combustion. The dioxin source is by no means only waste incineration.
>
> We lack risk-free disposal solutions and know too little to evaluate risks. But no evidence so far suggests that we stop planning incinerators. We should, however, move cautiously, fully informing the public, installing pollution controls as needed, and searching for better solutions.

Notice that the essential message is still intact; related ideas are again combined and fewer supporting details are included. Clearly, length is adjustable according to your audience and purpose.

## SUMMARIES AS INFORMATIVE OR DESCRIPTIVE ABSTRACTS

In long reports and proposals, summaries take the form of *informative abstracts* or *descriptive abstracts*. An informative abstract reflects *what the original contains* (like the summaries on pages 415, 550). A descriptive abstract reflects *what the original is about*. These differences can be clarified by an analogy. Imagine you are describing your summer travels to a friend. You have two options: (1) you might simply mention the places you visited, in chronological order; or (2) in addition to describing your itinerary, you might describe experiences in each place. Option 1 is like a roadmap, an overview of the areas traveled. This option is analogous to a descriptive abstract. Option 2, on the other hand, is expanded to include the significant experiences. This second option is analogous to an informative abstract, which gives the major facts from the original: key findings, conclusions, solutions, and recommendations.

A descriptive abstract, then, merely conveys the nature and extent of the document. It presents the broadest view, and offers no major facts from the original. Whereas the informative abstract contains the meat of the original, the descriptive abstract has only its skeletal structure; a descriptive abstract is a summary of a summary.

### INCINERATING TRASH: A HOT ISSUE GETTING HOTTER
As an alternative to landfill dumps, municipal trash incinerators offer environmental benefits, but they also pose the danger of dioxin emissions.

Because it merely previews the original, a descriptive abstract is always brief—usually no longer than a short paragraph. One- or two-sentence abstracts often follow article titles in journal or magazine tables of contents; they give a bird's-eye view.

On the job, you might write informative abstracts for a boss who needs the information, but has no time to read the original. Or you might write descriptive abstracts to accompany a bibliography of works that you recommend to colleagues or clients (an *annotated bibliography*).

## PLACING ABSTRACTS IN REPORTS

If your reader asks for a descriptive abstract of your report, place it single-spaced on a separate page right after your table of contents or on the title page. Usually you will place your informative abstract in front, instead of writing a descriptive abstract. Practice can vary from company to company. In general, however, a summary at the beginning helps readers decide whether to read the document, and it gives them a framework for understanding the document. A summary at the end helps readers remember what they have read.

(Revision Checklist for Summaries follows.)

## REVISION CHECKLIST FOR SUMMARIES

Use this checklist to refine your summaries. (Page numbers in parentheses refer to the first page of discussion.)

### Content

☐ Does the summary contain only the essential message? (141)

☐ Does it make sense as an independent piece? (142)

☐ Is the summary accurate when checked against the original? (143)

☐ Is it free of personal comments or additions to the original? (142)

☐ Is it free of needless details? (142)

☐ Is it economical yet clear and comprehensive? (142)

☐ Is the source documented? (143)

☐ Does the descriptive abstract tell what the original is about? (148)

### Organization

☐ Is the summary coherent? (147)

☐ Are there enough transitions to reveal the line of thought? (143)

☐ Does it have an introduction-body-conclusion structure? (142)

### Style

☐ Is it at the best level of technicality for *all* anticipated readers? (142)

☐ Is it free of needless words? (143)

☐ Are all sentences clear, concise, and fluent? (199)

☐ Is it written in correct English? (Appendix A)

## EXERCISES

**1.** Read each of these two paragraphs, and list the significant ideas comprising each essential message. Write a summary of each paragraph.

> In recent years, ski-binding manufacturers, in line with consumer demand, have redesigned their bindings several times in an effort to achieve a noncompromising synthesis between performance and safety. Such a synthesis depends on what appear to be divergent goals. Performance, in essence, is a function of the binding's ability to hold the boot firmly to the ski, thus enabling the skier to change rapidly the position of his or her skis without being hampered by a loose or wobbling connection. Safety, on the other hand, is a function of the binding's ability both to release the boot when the skier falls, and to retain the boot when subjected to the normal shocks of skiing. If achieved, this synthesis of performance and safety will greatly increase skiing pleasure while decreasing accidents.

Contrary to public belief, sewage-treatment plants do not fully purify sewage. The product that leaves the plant to be dumped into the leaching (sievelike drainage) fields is secondary sewage containing toxic contaminants such as phosphates, nitrates, chloride, and heavy metals. As the secondary sewage filters into the ground, this conglomeration is carried along. Under the leaching area develops a contaminated mound through which groundwater flows, spreading the waste products over great distances. If this leachate reaches the outer limits of a well's drawing radius, the water supply becomes polluted. And because all water flows essentially toward the sea, more pollution is added to the coastal regions by this secondary sewage.

**2.** Attend a campus lecture on a topic of interest and take notes on the significant points. Write a summary of the lecture's essential message.

**3.** Find an article about your major field or area of interest and write both an informative abstract and a descriptive abstract of the article.

**4.** Select a long paper you have written for one of your courses; write an informative abstract and a descriptive abstract of the paper.

**5.** Read the following article and write both a descriptive abstract and a summary (or informative abstract), using the steps under The Summary Process as a guide. Define a specific audience and use for your material. (A possible scenario: You are assistant shipping manager for a large wholesaler of roofing and construction supplies to hardware stores and lumber yards. Company executives have asked all managers to provide them with summaries of any news relevant to the building-supply business. After coming across this article, you decide to summarize it for your bosses. As far as you can tell, your intended readers have little or no background on the asbestos issue. Give them the information they need.) Bring your summary to class and exchange it with a classmate for editing according to the revision checklist. Revise your edited copy before submitting it to your instructor.

### MOVING TO RID AMERICA OF ASBESTOS

On January 19, EPA published a proposal in the *Federal Register* to rid the United States of the "miracle" fiber called asbestos.

In the proposed rule, issued under authority of the Toxic Substances Control Act (TSCA), EPA invites public opinion on its intent to immediately ban five major asbestos products and phase out all remaining uses of the substance over the next 10 years. Why is EPA proposing such a measure for a product long considered so commercially important, and still so pervasive throughout American society?

Asbestos is really a common name for a group of natural minerals—silicates—that separate into thin but strong fibers. The fibers are chemically inert and heat-resistant, and they cannot be destroyed or degraded easily.

Since 1900, over 30 million tons of asbestos have been used in hundreds of products. Much of it was sprayed on ceilings and other parts of schools and public and private buildings for fireproofing, sounddeadening, insulation, or decoration.

Unfortunately, some of the characteristics that make this mineral fiber so useful commercially—such as its great stability—also help make it a dangerous killer when it is breathed in. Unless completely sealed into a product, asbestos can easily break into a dust or into tiny fibers. These fibers can then float and be inhaled. Once asbestos gets into the body, it can remain there for many years.

"There can be no debate about the health risks of asbestos," says EPA Administrator Les Thomas. A well-documented cause of lung and other cancers in humans, including mesothelioma (a cancer of the chest and abdominal lining), asbestos is now generating up to 12,000 cancer cases a year in the United States, almost all of which are fatal. Aside from the cancer threat, about 65,000 persons in this country are currently suffering from asbestosis, a chronic scarring of the lungs which makes breathing more and more difficult and eventually causes death. Cigarette smokers exposed to asbestos face extra risk, having a much higher chance of getting lung cancer than exposed nonsmokers.

Asbestos-caused cancers can remain latent and not occur for 15 to 40 years after the first exposure. EPA also believes that even small amounts of asbestos in the air are dangerous.

Asbestos is released into the air throughout its entire life cycle: manufacturing, use, destruction and disposal. Since substitutes are, or will soon be, available for nearly all uses of the fiber, EPA has no feasible alternative but to phase out asbestos and all its products. This is what the January proposal sets out to do.

Prohibited would be the importing, manufacture, and processing of five products that account for as much as one half of United States asbestos consumption. The bulk of these products are used mainly in the construction and renovation industry. They are:

- *Saturated and unsaturated roofing felt* — This is a product made from paper felt and intended to cover or lie under other roof coverings. Its purpose is to insulate and help prevent corrosion.

- *Flooring felt and asbestos felt-backed sheet flooring* — Used as an underside backing for vinyl sheet flooring, this felt helps maintain original product shape and helps prolong floor life, especially when moisture from below the surface is a problem.

- *Vinyl-asbestos floor tile* — Especially popular for use in heavy traffic areas such as in stores, kitchens, and entry ways.

- *Asbestos-cement pipe and fittings* — This is used primarily to carry water or sewage, and to a lesser extent, as conduit pipe for the protection of electrical or telephone cable or for air ducts.

- *Asbestos clothing* — Not street clothes, but special occupational garments worn by those needing protection from extreme heat, such as firefighters.

What about the rest of the asbestos products in use? EPA is proposing to get rid of the asbestos in these products indirectly by phasing down all domestic mining and importing of asbestos by a certain percentage each year over the next 10 years. This phasedown would be carried out by allowing a company to mine or import an annually decreasing percentage of the amount of asbestos it mined or imported during the years 1981–1983.

EPA estimates that as a result of what it is proposing, about 1,900 cancer deaths from asbestos will be avoided.

EPA intends that labels be put on all products not immediately banned, warning users that they contain asbestos. EPA hopes that these warnings would encourage users to take steps to reduce their exposure.

For the five products banned under the proposal, EPA is convinced that industry has adequate, readily available substitutes which should minimize the economic impact of this action. The 10-year phaseout should give industry time to develop good alternatives for all remaining asbestos products.

(Public hearings on EPA's proposed rules are tentatively scheduled for mid-May; written public comments must be submitted by April 29.)

*Source:* Adapted from Dave Ryan, "Moving to Rid America of Asbestos," *EPA Journal* 12.2 (1986): 7–8.

## TEAM PROJECTS

**1.** Organize into groups of four or five and choose a topic for group discussion: an employment problem, a campus problem, plans for an event, suggestions for energy conservation, or the like. (A possible topic: Should employers have the right to require lie detector tests, drug tests, or AIDS tests for their employees?) Discuss the topic for a full class period, taking notes on significant points and conclusions. Afterward, organize and edit your notes in line with the directions for writing summaries. Next, write a summary of the group discussion in no more than 200 words. Finally, as a group, compare your individual summaries for accuracy, emphasis, conciseness, and clarity.

**2.** Your instructor will read a brief article to the class, *once* only. Take any notes you consider appropriate, and compose an informative abstract of the piece. Type and revise your abstract at home, paying careful attention to accuracy, completeness, and emphasis. In class, compare your version with those of your editing group. Working together, compose the final version as a team project for submission to the instructor.

# 8

# Defining Your Terms

Purpose of Definitions

Elements of a Definition

Types of Definition

Expanding Your Definition

Applying the Principles

Placing Definitions in Reports

To define a term is to give the precise meaning you intend when you use it. *Clarity* is vital in any document. Clear writing depends on definitions that both reader and writer understand. Unless you are sure the reader knows the exact meaning you intend, always define something before you discuss it.

## PURPOSE OF DEFINITIONS

Every specialty has its own technical "language." Engineers, architects, and programmers talk about "prestressed concrete," "tolerances," or "microprocessors"; lawyers, real estate brokers, and investment counselors discuss "easements," "liens," "amortization," or "escrow accounts"—and so on. Whenever such terms are unfamiliar to a nonspecialized audience, they need to be defined.

For colleagues, rarely do you have to define specialized terms in your field (unless the term is new), but reports often are written for the layperson—the client or some other general reader. When you write for nonspecialists, think about their needs. Don't force readers to a dictionary or encyclopedia to make sense of your message. Clarify your meaning with definitions.

Most of the specialized expressions previously mentioned are concrete and specific. Once a term such as "microprocessor" has been defined in enough detail to suit the reader's needs, its meaning will not differ appreciably in another context. And when a term is highly technical, a writer can easily figure out that it should be defined for some readers. Any nonspecialist knows that he or she has no idea what "prestressed concrete" or "diffraction" means. Readers are less likely to be aware, however, that more familiar terms like "disability," "guarantee," "tenant," "lease," or "mortgage" acquire very specialized meanings in specialized contexts. Here definition (by all parties) becomes crucial. What "guarantee" means in one situation is not necessarily what it will mean in another. That's why a contract is a detailed (and legal) definition of the subject of the contract.

Let's assume you're shopping for disability insurance to provide a steady income in case injury or illness should cause you to lose your job. Besides comparing prices of various policies, you will want each company to define "physical disability." Although Company A offers the least expensive policy, it might define physical disability as your inability to work at *any* job whatsoever. Therefore, if a neurological disease should prevent you from continuing your work as a designer of delicate electronic devices, without disabling you for work as a salesperson or clerk, you might not qualify as "disabled," according to Company A's definition. In contrast, Company B's policy, which is more expensive, might define physical disability as your inability to work at your *specific* job. Even though both companies use the term "physical disability," each defines the term differently. Because you are legally responsible for all documents bearing your signature, you need to understand the importance and technique of clear definition.

Growth in technology and specialization will continue to make definitions vital to communication. Use definitions, however, only when the audience needs them. Know for whom you're writing, and why. If unable to pinpoint your audience, assume a general readership and define generously.

## ELEMENTS OF A DEFINITION

For all definitions, use these guidelines:

### Plain English

Your purpose is to clarify meaning, not muddy it. Use language your readers will understand.

| | |
|---|---|
| Incorrect | A tumor is a neoplasm. |
| Correct | A tumor is a growth of cells that occurs independently of surrounding tissue, and serves no useful function. |
| Incorrect | A solenoid is an inductance coil that serves as a tractive electromagnet. *(A definition appropriate for an engineering manual, but too specialized for general readers.)* |
| Correct | A solenoid is an electrically energized coil that converts electrical energy to magnetic energy capable of performing mechanical functions. |

### Basic Properties

Any item has characteristics that differentiate it from all others. Its definition should express these basic properties. A thermometer has a singular function:

it measures temperature; this is the essential information a reader needs. All other data about thermometers, such as types, special uses, materials used in construction, and cost, are secondary. A book, on the other hand, cannot be defined in functional terms because books can have several functions. A book can be used to write in or to display pictures, to record financial transactions, to read, and so on. Also, other items (individual sheets of paper, posters, newspapers, picture frames) serve the same functions. The basic property of a book is physical: it is a bound volume of pages. To be called a book, an item must have that property; it is what your reader would have to know *first,* to understand what a book is.

## Objectivity

Leave your opinions out of a definition. "Bomb" is defined as "an explosive weapon detonated by impact, proximity to an object, a timing mechanism, or other predetermined means." If instead you define a bomb as "a weapon devised and perfected by hawkish idiots to blow up the world," you are editorializing; furthermore, you are ignoring a bomb's basic property.

Likewise, in defining "diesel engine," simply tell your reader what it is and how it works. You might think that diesels are too noisy and sluggish for automobiles, but omit these judgments unless your reader has asked for them (say, in a report recommending gasoline powered minibuses for your town's public transportation system).

## TYPES OF DEFINITION

Depending on your subject, purpose, and audience, definitions vary greatly in length. Often you will need only a *parenthetical definition*—a few words or a synonym in parentheses after the term. Sometimes your definition will require one or more complete sentences. Other terms require a definition of greater length—an *expanded definition.*

Your choice of definition type depends on what information readers need, and that, in turn, depends on why they need it. "Carburetor," for instance, could be defined in one sentence, briefly telling readers what it is and how it works. But this definition should be expanded for the student mechanic who needs to know the origin of the term, how the device was developed, what it looks like, how it is used, and how its parts interact. Your audience's needs should guide your choice.

### Parenthetical Definition

A parenthetical definition explains the terms in a word or phrase, often as a synonym in parentheses following the term:

The effervescent (bubbling) mixture is highly toxic.

The leaching field (sievelike drainage area) requires 15 inches of crushed stone.

Another option is to express your definition as a clarifying phrase:

The trees on the site are mostly deciduous; that is, they shed their foliage at season's end.

Use parenthetical definitions to give readers a general understanding of specialized terms so they can follow the discussion where these terms are used. A parenthetical definition of "leaching field" might be adequate in a progress report to a client whose house you are building. But a public-health report titled "Groundwater Contamination from Leaching Fields" would call for expanded definition.

## Sentence Definition

Often, a definition requires one or more whole sentences. Sentence definitions have this structure: (1) the term being defined, (2) the class (specific group) to which the term belongs, and (3) the features that differentiate the term from all others in its class.

| Term | Class | Distinguishing features |
|---|---|---|
| carburetor | a mixing device | in gasoline engines that blends air and fuel into a vapor for combustion within the cylinders |
| transit | a surveying instrument | that measures horizontal and vertical angles |
| diabetes | a metabolic disease | caused by a disorder of the pituitary gland or pancreas, and characterized by excessive unination, persistent thirst, and decreased ability to metabolize sugar |
| liberalism | a political concept | based on belief in progress, the essential goodness of man, and the autonomy of the individual, and standing for the protection of political and civil liberties |
| brief | a legal document | containing all the facts and points of law pertinent to a specific case, and filed by an attorney before the case is argued in court |
| stress | an applied force | that strains or deforms a body |
| laser | an electronic device | that converts electrical energy to light energy, producing a bright, intensely hot, and narrow beam of light |
| fiber optics | a technology | that uses light energy to transmit voices, video images, and data through hair-thin glass fibers |

In this presentation, these elements are combined into one or more complete sentences.

> Diabetes is a metabolic disease caused by a disorder of the pituitary gland or pancreas. This disease is characterized by excessive urination, persistent thirst, and decreased ability to metabolize sugar.

Sentence definition is especially useful if you need to stipulate the precise working definition of a term that has several possible meanings. State your working definitions at the beginning of your report:

> Throughout this report, the term "disadvantaged student" means . . . .

**Classifying the Term.**  Be specific and precise in your classification. The narrower your class, the more specific your meaning. "Transit" is correctly classified as a "surveying instrument," not as a "thing" or an "instrument." "Stress" is classified as "an applied force"; to say that stress "takes place when . . ." or "is something that . . ." fails to reflect a specific classification. Be sure to select precise terms of classification: "Diabetes" is precisely classified as "a metabolic disease," not as "a medical term."

**Differentiating the Term.**  Differentiate the term by separating the item it names from every other item in its class. If you can apply the distinguishing features to more than one item, your definition is imprecise. Make these features narrow enough to pinpoint the item's unique identity and meaning, yet broad enough to be inclusive. A definition of "brief" as "a legal document introduced in a courtroom" is too broad because the definition doesn't differentiate "brief" from all other legal documents (wills, written confessions, etc.). Conversely, differentiating "carburetor" as "a mixing device used in automobile engines" is too narrow because it ignores the carburetor's use in all other gasoline engines.

Also, avoid circular definitions (repeating, as part of the distinguishing features, the word you are defining). Thus "stress" should not be defined as "an applied force that places stress on a body." The class and distinguishing features must express the item's basic property ("an applied force that strains or deforms a body").

## Expanded Definition

An expanded definition can include parenthetical and sentence definitions, but it provides greater detail for readers who need it. The sentence definition of "solenoid" on page 155 is good for a general reader who simply needs to know what a solenoid is. But a manual for mechanics or mechanical engineers would define this item in detail (as on pages 165–166); these readers need to know what a solenoid is, how it works, and how to use it.

The problem with defining an abstract and general word, such as "condominium" or "loan," is different. "Condominium" is a vaguer term than "solenoid" (solenoid A is pretty much like solenoid B) because the former refers to many types of ownership agreements.

Concrete, specific terms such as "diabetes," "transit," and "solenoid" often can be defined in a sentence, and require expanded definition only for certain audiences. But terms such as "disability" and "condominium" require expanded definition for almost any audience. The more general or abstract the term, the more likely the need for an expanded definition.

An expanded definition may be a single paragraph (as for a simple tool) or may extend to many pages (as for a digital dosimeter—a device for measuring radiation exposure); sometimes the definition itself *is* the whole report.

This excerpt from an automobile insurance policy defines the coverage for "bodily injury to others." Its style and detail make this definition clear to general readers. Instead of the fine-print "legalese" seen in many policies, this policy is written in plain English.

**Part I. Bodily Injury to Others**

Under this Part, we will pay damages to people injured or killed by your auto in Massachusetts accidents. Damages are the amounts an injured person is legally entitled to collect for bodily injury through a court judgment or settlement. We will pay only if you or someone else using your auto with your consent is legally responsible for the accident. The most we will pay for injuries to any one person as a result of any one accident is $5,000. The most we will pay for injuries to two or more people as a result of any one accident is a total of $10,000. This is the most we will pay as the result of a single accident no matter how many autos or premiums are shown on the Coverage Selections page.

We will *not* pay:

1. For injuries to guest occupants of your auto.
2. For accidents outside of Massachusetts or in places in Massachusetts where the public has no right of access.
3. For injuries to any employees of the legally responsible person if they are entitled to Massachusetts workers' compensation benefits.

## EXPANDING YOUR DEFINITION

How you expand a definition depends on the questions you think readers need answered, as shown in Figure 8.1.

Choose from among the following strategies as you plan, draft, and revise your expanded definition. Always begin an expanded definition with a sentence definition, and then use only those expansion strategies which serve your reader's needs.

### Etymology

A word's origin (its development and changing meanings) can help clarify its definition. *Biological control* of insects is derived from the Greek "bio," mean-

**FIGURE 8.1** Directions in Which a Definition Can Be Expanded

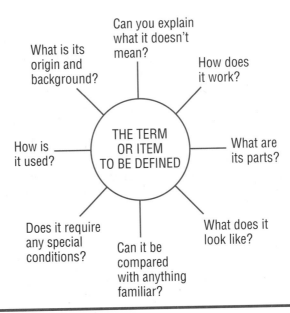

ing *life* or *living organism,* and the Latin "contra," meaning *against* or *opposite*. Biological control, then, is the use of living organisms against insects. College dictionaries contain etymological information, but your best bet is *The Oxford English Dictionary* and encyclopedic dictionaries of science, technology, and business.

Some technical terms are acronyms, derived from the first letters or parts of several words. *Laser* is an acronym for *light amplification by stimulated emission of radiation.*

Sometimes a term's origin can be colorful as well as informative. *Bug* (jargon for *programming error*), is said to derive from an early computer at Harvard that malfunctioned because of a dead bug blocking the contacts of an electrical relay. Because programmers, like most of us, hate to admit mistakes, the term became a euphemism for *error.* Correspondingly, *debugging* is the correcting of errors in a program.

### History and Background

The meaning of specialized terms like "radar," "bacteriophage," "silicon chips," or "X ray" can often be clarified through a background discussion: discovery or history of the concept, development, method of production, applications, etc. Specialized encyclopedias are a good background source.

> The idea of lasers . . . dates back as far as 212 B.C., when Archimedes used a [magnifying] glass to set fire to Roman ships during the siege of Syracuse. (Gartaganis 22).

The early researchers in fiber optic communications were hampered by two principal difficulties—the lack of a sufficiently intense source of light and the absence of a medium which could transmit this light free from interference and with a minimum signal loss. Lasers emit a narrow beam of intense light, so their invention in 1960 solved the first problem. The development of a means to convey this signal was longer in coming, but scientists succeeded in developing the first communications-grade optical fiber of almost pure silica glass in 1970 (Stanton 28).

## Negation

Readers can better grasp some meanings by understanding clearly what the term *does not* mean. The insurance policy on page 159 defines coverage for "bodily injury to others" partly by using negation: "We will *not* pay: 1. For injuries to guest occupants of your auto . . . ." In this next example, negation helps clarify the definition of "lasers": "Lasers are not merely weapons from science fiction."

## Basic Operating Principle

Most items work according to a basic operating principle, whose explanation should be part of your definition:

> A clinical thermometer works on the principle of heat expansion: as the temperature of the bulb increases, the mercury inside expands, forcing a mercury thread up into the hollow stem.

> Air-to-air solar heating involves circulating cool air, from inside the home, across a collector plate (heated by sunlight) on the roof. This warmed air is then circulated back into the home.

> Basically, a laser [uses electrical energy to produce] coherent light, light in which all the waves are in phase with each other, making the light hotter and more intense (Gartaganis 23).

> [A fiber optics] system works as follows: An electrical charge activates the laser . . . , and the resulting light . . . energy passes through the optical fiber. At the other end of the fiber, a . . . receiver . . . converts this light signal back into electrical impulses (Stanton 28).

Even abstract concepts or processes can be explained through their operating principle:

> Economic inflation functions according to the principle of supply and demand: If an item or service is in short supply, its price increases in proportion to its demand.

## Analysis of Parts

When your subject can be divided into parts, identify and explain them:

The standard frame of a pitched-roof wooden dwelling consists of floor joists, wall studs, roof rafters, and collar ties.

Psychoanalysis is an analytic and therapeutic technique consisting of four parts: (1) free association, (2) dream interpretation, (3) analysis of repression and resistance, and (4) analysis of transference.

In discussing each part, of course, you would further define specialized terms such as "floor joists" and "repression."

Analysis of parts is particularly useful for helping nontechnical readers understand a highly technical subject. Notice how this next analysis helps explain the physics of lasing by dividing the process into three discrete parts:

1. [Lasers require] a source of energy, [such as] electric currents or even other lasers.
2. A resonant circuit . . . contains the lasing medium and has one fully reflecting end and one partially reflecting end. The medium — which can be a solid, liquid, or gas — absorbs the energy and releases it as a stream of photons [electromagnetic particles that emit light]. The photons . . . vibrate between the fully and partially reflecting ends of the resonant circuit, constantly accumulating energy — that is, they are amplified. After attaining a prescribed level of energy, the photons can pass through the partially reflecting surface as a beam of coherent light and encounter the optical elements.
3. Optical elements — lenses, prisms, and mirrors — modify size, shape, and other characteristics of the laser beam and direct it to its target (Gartaganis 23).

Figure 1 shows the three parts of a laser.

FIGURE 1  Description of a Simple Laser

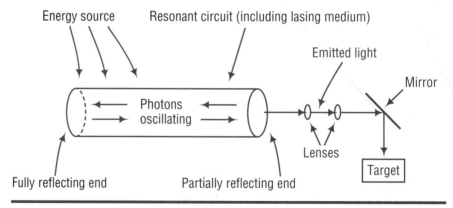

## Visuals

Well-labeled visuals (such as Figure 1 in the previous section) are excellent for clarifying definitions. Always introduce your visual, and explain it. If your visual is borrowed, credit your source at the bottom left. Unless the visual takes up one whole page or more, do not place it on a separate page. Include the visual near its discussion in your definition.

## Comparison and Contrast

Comparisons and contrasts help readers understand. Analogies (a type of comparison) to something familiar can help explain the unfamiliar:

> To visualize how a simplified earthquake starts, imagine an enormous block of gelatin with a vertical knife slit through the middle of its lower half. Gigantic hands are slowly pushing the right side forward and pulling the left side back along the slit, creating a strain on the upper half of the block that eventually splits it. When the split reaches the upper surface, the two halves of the block spring apart and jiggle back and forth before settling into a new alignment. Inhabitants on the upper surface would interpret the shaking as an earthquake ("Earthquake Hazard Analysis" 8).

> The average diameter of an optical cable is around two-thousandths of an inch, making it about as fine as a hair on a baby's head (Stanton 29–30).

Here is a contrast between optical fiber and conventional copper cable:

> Beams of laser light coursing through optical fibers of the purest glass can transmit many times more information than the present communications systems. . . . A pair of optical fibers has the capacity to carry more than 10,000 times as many signals as conventional copper cable. A ½-inch optical cable can carry as much information as a copper cable as thick as a person's arm. . . .
>     Not only does fiber optics produce a better signal, [but] the signal travels farther as well. All communications signals experience a loss of power, or attenuation, as they move along a cable. This power loss necessitates placement of repeaters at one- or two-mile intervals of copper cable in order to regenerate the signal. With fiber, repeaters are necessary about every thirty or forty miles, and this distance is increasing with every generation of fiber (Stanton 27–28).

Here is a combined comparison and contrast:

> Fiber optics technology results from the superior capacity of lightwaves to carry a communications signal. Sound waves, radio waves, and light waves can all carry signals; their capacity increases with their frequency. Voice frequencies carried by telephone operate at 1000 cycles per second, or hertz. Television signals transmit at about 50 million hertz. Light waves, however, operate at frequencies in the hundreds of trillions of hertz (Stanton 28).

## Special Materials or Conditions Required

Some items or processes need special materials and handling, or they may have other requirements or restrictions. An expanded definition should include this important information.

Besides training in engineering, physics, or chemistry, careers in laser technology require a strong background in optics (study of the generation, transmission, and manipulation of light).

Abstract concepts might also be defined in terms of special conditions:

To be held guilty of libel, a person must have defamed someone's character through written or pictorial statements.

### Example

Familiar examples showing types or uses of an item can help clarify your definition. This example shows how laser light is used as a heat-generating device:

Lasers are increasingly used to treat health problems. Thousands of eye operations involving cataracts and detached retinas are performed every year by ophthalmologists. . . . Dermatologists treat skin problems. . . . gynecologists treat problems of the reproductive system, and neurosurgeons even perform brain surgery—all using lasers transmitted through optical fibers (Gartaganis 24–25).

The next example shows how laser light is used to carry information:

The use of lasers in the calculating and memory units of computers, for example, permits storage and rapid manipulation of large amounts of data. And audiodisc players use lasers to improve the quality of the sound they reproduce. The use of optical cable to transmit data also relies on lasers (Gartaganis 25).

And this next example shows how optical fiber can relay a video signal:

Acting, in essence, as tiny cameras, optical fibers can be inserted into the body and relay an image to an outside screen (Stanton 28).

Examples are one of your most powerful communications tools—as long as you tailor the examples to your readers' level of specialization. To be informative, an example has to be understood.

Whichever combination of expansion strategies you decide to use, be sure to document fully your sources of information. (See Chapter 6 for documentation formats.) And place directly quoted statements within quotation marks.

---

## APPLYING THE PRINCIPLES

The following definitions employ expansion strategies appropriate to their audiences' needs. Specific strategies are labeled in the margin. Each definition, like a good essay, is unified and coherent: each paragraph is developed around a main idea and logically connected to other paragraphs. Visuals are incorporated. Transitions emphasize the connection between ideas. Each definition is at a level of technicality that connects with the intended audience.

To illustrate the importance of audience analysis in a writer's decision about "How much is enough?" this example, like many throughout the text, is preceded by an audience-and-use profile based on the worksheet on page 33.

---

**Audience-and-Use Profile**

The intended readers of this material are beginning student mechanics. Before they can repair a solenoid, they will need to know where the term comes from, what a solenoid looks like, how it works, how its parts operate, and how it is used. This definition is designed as merely an *introduction,* and so it offers only a general (but comprehensive) view of the mechanism.

Because the intended readers are not engineering students, they do *not* need details about electromagnetic or mechanical theory (e.g., equations or graphs illustrating voltage magnitudes, joules, lines of force).

**EXPANDED DEFINITION: "SOLENOID"**

A solenoid is an electrically energized coil that forms an electromagnet capable of performing many mechanical functions. The term "solenoid" is derived from the word "sole," which in reference to electrical equipment means "a part of," or "contained inside, or with, other electrical equipment." The Greek word *solenoides* means "channel," or "shaped like a pipe."

**Formal sentence definition**

**Etymology**

A simple plunger-type solenoid consists of a coil of wire attached to an electrical source, and an iron rod that passes in and out of the coil along the axis of the spiral. A spring holds the rod outside the coil when the current is de-energized, as shown in Figure 1.

**Description and analysis of parts**

**FIGURE 1** Side View of a Plunger-type Solenoid

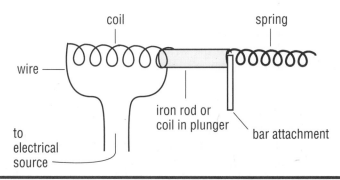

When the coil receives electric current, it becomes a magnet and thus draws the iron rod inside, along the length of its cylindrical center. With a lever attached to its end, the rod can transform electrical energy into mechanical force. The amount of mechanical force produced is determined by the product of the number of turns in the coil, the strength of the exciting current, and the magnetic conductivity of the iron rod.

**Special conditions and principle of operation**

The plunger-type solenoid in Figure 1 is commonly used in the starter motor of an automobile engine. This type is 4½ inches long and 2 inches in diameter, with a steel casing attached to the casing of the starter motor. A linkage (pivoting lever) is attached at one end to the iron rod of the solenoid, and at the other end to the drive gear of the starter, as shown in Figure 2. When the ignition key is turned, current from the battery is supplied to the solenoid coil, and the iron rod is drawn inside the coil, thereby shifting the attached linkage. The linkage, in turn, engages the drive gear, activated by the starter motor, with the flywheel (the main rotating gear of the engine).

**FIGURE 2**  Side View of Solenoid and Starter Motor

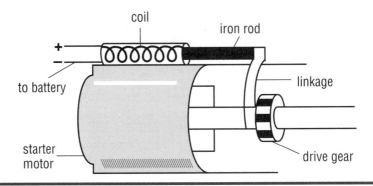

Because of the solenoid's many uses, its size varies according to the work it must do. A small solenoid will have a small wire coil, hence a weak magnetic field. The larger the coil, the stronger the magnetic field; in this case, the rod in the solenoid can do harder work. An electronic lock for a standard door would require a much smaller solenoid than one for a bank vault.

———

The audience for the following definition (an entire community) is too diverse to define precisely, and so the writer wisely addresses the lowest level of technicality—to ensure that all readers will understand.

———

### Audience-and-Use Profile

The following definition is written for members of a community whose water supply (all obtained from wells, because the town has no reservoir) is doubly threatened: (1) by chemical seepage from a recently discovered toxic dump site, and (2) by a two-year drought that has severely depleted the water table. This definition forms part of a report analyzing the severity of the problems and exploring possible solutions.

To understand the problems, these readers first need to know what a water table is, how it is formed, what conditions affect its level and quality, and how it figures into town planning decisions. The concepts of *recharge* and *permeability* are vital to readers' understanding of the problem here, and so these terms are defined parenthetically. These readers have no interest in geological or hydrological (study of water resources) theory. They simply need a broad picture.

# EXPANDED DEFINITION: WATER TABLE

The water table is the level below the earth's surface at which the ground is saturated with water. Figure 1 shows a typical water table that might be found in the East. Wells driven into such a formation will have a water level identical to that of the water table.

**Formal sentence definition**

**Example**

**FIGURE 1** A Typical Water Table (Eastern United States)

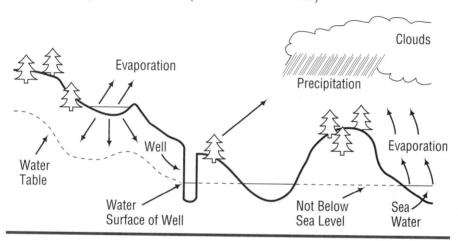

The world's supply of fresh water comes almost entirely as precipitation that begins with the evaporation of sea and lake water. This precipitation falls to earth, and follows one of three courses: it may fall directly onto bodies of water, such as rivers or lakes, where it is directly used by humans; it may fall onto land, and either evaporate or run over the ground to the rivers, etc.; or it may fall onto land, be contained, and seep into the earth. The latter precipitation makes up the water table.

**Basic operating principle**

Similar in contour to the earth's surface above, the water table generally has a level that reflects such features as hills and valleys. Where the water table intersects the ground surface, a stream or pond results.

**Comparison**

A water table's level, however, will vary, depending on the rate of recharge (replacement of water). The recharge rate is affected by rainfall or soil permeability (the ease with which water flows through the soil). A water table then is never static; rather it is the surface of a body of water striving to maintain a balance between the forces which deplete it and those which replenish it. In areas of Florida and some western states where the water table is depleted, the earth collapses, leaving sinkholes.

**Basic operating principle**

**Example**

The water table's depth below ground is vital in water resources engineering and planning. It determines an area's suitability for wastewater disposal, or a building lot's ability to handle sewage. A high water table could become contaminated by a septic system. Also, bacteria and chemicals seeping into a water table can pollute an entire town's water supply. Another consideration in water-table depth is

**Special conditions and examples**

the cost of drilling wells. These conditions obviously affect an industry's or homeowner's decision on where to locate.

Special conditions

The rising and falling of the water table give an indication of the pumping rate's effect on a water supply (drawn from wells), and of the sufficiency of the recharge rate in meeting demand. This kind of information helps water resources planners decide when new sources of water must be made available.

## PLACING DEFINITIONS IN REPORTS

Poorly placed definitions interrupt the information flow. If you have only a few parenthetical definitions, place them in parentheses after the terms. Any more than a few definitions per page will be disruptive. Rewrite them as sentence definitions and place them in a "Definitions" section of the introduction to your report, or in a glossary (Chapter 18).

If your sentence definitions are few, place them in a "Definitions" section of your introduction; otherwise, in a glossary. Definitions of terms in the report's title belong in your introduction.

Place expanded definitions in one of three locations:

1. If the definition is essential to the reader's understanding of the *entire* report, place it in the introduction. A report titled "The Effects of Aerosol on the Earth's Ozone Shield" would require expanded definitions of "aerosol" and "ozone shield" early in the report.

2. When the definition clarifies a major part of your discussion, place it in that section of your report. In a report titled "How Advertising Influences Consumer Habits," "operant conditioning" might be defined early in the appropriate section. Too many expanded definitions *within* a report, however, can be disruptive.

3. If the definition serves only as a reference, place it in an appendix (Chapter 18). A report on fire safety in a public building might have an expanded definition of "smoke detectors" in an appendix.

## REVISION CHECKLIST FOR DEFINITIONS

Use this list to revise your definition. (Numbers in parentheses refer to the first page of discussion.)

### Content

☐ Is the type of definition (parenthetical, sentence, expanded) suited to your purpose and readers' needs? (156)

☐ Does the definition express the basic property of the item? (155)

☐ Is the definition objective? (156)

☐ Is your expanded definition adequately developed? (159)

☐ Have you documented all data sources? (164)

☐ Have you used visuals adequately and appropriately? (162)

## Arrangement

☐ Does the sentence definition describe features that distinguish the term from other items in the same class? (157)

☐ Is the expanded definition unified and coherent (like an essay)? (164)

☐ Do you provide adequate transition between ideas? (164)

☐ Have you placed your definition in the appropriate location? (168)

## Style

☐ Is your definition in plain English? (155)

☐ Will its level of technicality connect with the audience? (155)

☐ Are all sentences clear, concise, and fluent? (199)

☐ Is the word choice precise? (221)

☐ Is it written in correct English? (Appendix A)

Now revise all material that has not been checked off.

## EXERCISES

**1.** Sentence definitions require precise classification and differentiation. Is each of these definitions adequate for a general reader? Rewrite those which seem inadequate. Consult dictionaries and encyclopedia as needed.

    *a.* A bicycle is a vehicle with two wheels.

    *b.* A transistor is a device used in transistorized electronic equipment.

    *c.* Surfing is when one rides a wave to shore while standing on a board specifically designed for buoyancy and balance.

    *d.* Bubonic plague is caused by an organism known as *pasteurella pestis*.

    *e.* Mace is a chemical aerosol spray used by the police.

    *f.* A Geiger counter measures radioactivity.

    *g.* A cactus is a succulent.

    *h.* In law, an indictment is a criminal charge against a defendant.

    *i.* A prune is a kind of plum.

    *j.* Friction is a force between two bodies.

    *k.* Luffing is what happens when one sails into the wind.

    *l.* A frame is an important part of a bicycle.

    *m.* Hypoglycemia is a medical term.

    *n.* An hourglass is a device used for measuring intervals of time.

    *o.* A computer is a machine that handles information with amazing speed.

    *p.* A Ferrari is the best car in the world.

    *q.* To meditate is to exercise mental faculties in thought.

**2.** Standard dictionaries define for the general reader, whereas specialized reference books define for the specialist. Choose an item in your field and copy the definition (1) from a standard dictionary, and (2) from a technical reference book. For the technical definition, label each expansion strategy. Rewrite the specialized definition for a general reader.

**3.** Using reference books as necessary, write sentence definitions for these terms or for terms from your field.

| | | |
|---|---|---|
| biological insect control | economic inflation | estuary |
| generator | anorexia nervosa | acid rain |
| dewpoint | low-impact camping | classical conditioning |
| microprocessor | hemodialysis | hypothermia |
| capitalism | gyroscope | thermistor |
| economic recession | coronary bypass | aquaculture |
| marsh | oil shale | nuclear fission |
| artificial intelligence | chemotherapy | modem |

**4.** Select an item from the list in exercise 3 or from an area of interest. Identify an audience and purpose. Complete an audience-and-use profile sheet (page 33). Begin with a sentence definition of the term. Then write an expanded definition for a first-year student in that field. Next, write the same definition for a layperson (client, patient, or other interested party). Leave a margin at the left side of your page to list expansion strategies (use at least four in each version, and document your sources). Submit, with your two versions, an explanation of your changes from the first version to the second.

**5.** Figure 8.3 on page 172 shows a page from a brochure titled *Congeneration.* The brochure provides an expanded definition for potential users of fuel conservation systems engineered and packaged by Ewing Power Systems. The intended readers are plant engineers and other technical experts unfamiliar with cogeneration.

One page of the brochure is designed in a question/answer format. Figure 8.2 shows parts of that page.

Identify the specific expansion strategies in Figure 8.2 and 8.3. Is the definition appropriate for a technical audience? Why or why not? Be prepared to discuss your analysis and evaluation in class.

## TEAM PROJECT

**First Class Meeting.**  Divide into small groups on the basis of academic majors or interests. Appoint one person as group manager. Decide on an item, concept, or process that would require expanded definition for a layperson.

### Examples
From computer science: an algorithm, an applications program, artificial intelligence, binary coding, top-down procedural thinking, or systems analysis

From nursing: a pacemaker, coronary bypass surgery, or natural childbirth

Complete an audience-and-use profile (page 33).

FIGURE 8.2  Expanded Definition in a Technical Brochure

**Q. What is cogeneration?**

A. It is the simultaneous production of electricity *and* useful thermal energy. This means that you can generate electricity with the same steam you are now using for heating or process. *You can use the same steam twice.* In modern usage cogeneration has also come to mean using waste fuel for in-plant electricity generation.

**Q. How does it save money?**

A. Cogeneration saves money by allowing you to produce your own electricity for a fraction of the cost of utility power. Cogenerated power is cheaper because cogeneration systems are much more efficient than central utility plants. By using the same steam twice, cogeneration systems can achieve efficiencies of up to 80%, whereas the best utilities can do is about 40%.

**Q. Are there other benefits to cogeneration?**

A. Yes. For companies using waste fuel, elimination of waste disposal costs can be a very important benefit. Depending upon design, a cogeneration system can provide emergency standby power and can smooth out boiler load swings.

**Q. Is cogeneration new?**

A. No. It's been done ever since the beginning of the electrification of industrial America. Originally, most electric power was cogenerated by individual manufacturers, not the utilities. In the 1920s and '30s, as cheaper utility electricity became available, cogeneration waned. With cheap oil available, power rates continued to decline through the 1950s and '60s. Then came the 1973–74 Arab Oil Embargo. Everything changed abruptly. Since then electricity prices have risen. Further upward pressure was produced by some utilities' nuclear power plant building programs. Now many companies are getting back to their original source of power: cogeneration.

Once your group has selected a term and decided on the appropriate expansion strategies (etymology, negation, etc.), the group manager will assign each member to work on one or two specific strategies as part of the definition.

**Second Class Meeting.**  As a group, edit and incorporate the collected material into an expanded definition, revising as often as needed.

**Third Class Meeting.**  The group manager will assign one member to present the definition in class, using either opaque or overhead projection, a large-screen monitor, or mimeographed copies.

(Figure 8.3 follows.)

**FIGURE 8.3** Expanded Definition in a Technical Brochure

# TECHNICAL CONSIDERATIONS

Turbine generator sets make electricity by converting a steam pressure drop into mechanical power to spin the generator. Conceptually, steam turbines work much the same way as water turbines. Just as water turbines take the energy from water as it flows from a high elevation to a lower elevation, steam turbines take the energy from steam as it flows from high pressure to low pressure. The amount of energy that can be converted to electricity is determined by the difference between the inlet pressure and the exhaust pressure (pressure drop) and the volume of steam flowing through the turbine.

Steam turbines have been used in industry in a variety of applications for decades and are the most common way utilities generate electricity. Exactly how a steam turbine generator can be used in your plant depends upon your circumstances.

## IF YOU USE WASTE AS A BOILER FUEL

If you use wood waste or incinerator waste as a boiler fuel you can afford to condense turbine exhaust steam in a condenser. This allows you to convert waste fuel into electricity.

The simplest form of a condensing turbine generator set is the Ewing Power Systems C Series. All surplus steam enters the turbine at high pressure and exhausts to a condenser at a very low pressure, usually a vacuum. Because of the very low exhaust pressure, the pressure drop through the turbine is greater and more energy is extracted from each pound of steam. This is the same basic design as utilities use to produce power. The condenser can be either air or water cooled. In water cooled systems the "cooling" water can be hot enough for use as process hot water or for space heat.

In situations where there is surplus fuel and also a need for low pressure process steam, the Ewing Power Systems CX Series is the system of choice. This arrangement includes a back pressure turbine and a condensing turbine connected to a common generator. Low pressure process or space heating loads are met with the back pressure turbine while surplus steam is directed to the condensing turbine to maximize power production.

## IF YOU PURCHASE BOILER FUEL SUCH AS OIL OR GAS

If oil or gas is used as boiler fuel the best use of a turbine is as a replacement for a steam pressure reducing valve. Many plants produce steam at high pressure and then use some or all of the steam at low pressure after passing it through a pressure reducing valve. Other plants have high pressure boilers but run them at low pressure because they do not need high pressure steam for their process. In either case, a turbine generator can turn the pressure drop energy potential into electricity.

The Ewing Power Systems BP Series turbine generator sets are designed for pressure reducing (back pressure) applications. Very little energy is consumed by the turbine, so most of the inlet steam is available for process. The turbine generator uses about 3631 BTU's per hour for each kilowatt-hour produced. At 40 cents per gallon for No. 6 fuel oil and 85% boiler efficiency, it will cost about 1.1 cents per kilowatt-hour to generate your own power with a BP Series turbine. Generating costs for gas-fired boilers are similar.

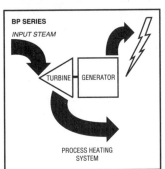

To generate power at this very low cost, all the exhaust steam must be used productively. Generator output is therefore completely governed by process steam demand. For example, if steam is used for space heating you will make more electricity on cold days than on warmer days because more steam will flow through the turbine.

## ELECTRICAL CONSIDERATIONS

In most cases the generator will be connected to your plant electrical system and to your utility. This means that you will not give up the security of utility power. It also means that you do not have to generate *all* your power; most cogeneration systems provide only part of the plant load. The more power the generator is making, the less you buy from the utility. If you make more power than you use, you will be able to sell the excess to the utility. If your generator is off-line for any reason, you will be able to buy power from the utility, just as you do now.

There are two primary generator designs: induction and synchronous. Induction generators are similar to induction motors and are much simpler than synchronous generators. Synchronous generators require more elaborate controls and are usually more expensive but offer the advantage of stand-alone capability. Whereas induction generators cannot operate unless they are connected to a utility grid, synchronous sets can be operated in isolation as emergency units or when it is economically advantageous to avoid interconnection with the utility.

All turbine generator sets from Ewing Power Systems include a complete electrical control panel. Our standard panels meet most utility interconnection requirements and we will customize the panel to meet unusual requirements. Synchronous panels can be built for full utility paralleling, stand-alone capability or both.

Courtesy of Ewing Power Systems, So. Deerfield, MA 01373

# Sequence, Shape, and Style in Your Document

■

# Outlining

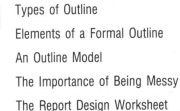

Types of Outline
Elements of a Formal Outline
An Outline Model
The Importance of Being Messy
The Report Design Worksheet

*Successful writers spend more time thinking and planning than writing.* And a major part of any writer's challenge is to get organized. A 1982 survey of workers, full-time students, and students working part-time found that all seventy respondents used outlines to organize their writing, especially in preparing long documents (Roundy and Mair 91).

Last-minute changes in building a house are more easily made on the blueprints than by tearing down walls or moving a bedroom wing. Likewise, an outline is easier to alter than a written report. Like the blueprint, the outline is not a set of commandments, but simply a tool — an orienting device. The computer is especially useful for arranging outlines until they reflect the sequence in which you expect readers to approach your message.

An outline enables you to move from a random listing of items, as they occurred to you, to a deliberate information map that enables readers to locate, understand, and remember your information.

## TYPES OF OUTLINE

### The Informal Outline

An informal outline is a simple list, probably all you need for a short report. For a longer report, an informal outline can serve as a tentative (or working) outline, because it keeps you on track and allows for alterations as you work.

Here is an informal outline for a report titled "An Analysis of the Advisability of Converting Our Office Building from Oil to Gas Heat." The writer begins by formulating her statement of purpose.

> The president of Abco Engineering Consultants has requested a report on the advisability of converting our office building from oil to gas heat.

As the writer brainstorms, she produces this list of topic headings, which she arranges in chronological sequence.

1. Description of Our Present Heating System
2. Removal of the Oil Burner and Tank
3. Installation of a Gas Pipe from the Street to the Building
4. Installation of a Gas Burner
5. Estimation of Gas Heating Costs

Beginning with a description of the present system, the topics then follow the sequence of the actual conversion process. Our writer now has a general plan for gathering data and writing a first draft. She can expand this outline and make it more specific by adding subtopics as shown below. Before writing her final draft, she will develop a formal outline, to check the organization of her report, and to reveal that organization to her readers.

## The Formal Topic Outline

The formal topic outline is detailed and systematic, using notation (numbers and letters) to mark divisions and to show how parts relate to parts and to the whole. Choose between two systems of notation: the roman numeral–letter–arabic numeral system or the decimal system.

**Roman Numeral–Letter–Arabic Numeral Notation.** Here are the five topic headings from the earlier informal outline developed into a formal outline, using roman numeral–letter–arabic numeral notation:

II. REPORT BODY (or COLLECTED DATA)
   A. Description of Our Present Heating System
      1. Physical condition
      2. Required yearly maintenance
      3. Fuel supply problems
         a. Overworked distributor
         b. Varying local supply[1]
      4. Cost of operation
   B. Removal of the Oil Burner and Tank
      1. Data from the oil company
      2. Data from the salvage company
         a. Procedure
         b. Cost
      3. Possibility of private sale
   C. Installation of a Gas Pipe from the Street to the Building
      1. Procedure
      2. Cost of installation
      3. Cost of landscaping

[1]Any division must yield at least two subparts. You could not logically divide "Types of Strip Mining" into "1. Contour Mining," without other subparts. If you can't divide your major topic into at least two subtopics, change your original heading.

    D. Installation of a Gas Burner
       1. Procedure
       2. Cost of plumber's labor and materials
    E. Estimation of Gas Heating Costs
       1. Rate determination
       2. Required yearly maintenance
       3. Cost data from neighboring facility
       4. Overall cost of operation
          a. Cost of conversion
          b. Cost of maintenance
          c. Cost of gas supply

The writer now adds the introduction and conclusion.

  I. INTRODUCTION
    A. Background
    B. Purpose of the Report
    C. Intended Audience (usually for in-school reports only)
    D. Information Sources
    E. Limitations of the Report
    F. Scope (list of major topics)
 II. BODY (as shown earlier)
III. CONCLUSION
    A. Summary of Findings
    B. Comprehensive Interpretation of Findings
    C. Recommendations

In short reports, the introduction and conclusion are still included, but usually shortened to one or two sentences.

A formal outline easily converts into a table of contents for the finished report, as shown in Chapter 18. (Because they serve mainly to guide the *writer*, many minor headings from an outline are omitted from the table of contents or in the report itself. Needless headings can make a document seem fragmented.)

**Decimal Notation.** Here is a partial outline in decimal notation:

2.0 Collected Data
  2.1 Description of Our Present Heating System
    2.1.1 Physical condition
    2.1.2 Required yearly maintenance
    2.1.3 Fuel supply problems
        2.1.3.1 Overworked distributor
        2.1.3.2 Varying local supply
    2.1.4 Cost of operation
  2.2 Removal of the Oil Burner and Tank
    2.2.1 Data from the oil company
    2.2.2 (and so on)

The decimal outline makes it easier to refer readers to various sections. But both systems achieve the same organizing objective. Unless readers express a preference, use whichever system you prefer.

## The Formal Sentence Outline

The previous outline is a *topic outline* because each division is expressed as a topic phrase. A topic outline may be expanded into a *sentence outline*.

> II. COLLECTED DATA
>   A. Our present heating needs are supplied by circulating hot air generated by a Model A-12, electrically fired Zippo oil burner fed by a 275-gallon fuel tank. Both are twelve years old.
>     1. Both burner and tank are in good working order and physical condition, for they have been carefully maintained.
>     2. The oil burner and associated components require cleaning once yearly at a service charge of $38. The air filter, costing $6.25, is replaced three times yearly at a total cost of $18.75.
>     3. (and so on)

Each sentence in turn serves as a topic sentence for a paragraph in the report. Sentence outlines are used mainly in large projects in which various team members prepare different sections of a long document.

## ELEMENTS OF A FORMAL OUTLINE

### Parallel Construction for Parallel Levels

Make all items at the same level parallel, or grammatically equal, to emphasize connections among related ideas.

> **Not parallel** E. Estimation of Gas Heating Costs
>     1. Rate determination
>     2. The system requires yearly maintenance
>     3. Cost data were obtained from a neighboring facility
>     4. Overall cost of operation
>
> **Parallel** E. Estimation of Gas Heating Costs
>     1. Rate determination
>     2. Required yearly maintenance
>     3. Cost data from neighboring facility
>     4. Overall cost of operation

For a full discussion of parallelism, see Appendix A.

## Clear and Informative Headings

Be sure topic headings contain informative words. Under "Description of Our Present Heating System," a heading titled "Fuel" is less informative than "Fuel supply problems."

Avoid repetitious headings that add no information.

| | |
|---|---|
| **Not informative** | C. Environmental Effects of Strip Mining |
| |    1. Effects on land |
| |    2. Effects on erosion |
| |    3. Effects on water |
| |    4. Effects on flooding |
| **Informative** | C. Environmental Effects of Strip Mining |
| |    1. Permanent land scarring |
| |    2. Increased erosion |
| |    3. Water pollution |
| |    4. Increased flood hazards |

Each revised heading has a key phrase that summarizes the message.

## Parts in a Sequence That Readers Will Find Logical

A sequence that is logical to your readers will enable them to follow your reasoning, to see the relationships. Does your subject itself suggest a sequence? How should you organize to make your material logical—from your audience's point of view? Specialists in document design suggest that you list the questions you could expect from readers, and order these questions in the sequence you would expect readers to ask their questions (Redish et al. 142). Before deciding on a sequence for your material, discover all you can about your audience by completing an audience-and-use profile sheet (page 33). Here are some possible sequences.

**Chronological Sequence.**   In a chronological sequence, follow the order of events (as in the sequence in a set of instructions). Also, follow this sequence to explain how the parts of a mechanism operate (as in how the heart pumps blood). Begin with the first step, and end with the last.

**Spatial Sequence.**   In a spatial sequence, follow the arrangement of parts (left to right, top to bottom, front to rear), as when you describe how an office will be remodeled to accommodate automated equipment.

**Reasons For and Against.**   In giving reasons for and against something, follow the sequence in which both sides of an issue are argued—first one side, then the other—as in an analysis of the value and danger of a proposed flu-vaccination program.

**Problem-Causes-Solution.**   Follow the sequence of the problem-solving process, from a description of the problem, through diagnosis, to a solution. An analysis of the rising rate of business failures in your area would require this sequence.

**Cause and Effect.** In a cause-and-effect sequence, follow actions to their results, as in analyzing the therapeutic benefits of transcendental meditation.

**Comparison and Contrast.** In evaluating two items, first discuss their similarities and then their differences. An example is an item-by-item comparison of two proposed sites for dumping low-level nuclear wastes.

**Simple to Complex.** Explain a complex subject by beginning with its most familiar or simplest parts. An explanation of satellite transmission should follow the logic of the learning process by describing what we see on our television screens *before* explaining how the video signal is transmitted.

**Sequence of Priorities.** Sometimes you will place items in sequence according to their relative importance, as in a proposal for increasing your town's budget.

Many reports require a combination of sequences. The heating conversion outline fuses the chronological sequence with comparison-contrast and cause-and-effect. A single paragraph generally follows *one* sequence, as shown on pages 192–196.

Research demonstrates that material presented in an organized sequence is easier for readers to understand and remember than unorganized material (Felker et al. 11). Organize in the sequence you expect readers to approach the material.

## AN OUTLINE MODEL

The following model is adaptable to most reports directed toward reaching a decision. (For information-type reports, the Conclusion becomes simply a Summary.)

### GENERAL OUTLINE MODEL

I. INTRODUCTION
    A. Definition, Description, and History
    B. Statement of Purpose
    C. Target Audience (omitted for workplace audiences)
    D. Information Sources (including research methods and materials)
    E. Working Definitions
    F. Limitations of the Report
    G. Scope of Coverage (sequence of major topics in the body)

II. BODY
    A. First Major Topic

1. First subtopic of A
2. Second subtopic of A
   a. First subtopic of 2
   b. Second subtopic of 2
      (and so on—subdivision carried as far as necessary)
    B. Second Major Topic
    (and so on)
III. CONCLUSION (where everything is tied together)
    A. Summary of Information in II
    B. Overall Interpretation of Information in II
    C. Recommendations Based on Information in II

Suggestions for developing each section follow.

## Introduction

A typical introduction has some combination of these sections:

1. *Definition, description, and history.* Begin by giving readers a background on the origins and significance of your topic.

2. *Statement of purpose.* Tell why you are writing the report, and what you plan to achieve.

3. *Target audience.* Identify your audience and its intended use of your information. (Omit this section if you submit an audience-and-use analysis or if your report is for a workplace audience.)

4. *Information sources.* If your data come from outside sources, identify the sources briefly here. (You will document them fully later.)

5. *Working definitions.* If needed, define specialized terms or general terms that have special meanings. If you must define more than five terms, place your definitions in a glossary at the end.

6. *Report limitations.* Explain any incomplete information. Perhaps you were unable to locate a key book or interview a key person. Or perhaps your study must be titled "preliminary" instead of "definitive" (the final word). Or perhaps your report treats only one side of a controversial issue, as in studying the *negative* effects of automation on employee morale.

7. *Scope.* Preview the territory your report will cover by listing major topics to be discussed in the body section.

Not all introductions require each of these sections. Keep your introduction as brief as you can—without compromising its informative value. Reports too often can waste readers' time with needless background. Know your readers, and give them only what they need.

## Body

In the body you present your evidence and explanations. "Show me!" is any reader's implied demand. Show step by step how you move from introduction to conclusion.

Give your body section an informative title. For a mechanism description, you might title the body "Description and Function of Parts." For instructions, "Required Steps." For a problem-solving report, "Collected Data." (See section titles in documents in this book.)

## Conclusion

The concluding section of a report has many purposes: it might evaluate the significance of the report, take a position, predict an outcome, offer a solution, or suggest further study. Whatever the conclusion's specific purpose, readers expect a report to conclude with a clear perspective on the whole report. Conclusions vary with the document. In a mechanism description, you might simply review the mechanism's major parts, then briefly describe one operating cycle. Here are the possible types of material in a conclusion:

1. *Summary of the body.* When your body section is long and involved or your major findings need reemphasis, summarize them briefly.

2. *Overall interpretation of the body.* Tell the readers what your data mean—even when the report does not call for recommendations.

3. *Recommendations.* Base recommendations directly on your findings and interpretations.

A good beginning, middle, and ending are indispensable, but alter your own outline as you see fit. No one model should be followed slavishly by any writer. *The organization of any document ultimately is determined by what your readers need.*

## THE IMPORTANCE OF BEING MESSY

Keep in mind that the neat and ordered outlines shown earlier represent the *products* of outlining, not the *process.* Beneath any finished outline (or any finished document, for that matter) lie pages and pages of scribbling and things crossed out, lists, arrows, and fragments of ideas. Writing begins in disorder. Messiness is a natural and often essential part of writing in its early stages.

In a recent survey to assess how computers influence workers' writing, traditional formal outlining was found to be giving way to outlining that took the form of "notes on audience, purpose, direction, key content points, tone." These outlines were "flexible, sketchy, punctuated by arrows, numbers or exclamation points; they looked more like lists . . ." (Halpern 179). Outlines needn't be pretty, so long as they help you control your material.

Not until you assemble your final draft of a long document do you compose the finished outline, which will serve as a model for your table of contents. At this final stage, a finished outline serves as a quality-control check on your reasoning, and as a way of revealing to your readers a logical line of thinking.

## THE REPORT DESIGN WORKSHEET

As an alternative to the audience-and-use profile sheet (page 33), the worksheet in Figure 9.1 can supplement your outline, to help you zero in on your audience and purpose. This version is based on a worksheet model developed by Professor John S. Harris of Brigham Young University. The figure has been completed for the heating-conversion report outlined earlier.

## EXERCISE

For each of the following documents, indicate the best sequence. (For example, a description of a proposed computer lab would follow a spatial sequence.)

- a set of instructions for operating a power tool
- a campaign report describing your progress in political fund-raising
- a report analyzing the weakest parts in a piece of industrial machinery
- a report analyzing the desirability of a proposed oil refinery in your area
- a detailed breakdown of your monthly budget to trim excess spending
- a report investigating the reasons for student apathy on your campus
- a report investigating the effects of the ban on DDT in insect control
- a report on any highly technical subject, written for a general reader
- a report investigating the success of a no-grade policy at other colleges
- a proposal for a no-grade policy at your college

## TEAM PROJECTS

1. Organize into small groups. Choose *one* of these topics, formulate a statement of purpose, and brainstorm to develop a formal outline for the body section of a report. One representative from your group can write the final draft on the board for class revision.

- job opportunities in your career field
- a physical description of the ideal classroom
- how to organize an effective job search
- how the quality of your higher educational experience can be improved
- arguments for and against a formal grading system
- arguments for and against a college-wide computer literacy requirement

2. Assume your group is preparing a report titled "The Negative Effects of Strip Mining on the Cumberland Plateau Region of Kentucky." After brainstorming and researching your subject, you all settle on these four major topics:

**FIGURE 9.1** Completed Report Design Worksheet

## REPORT DESIGN WORKSHEET

Preliminary Information

What is to be done?  *A report on the feasibility of converting our home office from oil to gas heat*

Whom is it to be presented to, and when?  *Charles Jones, company president: April 1*

| Audience Analysis | Primary Reader(s) | Secondary Reader(s) |
|---|---|---|
| Position and title: | *President, Abco Engineering consultants* | *Company officers engineering staff* |
| Relationship to author or organization: | *Employer* | *supervisors, colleagues, junior members* |
| Technical expertise: | *nontechnical (for this subject)* | *nontechnical* |
| Personal characteristics: | *highly efficient; demands quality and economy* | *all serious-minded professionals* |
| Attitude toward author or organization: | *is considering me for promotion to assistant V.P.* | *friendly and respectful; officers will vote on my promotion* |
| Attitude toward subject: | *highly interested because of last winter's inconvenience* | *interested* |
| Effect of report on readers or organization: | *will be read closely and acted upon* | *will be read and discussed at our next staff meeting* |

Reader's Purpose

| | | |
|---|---|---|
| Why has reader requested it? | *wants to make a practical decision* | _____ |
| What does reader plan to do with it? | *use the data to make the best choice* | *confer with the president about the choice* |
| What should reader know beforehand to understand it as written? | *nothing special; history of problem is reviewed in report* | *same* |
| What does reader already know? | *remembers last winter's problems* | *same* |
| What amount and kinds of detail will reader find significant? | *brief description of conversion procedures and detailed cost analysis* | *same* |
| What should reader know and/or be able to do after reading it? | *make an educated decision* | *advise the president about his decision* |

**FIGURE 9.1** Completed Report Design Worksheet  *Continued*

Writer's Purpose

Why am I writing it?   *to communicate my research findings clearly*

What effect(s) do I wish to achieve?   *to have my readers conclude that conversion is not economically feasible; to persuade them to accept my recommendation of an alternative to conversion*

Design Specifications

Sources of data:   *gas company, our oil company representative; Tubo Plumbing Corp., Jumbo Salvage Co., Watt Electronics, Inc.*

Tone:   *semiformal*

Point of view:   *first- and second-person*

Needed visuals and supplements:   *title page, letter of transmittal, table of contents, informative abstract, data sheet appendix reviewing the procedure for cost analysis*

Appropriate format (letter, memo, etc.):   *formal report format with full heading system*

Rhetorical mode (description, definition, classification, etc.—or some combination):   *primary mode: analysis; secondary modes: description, process narration*

Basic organization (problem-causes-solution, intro-instructions-summary, etc.):   *questions-answers-conclusions and recommendations*

Main points in introduction:   *Background*
*Purpose*
*Intended Audience*
*Data Sources*
*Limitatioins Scope*

Main points in body:   *Description of Present System*
*Removal of Oil Burner and Tank*
*Installation of Oil Pipe*
*Installation of Gas Burner*
*Estimation of Gas Heating Costs*

Main points in conclusion:   *Summary of Findings*
*Interpretation of Findings*
*Recommendation*

Other Considerations   *no frills: these readers are all engineers interested in hard facts.*

- economic and social effects of strip mining
- description of the strip-mining process
- environmental effects of strip mining
- description of the Cumberland Plateau

Arrange these topics in the most sensible sequence.

When your topics are arranged, assume that subsequent research and further brainstorming produce this list of subtopics:

- method of strip mining used in the Cumberland Plateau region
- location of the region
- permanent land damage
- water pollution
- lack of educational progress
- geological formation of the region
- open-pit mining
- unemployment
- increased erosion
- auger mining
- natural resources of the region
- types of strip mining
- increased flood hazards
- depopulation
- contour mining

Arrange each subtopic (and perhaps some sub-sub-topics) under its appropriate topic headings. Use effective notation to create the body section of a formal outline. Appoint one group member to present the outline in class.

*Hint:* Assume that this is your thesis sentence: "The federal government needs to acknowledge that decades of strip mining (without reclamation) in the Cumberland Plateau have devastated this region's environment, economy, and social structure."

# 10

# Shaping the Paragraphs

 The Standard Paragraph

Paragraph Unity

Paragraph Coherence

Sequence in a Paragraph

Paragraph Length

To follow a writer's line of thinking, readers need a message that is sensibly organized. But thinking rarely occurs in a neat, predictable sequence, meaning that writers cannot merely record their thoughts in the original order. The material must instead be *shaped* into an organized unit of meaning. In setting out to organize a message, writers face deliberate decisions:

- What do I want to emphasize?
- What do I say first?
- What comes after that?
- How do I end?

Useful messages of any length—whether in the form of a book, chapter, news article, letter, or memo—typically follow a common organizing pattern:

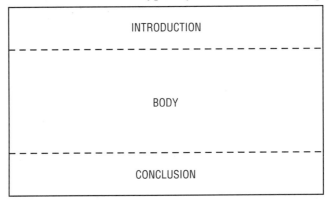

INTRODUCTION

BODY

CONCLUSION

This pattern is best illustrated in the form of a standard paragraph.

Whereas outlines help us shape the whole message into this logical pattern, paragraphs help us shape its parts.

## THE STANDARD PARAGRAPH

A standard paragraph is a group of sentences focused on one main organizing point. The main point is expressed as a *topic sentence*:

Computer literacy soon will be a requirement for all "educated" people.

A video display terminal can endanger the operator's health.

Chemical pesticides and herbicides are both ineffective and hazardous.

Each of these topic sentences introduces only one way of seeing a subject. Without explanations, we could not possibly grasp the writer's exact meaning. Consider the third statement:

Chemical pesticides and herbicides are both ineffective and hazardous.

Imagine that you are a researcher for the Epson Electric Light Company, and you have been assigned this task: determine whether the company (1) should begin spraying pesticides and herbicides under its power lines, as many other utilities are doing, or (2) should continue with its manual (and nonpolluting) ways of minimizing foliage and insect damage to lines and poles. If you simply responded with the preceding statement, your boss would have questions:

- Why, exactly, are chemical pesticides and herbicides ineffective and hazardous?
- What are the problems?
- Can you explain?

By answering these questions in a fully developed paragraph, you provide the necessary supporting details:

[1]Chemical pesticides and herbicides are both ineffective and hazardous. [2]Because none of these chemicals has permanent effects, pest populations invariably recover and require respraying. [3]Repeated applications cause pests to develop immunity to the chemicals. [4]Furthermore, most of these products attack species other than the intended pest, killing off its natural predators, and thereby actually increasing the pest population. [5]Above all, chemical residues survive in the environment (and living tissue) for years, often carried hundreds of miles by wind and water. [6]This toxic legacy includes such biological disasters as birth deformities, reproductive failures, brain damage, and cancer. [7]The ultimate victims of these chemicals would be our customers. [8]I therefore recommend we continue our present control methods.

Introduction

Body (2-6)

Conclusion (7-8)

Most paragraphs in technical writing have an introduction-body-conclusion structure. They begin with a clear topic (or orienting) sentence stating a generalization. Details in the body support the generalization.

## Writing the Topic Sentence

Readers look to the first one or two sentences in a paragraph to orient themselves, to build clear expectations. A topic sentence gives readers a framework for understanding your message. Without this framework, readers cannot possibly grasp your exact meaning. Consider this paragraph, whose topic sentence has been left out (read it *once* only):

> Besides containing several toxic metals, it percolates through the soil, leaching out naturally present metals. Pollutants such as mercury invade surface water, accumulating in fish tissues. Any organism eating the fish — or drinking the water — in turn faces the risk of heavy metal poisoning. Moreover, acidified water can release heavy concentrations of lead, copper, and aluminum from metal plumbing, making ordinary tap water hazardous.

After one reading, are you able to provide a main point for the paragraph? Could you restate the message accurately in your own words? Probably not, even after a second reading. Without the orientation given by a topic sentence, you have no framework for understanding the information's larger meaning. And because you don't know what to look for, you can't figure out where to place the emphasis: on polluted fish, on metal poisoning, on tap water?

Now, after inserting this sentence at the beginning, reread the paragraph:

> Acid rain indirectly threatens human health.

In light of this organizing point, the exact meaning of the message becomes obvious. The topic sentence gives us a framework by:

1. Naming the subject of the message (acid rain).
2. Stating the topic — the writer's specific viewpoint on the subject (that acid rain threatens human health).
3. Forecasting how the message will be developed (by explaining the process) in response to the reader's main question: *How exactly does acid rain threaten human health?*

In most standard paragraphs, readers expect the topic sentence to serve as the key to understanding the whole. Your topic sentence therefore should appear *first* in the paragraph, unless you have good reason to place it elsewhere. Think of your topic sentence as the one sentence you would keep if you could keep only one (U.S. Air Force Academy 11).

Let your readers know immediately what to expect. Don't write *Some pesticides are less hazardous and often more effective than others* when you mean *Organic pesticides are less hazardous and often more effective than their chemical counterparts.* The first topic sentence leads everywhere and nowhere; the second helps us focus, tells us what to expect from the paragraph. Don't write *Acid rain poses a danger*, thereby imposing on readers the job of figuring our your meaning of *danger*. If you mean that *Acid rain is killing our lakes and polluting our water supplies*, say so. Uninformative topic sentences keep readers guessing.

## Writing the Body

The paragraph body contains the supporting details that explain and expand your main idea. This support material answers the questions you can expect from readers about your topic sentence:

- Says who?
- What proof do you have to support your claim?
- Can you give examples?

Here is how writer Tracy Kidder develops a paragraph to support his topic sentence (in boldface):

> **Obviously, computers have made differences.** They have fostered the development of spaceships—as well as a great increase in junk mail. The computer boom has brought the marvelous but expensive diagnostic device known as the CAT scanner, as well as a host of other medical equipment; it has given rise to machines that play good but rather boring chess, and also, on a larger game board, to a proliferation of remote-controlled weapons in the arsenals of nations. Computers have changed ideas about waging war and about pursuing science, too. It is hard to see how contemporary geophysics or meteorology or plasma physics can advance very far without them now. Computers have changed the nature of research in mathematics, though not every mathematician would say it is for the better. And computers have become a part of the ordinary conduct of businesses of all sorts (243–244).

With the topic sentence and supporting details on paper, the paragraph is ready for its conclusion.

## Writing the Conclusion

Your concluding statement signals that discussion of the main idea stated in your topic sentence is ending. In it you usually tie the paragraph together by summarizing, interpreting, or judging the facts. If the paragraph is part of a longer document, your conclusion can also prepare readers for a subsequent paragraph. Here is Tracy Kidder's conclusion for the preceding paragraph:

> They really help, in some cases.

This particular conclusion serves as transition to Kidder's next main idea:

> **Not always, though.** One student of the field has estimated that about forty percent of commercial applications of computers have proved uneconomical, in the sense that the job the computer was bought to perform winds up costing more

to do after the computer's arrival than it did before. Most computer companies have boasted that they aren't just selling machines, they're selling *productivity*. ("We're not in competition with each other," said a PR man. "We're in competition with labor.") But that clearly isn't always true. Sometimes they're selling paper-producers that require new legions of workers to push that paper around (244).[1]

An introduction-body-conclusion structure should serve most of your paragraph needs in technical writing.[2] Begin each support paragraph with a solid topic sentence and you will stay on target.

## Structural Variation

Sometimes a main idea that needs detailed definition calls for a topic statement of two or more sentences (as shown in the needle description paragraph, on page 193). Or your main idea might have several distinct parts, which would result in an excessively long paragraph. You might break up the paragraph, letting your topic statement stand as a brief introductory paragraph and serve the separate subparts, which are set off as independent paragraphs for the readers' convenience.

> **The most common types of strip-mining procedures are open-pit mining, contour mining, and auger mining. The specific type employed will depend on the type of terrain covering the coal.**
>
> Open-pit mining is employed in the relatively flat lands in western Kentucky, Oklahoma, and Kansas. Here, draglines and scoops operate directly on the coal seams, producing long, parallel rows of packed spoil banks, ten to thirty feet high, with steep slopes. Between the spoil banks are large pits that fill with water to create pollution and flood hazards.
>
> Contour mining is most widely practiced in the mountainous terrain of the Cumberland Plateau and eastern Kentucky. Here, bulldozers and explosives cut and blast the earth and rock covering a coal seam. Wide bands are removed from the mountain's circumference to reach the embedded coal beneath. The cutting and blasting result in a shelf along with a manmade cliff some sixty feet high, at a right angle to the shelf. The blasted and churned earth is pushed over the shelf to form a massive and unstable spoil bank that creates a danger of mud slides.
>
> Auger mining is employed when the mountain has been cut so thin it can no longer be stripped. It is also used in other difficult-access terrain. Here, large augers bore parallel rows of holes into the hidden coal seams to extract the embedded coal. Among the three strip-mining methods, auger mining causes least damage to the surrounding landscape.

Each paragraph begins with a clear statement about the part of the subtopic to be discussed.

[1]From THE SOUL OF A NEW MACHINE by Tracy Kidder. Copyright © 1981 by John Tracy Kidder. By permission of Little, Brown and Co.
[2]Besides support paragraphs, of course, you will write introductory paragraphs, transitional paragraphs, and concluding paragraphs (as illustrated in the sample reports in later chapters).

## PARAGRAPH UNITY

A paragraph is unified when all its parts work toward the same end—when every word, phrase, and sentence explains, illustrates, and clarifies the idea expressed in the topic sentence.

Solar power offers an efficient, economical, and safe solution to the Northeast's energy problems. To begin with, solar power is highly efficient. Solar collectors installed on fewer than 30 percent of roofs in the Northeast would provide more than 70 percent of the area's heating and air-conditioning needs. Moreover, solar heat collectors are economical, operating for up to twenty years with little or no maintenance. These savings recoup the initial cost of installation within only ten years. Most important, solar power is safe. It can be transformed into electricity through photovoltaic cells (a type of storage battery) in a noiseless process that produces no air pollution—unlike coal, oil, and wood combustion. In sharp contrast to its nuclear counterpart, solar power produces no toxic waste and poses no catastrophic danger of meltdown. Thus, massive conversion to solar power would ensure abundant energy and a safe, clean environment for future generations.

*A unified paragraph*

One way to destroy unity in the paragraph above would be to discuss the differences between active and passive solar heating, or manufacturers of solar technology, or the advantages of solar power over wind power. Although these matters do *broadly* relate to the general issue of solar energy, none directly advances the meaning of *efficient, economical,* or *safe.*

Every topic sentence has a key word or phrase that carries the meaning. In the pesticide-herbicide paragraph (page 187), the key words are *ineffective* and *hazardous.* Anything that fails to advance their meaning throws the paragraph—and the readers—off track. Paragraph unity is destroyed when you drift from your stated purpose by adding irrelevant details.

## PARAGRAPH COHERENCE

A paragraph is coherent when it hangs together and flows smoothly in a clear direction—when all sentences are logically connected like links in a chain, leading toward a definite conclusion.

Three ways to damage paragraph coherence are (1) to use short, choppy sentences, (2) to place sentences in the wrong sequence, and (3) to use insufficient transitions and other connectors (pages 518–521) to link related ideas. Here is how the paragraph on solar energy could become incoherent:

Solar power offers an efficient, economical, and safe solution to the Northeast's energy problems. Unlike nuclear power, solar power produces no toxic waste and poses no danger of meltdown. Solar power is efficient. Solar collectors could be installed on fewer than 30 percent of roofs in the Northeast. These collectors would provide more than 70 percent of the area's heat-

*An incoherent paragraph*

ing and air-conditioning needs. Solar power is safe. It can be transformed into electricity. This transformation is made possible by photovoltaic cells (a type of storage battery). Solar heat collectors are economical. The photovoltaic process produces no air pollution.

The paragraph above (only part is shown) lacks coherence in three ways: the sentences are choppy; the sequence of sentences is chaotic; and transitions and connectors are missing.

Here again is the original, coherent paragraph. Notice how this version reveals a clear line of thought:

**A coherent paragraph**

[1]Solar power offers an efficient, economical, and safe solution to the Northeast's energy problems. [2]**To begin with,** solar power is highly efficient. [3]Solar collectors installed on fewer than 30 percent of roofs in the Northeast would provide more than 70 percent of the area's heating and air-conditioning needs. [4]**Moreover,** solar heat collectors are economical, operating for up to twenty years with little or no maintenance. [5]**These savings** recoup the initial cost of installation within only ten years. [6]**Most important,** solar power is safe. [7]**It** can be transformed into electricity through photovoltaic cells (a type of storage battery) in a noiseless process that produces no air pollution—unlike coal, oil, and wood combustion. [8]**In sharp contrast** to its nuclear counterpart, solar power produces no toxic waste and poses no danger of catastrophic meltdown. [9]**Thus,** massive conversion to solar power would ensure abundant energy and a safe, clean environment for future generations.

We can easily trace the sequence of thoughts in this paragraph.

  1. The topic sentence establishes a clear direction.

2–3. The first reason is given and then explained.

4–5. The second reason is given and explained.

6–8. The third and major reason is given and explained.

  9. The conclusion sums up and reemphasizes the main point.

Within this line of thinking, each sentence follows logically from the one before it. Readers know exactly where they are at any point in the paragraph. To reinforce the logical sequence, related ideas are combined in individual sentences, and transitions and connectors (in boldface) signal clear relationships. The whole paragraph sticks together.

## SEQUENCE IN A PARAGRAPH

In developing the sequence of a paragraph, you decide on which idea to discuss first, which second, and so on. The sequence you select will depend on your subject, purpose, and readers' needs. Some possibilities follow.

### Spatial Sequence

A spatial sequence begins at one location and ends at another. It is most useful in describing a physical item or a mechanism. Describe the parts in the order in which readers would actually view them: left or right, inside to outside. This description of a hypodermic needle proceeds from the needle's base (hub) to its point:

> A hypodermic needle is a slender, hollow steel instrument used to introduce medication into the body (usually through a vein or muscle). It is a single piece composed of three parts, all considered sterile: the hub, the cannula, and the point. The hub is the lower, larger part of the needle that attaches to the necklike opening on the syringe barrel. Next is the cannula (stem), the smooth and slender central portion. Last is the point, which consists of a beveled (slanted) opening, ending in a sharp tip. The diameter of a needle's cannula is indicated by a gauge number; commonly, a 24–25 gauge needle is used for subcutaneous injections. Needle lengths are varied to suit individual needs. Common lengths used for subcutaneous injections are ⅜, ½, ⅝, and ¾ inch. Regardless of length and diameter, all needles have the same functional design.

## Chronological Sequence

A paragraph describing a series of events or giving instructions is most effective when its details are arranged according to a strict time sequence: first step, second step, and so on.

> Most jogging injuries occur when inexperienced runners begin too quickly. Before taking a step, spend at least ten minutes stretching and warming up, using any exercises you find comfortable. (After your first week, consult a jogging book for specialized exercises.) When you've completed your warmup, set a brisk pace walking. Exaggerate the distance between steps, taking long strides and swinging your arms briskly and loosely. After roughly 100 yards at this brisk pace, you should feel ready to jog. Immediately break into a very slow trot: lean your torso forward and let one foot fall in front of the other (one foot barely leaving the ground while the other is on the pavement). Maintain the slowest pace possible, just above a walk. *Do not bolt out like a sprinter!* The biggest mistake is to start fast and injure yourself. While jogging, relax your body. Keep your shoulders straight and your head up, and enjoy the scenery—after all, it is one of the joys of jogging. Keep your arms low and slightly bent at your sides. Move your legs freely from the hips in an action that is easy, not forced. Make your feet perform a heel-to-toe action: land on the heel; rock forward; take off from the toe.

The paragraph explaining how acid rain endangers human health (page 188) is another example of chronological sequence.

## Effect-to-Cause Sequence

A paragraph that first identifies a problem and then discusses its causes is typically found in problem-solving reports.

> Modern whaling techniques have brought the whale population to the threshold of extinction. In the nineteenth century, invention of the steamboat increased hunters' speed and mobility. Shortly afterward, the grenade harpoon was invented

so that whales could be killed quickly and easily from the ship's deck. In 1904, a whaling station opened on Georgia Island in South America. This station became the gateway to Antarctic whaling for the nations of the world. In 1924, factory ships were designed that enabled round-the-clock whale tracking and processing. These ships could reduce a ninety-foot whale to its by-products in roughly thirty minutes. After World War II, more powerful boats with remote sensing devices gave a final boost to the whaling industry. The number of kills had now increased far beyond the whales' capacity to reproduce.

### Cause-to-Effect Sequence

In a cause-to-effect sequence, the topic sentence identifies the cause(s), and the remainder of the paragraph discusses its effects.

> Some of the most serious accidents involving gas water heaters occur when a flammable liquid is used in the vicinity. The heavier-than-air vapors of a flammable liquid such as gasoline can flow along the floor—even the length of a basement—and be explosively ignited by the flame of the water heater's pilot light or burner. Because the victim's clothing frequently ignites, the resulting burn injuries are commonly serious and extremely painful. They may require long hospitalization, and can result in disfigurement or death. *Never, under any circumstances, use a flammable liquid near a gas heater or any other open flame.* (Consumer Product Safety Commission)

### Emphatic Sequence

Paragraphs that provide detailed reasons to support a specific viewpoint or recommendation often appear in workplace writing, as in the pesticide-herbicide paragraph on page 187 or the solar energy paragraph on page 191. For emphasis, the reasons or examples usually are arranged in decreasing or increasing order of importance.

> Although strip mining is safer and cheaper than conventional mining, it is highly damaging to the surrounding landscape. Among its effects are scarred mountains, ruined land, and polluted waterways. Strip operations are altering our country's land at the rate of 5,000 acres per week. An estimated 10,500 miles of streams have been poisoned by silt drainage in Appalachia alone. If strip mining continues at its present rate, 16,000 square miles of U.S. land eventually will be stripped barren.

In this paragraph, the most dramatic example is saved for the end.

### Problem-Causes-Solution Sequence

The problem-causes-solution paragraph is commonly used in daily activity or progress reports. After outlining the cause of the problem, the writer of this paragraph explains how the problem has been solved:

> On all waterfront buildings, the unpainted wood exteriors had been severely damaged by the high winds and sandstorms of the previous winter. After repairing

the damage, we took protective steps against further storms. First, all joints, edges, and sashes were treated with water-repellent preservative to protect against water damage. Next, three coats of nonporous primer were applied to all exterior surfaces to prevent paint from blistering and peeling. Finally, two coats of wood-quality latex paint were applied over the nonporous primer. To keep coats of paint from future separation, the first coat was applied within two weeks of the priming coats, and the second within two weeks of the first. Two weeks after completion, no blistering, peeling, or separation has occurred.

## Comparison/Contrast Sequence

A paragraph discussing the similarities or differences (or both) between two or more items often is used in job-related writing.

The ski industry's quest for a binding that ensures good performance as well as safety has led to development of two basic types. Although both bindings improve performance and increase the safety margin, they have different release and retention mechanisms. The first type consists of two units (one at the toe, another at the heel) that are spring-loaded. These units apply their retention forces directly to the boot sole. Thus the friction of boot against ski allows for the kind of ankle movement needed at high speeds over rough terrain, without causing the boot to release. In contrast, the second type has one spring-loaded unit at either the toe or the heel. From this unit extends a boot plate that travels the length of the boot to a fixed receptacle on its opposite end. With this plate binding, the boot has no part in release or retention. Instead, retention force is applied directly to the boot plate, providing more stability for the recreational skier, but allowing for less ankle and boot movement before releasing. Overall, the double-unit binding performs better in racing, but the plate binding is safer.

For comparing and contrasting more specific data on these bindings, two lists would be most effective.

The Salomon 555 offers the following features:

1. upward release at the heel and lateral release at the toe (thus eliminating 80 percent of leg injuries)
2. lateral antishock capacity of 15 millimeters, with the highest available return-to-center force
3. two methods of reentry to the binding: for hard and deep powder conditions
4. five adjustments
5. (and so on)

The Americana offers these features:

1. upward release at the toe as well as upward and lateral release at the heel
2. lateral antishock capacity of 30 millimeters, with moderate return-to-center force
3. two methods of reentry to the binding
4. two adjustments, one for boot length and another for comprehensive adjustment for all angles of release and elasticity
5. (and so on)

Instead of this block structure (in which one binding is discussed and then the other), the writer might have chosen a point-by-point structure (in which points common to all items, such as "Reentry Methods," are listed together). The point-by-point comparison is particularly favored in feasibility and recommendation reports because it offers readers a meaningful comparison between common points rather than by cataloging items separately.

## PARAGRAPH LENGTH

Paragraph length depends on the writer's purpose and audience's needs. To guide your decision about length, answer this question:

> How much and what kind of support do I need, to convey my exact meaning to *these* readers?

When deciding about paragraph length, consider these guidelines:

1. Excessive use of short paragraphs makes a message seem choppy and poorly organized. A series of short paragraphs, however, *is* effective in step-by-step instructions.

2. Too many long paragraphs can be hard to follow. Important ideas can get buried in the middle of a long paragraph. Unless your paragraph is in list form (like this one), it should rarely be longer than fifteen lines (or about one-half of a double-spaced page).

3. A short paragraph (even a single-sentence paragraph) can attract attention to an important idea by setting it off.

   > We can prevent further damage from mud slides only by building a retaining wall behind the number 3 construction site immediately.

4. A combination of shorter and longer paragraphs is best for emphasis, if your subject allows this arrangement.

5. Avoid long paragraphs at the beginning or end of a document.

## EXERCISE

Here are assorted writing situations. Select one (or more, if assigned), and respond with a unified, coherent, and logically developed paragraph. Be sure to identify (on paper) your audience's needs, and its uses of your information. Begin with a definite topic sentence. List anticipated reader questions about your topic sentence. Brainstorm for worthwhile content.

Submit your final paragraph to your instructor, attaching your audience-and-use analysis, audience questions, brainstorming list, and early drafts.

   *a.* From the start, you've had major complaints about inadequate facilities in your dorm or apartment (leaking faucets, no parking space, or the like). You decide to take action by writing a one-paragraph note to the president of your dorm council (or the supervisor of your apartment building), spelling out the problems and demanding a solution.

b. The student senate has published a request for nominations for the Teacher-of-the-Year award. This request stipulates that all nominations be accompanied by a paragraph of no more than 200 words, showing why your candidate should win. You decide to nominate Professor X.

c. You've decided to apply for a scholarship offered by your college. Among other materials, the scholarship committee requests a paragraph explaining your reasons for attending this college. Write the paragraph.

d. Think about a place in your town or on campus that needs improvement. Describe the problem in one paragraph to a specified audience (mayor, dean, head of residence) who will use your information as a basis for action.

e. Assume it's time for end-of-semester course evaluations by students. Write a one-paragraph evaluation of your favorite course, to be read by the department chairperson.

f. Describe the job outlook in your chosen field. Write one paragraph for a high school senior interested in your major. Your paragraph will be included in the Career Handbook published by your college.

# Revising for a
# Readable Style

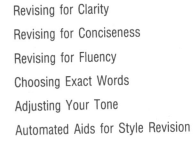

Revising for Clarity

Revising for Conciseness

Revising for Fluency

Choosing Exact Words

Adjusting Your Tone

Automated Aids for Style Revision

You might write for a diverse or specific audience, or for an expert or nonexpert audience. But no matter how technically appropriate your document, the audience's needs will not be served unless your style is readable.

But what is *writing style* and how does it influence reader response to a document? Your writing style is the product of

- the words you choose
- the way in which you put a sentence together
- the length of your sentences
- the way in which you connect sentences
- the tone you convey.

Efficient writing style is neither fancy nor impressive; instead it is straightforward, easy to follow and understand—in a word, *readable*.

Efficient style requires much more than correct grammar, punctuation, and spelling. Granted, basic mechanical errors do distract readers; but correctness alone is no guarantee that your style will be readable. This response to a job applicant is mechanically "correct" but highly inefficient:

> We are in receipt of your recent correspondence indicating your interest in securing the advertised position. Your correspondence has been duly forwarded to the office having employment candidate selection responsibility for consideration. You may expect to hear from the aforementioned office relative to your application as the selection process progresses. Your interest in the position is appreciated.

Notice how hard you have worked to extract information that could be expressed this simply:

> Your application for the advertised position has been received and forwarded to the office that will select candidates. At each stage in the selection, we will inform you of your status. Thank you for your interest.

Inefficient style makes readers work harder than they should.

Style can be inefficient for many reasons, but it is especially inept when it

- makes the writer's meaning impossible to interpret
- takes too long to make the point
- reads like a Dick-and-Jane story from primary school
- uses imprecise or needlessly big words
- sounds stuffy and impersonal

To help your audience spend less time reading, you must spend more time revising for a style that is *clear, concise, fluent, exact,* and *likable.*

## REVISING FOR CLARITY

A clear sentence conveys its exact meaning on the very first reading. It signals relationships among its parts, and it emphasizes the main idea. The strategies outlined here will help you revise for clarity.

### Avoid Ambiguous Phrasing

Any sentence in workplace writing should have *one* meaning only, should allow for *one* interpretation. For instance does one's "suspicious attitude" mean that one is "suspicious" or "suspect"? Make sure your meaning is absolutely clear.

> **Ambiguous**   All managers are not required to submit reports. (Are some or none required?)
>
> **Revised**   Managers are not all required to submit reports.
> *or*
> No Managers are required to submit reports.
>
> **Ambiguous**   Our two helicopters crashed, injuring eight people. (Did they crash separately or did they collide?)
>
> **Revised**   Each of our two helicopters crashed, injuring eight people.
> *or*
> Our two helicopters collided, injuring eight people.

Make sure your writing conveys the meaning you intend.

### Avoid Ambiguous Pronoun References

Whenever you use a pronoun (*he, she, it, their,* etc.), it must refer to one clearly identified noun (or referent); otherwise your message will be confusing.

| Ambiguous referent | Our patients enjoy the warm days while they last. (Are the patients or the warm days on the way out?) |

Depending on whether the referent for *they* is *patients* or *warm days*, the sentence can be clarified thus:

| Clear referent | While these warm days last, our patients enjoy them. |

*or*

Our terminal patients enjoy the warm days.

| Ambiguous referent | Jack resents his assistant because he is competitive. (Who's the competitive one—Jack or his assistant?) |
| Clear referent | Because his assistant is competitive, Jack resents him. |

*or*

Because Jack is competitive, he resents his assistant.

Be sure readers can identify the noun your pronoun replaces. (See pages 500–501 for a detailed discussion of pronoun references.)

## Avoid Ambiguous Punctuation

Too often, a missing comma, hyphen, or other punctuation mark can obscure your meaning.

| Missing hyphen | Replace the trailer's inner wheel bearings. (The inner-wheel bearings or the inner wheel-bearings?) |
| | Our president is a high fidelity fanatic. (Someone who likes drugs, but refuses to fool around?) |
| Missing comma | Does your company produce liquid hydrogen? If so, how[,] and where do you store it? (Notice how the meaning changes with a comma after "how.") |
| | Police surrounded the crowd[,] attacking the strikers. (Without the comma, the crowd appears to be attacking the strikers.) |

Although missing hyphens and commas are prime culprits, other omissions can cause ambiguity as well. A missing colon after *kill* yields the headline "Moose Kill 200." A missing apostrophe after *Myers* creates this gem: "Myers Remains Buried in Portland." Punctuation *does* affect meaning.

## Avoid Telegraphic Writing

*Function words* show relationships between the *content words* (nouns, adjectives, verbs, and adverbs) in a sentence. Some examples of function words:

- articles (*a, an, the*)
- prepositions (*in, of, to*)
- linking verbs (*is, has, looks*)
- relative pronouns (*who, which, that*)

Some writers mistakenly try to compress their writing by eliminating these function words.

| | |
|---|---|
| **Ambiguous** | Proposal to employ retirees almost dead. |
| **Revised** | The proposal to employ retirees **is** almost dead. |
| **Ambiguous** | Uninsulated end pipe ruptured. (What ruptured? The pipe or the end of the pipe?) |
| **Revised** | **The** uninsulated end **of the** pipe ruptured. |

*or*

**The** uninsulated pipe **on the** end ruptured.

| | |
|---|---|
| **Ambiguous** | The reactor operator told management several times he expected an accident. (Did he tell them once or several times?) |
| **Revised** | The reactor operator told management several times **that** he expected an accident. |

*or*

The reactor operator told management **that** several times he expected an accident.

## Avoid Ambiguous Modifiers

Modifiers explain, define, or add detail to other words or ideas. If a modifier is too far from the words it modifies, the message can be ambiguous.

| | |
|---|---|
| **Misplaced modifier** | **Only** press the red button in an emergency. (Does *only* modify *press* or *emergency?*) |
| **Revised** | Press **only** the red button in an emergency. |

*or*

Press the red button in an emergency **only**.

Another problem with ambiguity occurs when the modifier has no word to modify.

| | |
|---|---|
| **Dangling modifier** | **Being so well known in the computer industry, I** would appreciate your advice. |

The writer intended to say that the *reader* is well known, but with no word to modify, the modifying phrase dangles. We can repair the confusing message by adding a subject:

| | |
|---|---|
| **Revised** | Because **you** are so well known in the computer industry, I would appreciate your advice. |

## Unstack the Modifying Nouns

One noun can modify another noun (as in "software development"). But when two or more nouns modify a noun, the string of densely packed words becomes hard to read and ambiguous.

| | |
|---|---|
| **Stacked** | Be sure to leave enough time for a **training session participant** evaluation. (Evaluation of or by participants?) |

With no function words (articles, prepositions, verbs, relative pronouns) to break up the string of nouns, readers cannot see the relationships among the nouns. What modifies what?

Stacked nouns also deaden your style. Bring your style *and* your reader to life by using action verbs (*complete, prepare, reduce*) and prepositional phrases.

| | |
|---|---|
| **Revised** | Be sure to leave enough time **for** participants **to evaluate** the training session. |
| | *or* |
| | Be sure to leave enough time **to evaluate** participants **in the** training session. |

No such problem with ambiguity occurs when *adjectives* are stacked in front of a noun.

| | |
|---|---|
| **Clear** | He was a **nervous, angry, confused,** but **dedicated** employee. |

Readers can readily see that the adjectives modify *employee.*

## Arrange Words for Coherence and Emphasis

In a coherent document, everything sticks together; each sentence builds on the previous sentence, and looks ahead to the following sentence. Many sentences work best when the beginning looks back at familiar information and the end provides the new information. Here is how sentences progress from familiar to unfamiliar:

| *Familiar* | | *Unfamiliar* |
|---|---|---|
| My dog | has | fleas. |
| Our boss | just won | the lottery. |
| This company | is planning | a merger. |

Besides helping a message stick together, the familiar-to-unfamiliar structure emphasizes the new information. Just as every paragraph has a key sentence, every sentence has a key word or phrase that sums up the new information. That key word or phrase is best emphasized at the end of the sentence.

| | |
|---|---|
| **Faulty emphasis** | We expect a **refund** because of your error in our shipment. |
| **Correct** | Because of your error in our shipment, we expect a **refund**. |
| **Faulty emphasis** | In a business relationship, **trust** is a vital element. |
| **Correct** | A vital element in a business relationship is **trust.** |

One exception to placing key words last occurs with *instructions*. Each step in a list of instructions should contain an action verb (*insert, open, close, turn, remove, press*). To provide readers with a forecast, place the verb in that instruction at the beginning.

> **Correct**    **Insert** the diskette before activating the system.
>
> **Remove** the protective seal.

With the key word at the beginning of the instruction, readers know immediately the action they need to take.

## Use Active Voice Often, Passive Voice Selectively

The active voice ("I did it") is more direct, concise, and forceful than the passive voice ("It was done by me"). In the active voice, the agent performing the action serves as subject:

| | *Agent* | *Action* | *Recipient* |
|---|---|---|---|
| **Active** | Joe | lost | your report. |
| | *Subject* | *Verb* | *Object* |

The passive voice reverses the pattern, making the recipient of an action serve as subject.

| | *Recipient* | *Action* | *Agent* |
|---|---|---|---|
| **Passive** | Your report | was lost | by Joe. |
| | *Subject* | *Verb* | *Prepositional phrase* |

Sometimes the passive eliminates the agent altogether:

> **Passive**    Your report was lost. (Who lost it?)

Some writers mistakenly rely on the passive voice because they think it sounds more objective and important. But the passive voice often makes writing wordy or evasive and unethical. Consider the effect when the active statement below is recast in the passive voice:

> **Concise and direct (active)**    I underestimated labor costs for this project. (7 words)
>
> **Wordy and indirect (passive)**    Labor costs for this project were underestimated by me. (9 words)
>
> **Evasive (passive)**    Labor costs for this project were underestimated.

For economy, directness, and clarity, use the active voice in most of your writing. Do not evade responsibility by hiding behind the passive voice:

> **Passive "irresponsibles"**    A **mistake** was made in your shipment.
>
> **It** was decided not to hire you.
>
> A **layoff** is recommended.

Acknowledge responsibility for your actions:

> **Active**    **I** made a mistake in your shipment.
>
> **I** decided not to hire you.
>
> **Our committee** recommends a layoff.

In reporting errors or bad news, use the active voice. Readers appreciate clarity and sincerity.

But do use the passive voice if the person behind the action has reason for being protected.

| | |
|---|---|
| **Correct passive** | The criminal was identified. |
| | The embezzlement scheme was exposed. |

Here, the passive protects the innocent person who identified the criminal or who exposed the scheme.

Use the passive only when your audience does not need to know the agent.

| | |
|---|---|
| **Correct passive** | Mr. Jones was brought to the emergency room. |
| | The bank failure was publicized statewide. |

Readers do not need to know *who* brought Mr. Jones or *who* publicized the bank failure. Notice again how the passive voice focuses on the *recipient* rather than the *agent*.

Use the passive voice to focus on events or results when the agent is unknown, unapparent, or unimportant.

| | |
|---|---|
| **Correct passive** | All memos in the firm are filed in a database. |
| | Fred's article was published last week. |

The information that will interest readers here is *how* the memos are filed or *that* Fred's article was published.

Prefer the passive when you deliberately wish to be indirect or inoffensive (as in requesting the customer's payment or the employee's cooperation).

| | |
|---|---|
| **Active but offensive** | **You** have not paid your bill. |
| | **You** need to overhaul our filing system. |
| **Inoffensive passive** | **This bill** has not been paid. |
| | **Our filing system** needs overhauling. |

By focusing on the recipient rather than the agent, these passive versions help retain the audience's goodwill.

The passive voice is weaker than the active voice, and creates an impersonal tone.

| | |
|---|---|
| **Weak and impersonal** | An offer will be made by us next week. |
| **Strong and personal** | We will make an offer next week. |

Use the active voice when you want action or your statement will have no power.

| | |
|---|---|
| **Weak passive** | If my claim is not settled by May 15, the Better Business Bureau will be contacted, and their advice on legal action will be taken. |

This passive and tentative statement is unlikely to move readers to action: *Who* is making the claim? *Who* should settle the claim? *Who* will contact the Bureau? *Who* will take action? Nobody is here! Here is a more direct, concise, and forceful version, in the active voice:

**Strong active**   If you do not settle my claim by May 15, I will contact the Better Business Bureau for advice on legal action.

Notice how this active version emphasizes the new and significant information by placing it at the end.

Ordinarily, use the active voice for giving instructions.

**Faulty passive**   The bid should be sealed.

Care should be taken with the dynamite.

**Correct active**   **Seal** the bid.

**Be careful** with the dynamite.

Avoid shifts from active to passive voice in the same sentence.

**Faulty shift**   During the meeting, project members spoke and presentations were given.

**Correct**   During the meeting, project members spoke and gave presentations.

Unless you have a deliberate reason for choosing the passive voice, use the *active* voice in most of your writing.

## Avoid Overstuffed Sentences

Never force readers to digest too much information in one overstuffed sentence.

**Overstuffed**   Publicizing the records of a private meeting that took place three weeks ago to reveal the identity of a manager who criticized our company's promotion policy would be unethical.

Notice how the details are hard to remember, the relationships hard to identify. Here is an improved version:

**Revised**   In a private meeting three weeks ago, a manager criticized our company's policy on promotion. It would be unethical to reveal the manager's identity by publicizing the records of that meeting. (Other versions are possible here, depending on the writer's intended meaning.)

Give your readers only as much information as they can retain easily in one sentence.

## EXERCISES IN REVISING FOR CLARITY

**1.** Revise each sentence below to eliminate ambiguities in phrasing, pronoun reference, or punctuation.

| | |
|---|---|
| **Ambiguous phrasing** | Most city workers strike on Friday. |
| **Revised** | Most city workers **are planning to strike** Friday. |
| | *or* |
| | Most city workers **typically strike** on Friday. |
| **Ambiguous pronoun reference** | Fred reminded Joe of the contract he had signed. |
| **Revised** | Fred reminded Joe of the contract **Joe** had signed. |
| | *or* |
| | Fred reminded Joe of the contract that **Fred** had signed. |
| **Ambiguous punctuation** | Dial "10" to deactivate the system and sound the alarm. |
| **Revised** | Dial "10" to deactivate the system, and sound the alarm. |
| | *or* |
| | Dial "10" to deactivate the system and **to** sound the alarm. |

*a.* Call me any evening except Tuesday after 7 o'clock.
*b.* The benefits of this plan are hard to imagine.
*c.* I cannot recommend this candidate too highly.
*d.* Visiting colleagues can be tiring.
*e.* Janice dislikes working with Claire because she's impatient.
*f.* Despite his efforts, Joe misinterpreted Sam's message.
*g.* Our division needs more effective writers.
*h.* Tell the reactor operator to evacuate and sound a general alarm.
*i.* If you don't pass any section of the test, your flying days are over.

**2.** Revise each sentence below to replace missing function words or to clarify ambiguous modifiers.

| | |
|---|---|
| **Telegraphic writing** | Send response to our client advising immediate action. |
| **Revised** | Send a response to our client **who** advised immediate action. |
| | *or* |
| | Send a response to our client **to** advise immediate action. |
| **Ambiguous modifier** | As a microprocessor expert, I look forward to your reply. |
| **Revised** | Because you are a microprocessor expert, I look forward to your reply. |

a. The manager claims repeatedly he reported the danger.
b. I want the final Amex report written by your division.
c. Replace main booster rocket seal.
d. The president refused to believe any internal report was inaccurate.
e. Our client failed to understand our recent proposal was preliminary.
f. Only use this phone in a red alert.
g. After offending our best client, I am deeply annoyed with the new manager.
h. Send memo to programmer requesting explanation.
i. He failed completely to explain the malfunction.
j. Do not enter test area while contaminated.

**3.** Revise each sentence below to unstack modifying nouns or to rearrange the word order for clarity and emphasis.

**Stacked modifiers**   Sarah's job involves fault analysis systems troubleshooting handbook preparation.

**Revised**   Sarah prepares handbooks **for** troubleshooting fault analysis systems.

**Faulty word order**   We have a critical need for technical support.

**Revised**   Our need for technical support is **critical**.

a. Develop on-line editing system documentation.
b. We need to develop a unified construction automation design.
c. Install a hazardous materials dispersion monitor system.
d. I recommend the following management performance improvement incentives.
e. Our profits have doubled since we automated our assembly line.
f. Education enables us to recognize excellence and to achieve it.
g. In all writing, revision is required.

**4.** The sentences below are wordy, weak, or evasive because of passive voice. Revise each sentence as a concise, forceful, and direct expression in the active voice, to identify the person or agent performing the action.

**Wordy passive**   Our test results will be sent to you as soon as verification is completed.

**Revised**   We will send you our test results as soon as we verify them.

**Weak passive**   It is believed by us that this contract is faulty.

**Revised**   We believe that this contract is faulty.

**Evasive passive**   The decision was made that your request for promotion should be denied.

**Revised**   I decided to deny your request for promotion.

a. The evaluation was performed by us.
b. The report was written by our group.
c. Unless you pay me within three days, my lawyer will be contacted.
d. Hard hats should be worn at all times.

    *e.* It was decided to reject your offer.

    *f.* Gasoline was spilled on your Ferrari's leather seats.

    *g.* The manager was kissed.

**5.** The sentences below lack proper emphasis because of active voice. Revise each ineffective active as an appropriate passive, to emphasize the recipient rather than the actor.

| | |
|---|---|
| **Agent is unimportant** | The publisher has accepted Mary's story for publication. |
| **Revised** | Mary's story has been accepted for publication. |
| **Recipient needs emphasis** | A tornado destroyed the barn. |
| **Revised** | The barn was destroyed by a tornado. |
| **Active is too blunt** | You apparently failed to edit this report. |
| **Revised** | This report apparently was not edited. |

    *a.* Joe's company fired him.

    *b.* A rockslide buried the mine entrance.

    *c.* Someone on the maintenance crew has just discovered a crack in the nuclear-core containment unit.

    *d.* A power surge destroyed more than 2,000 lines of our new applications program.

    *e.* Your report confused me.

    *f.* You are paying inadequate attention to worker safety.

    *g.* You are checking temperatures too infrequently.

**6.** Unscramble this overstuffed sentence by breaking it into shorter, clearer sentences.

A smoke-filled room causes not only teary eyes and runny noses but also can alter people's hearing and vision, as well as creating dangerous levels of carbon monoxide, especially for people with heart and lung ailments, whose health is particularly threatened by "second-hand" smoke.

## REVISING FOR CONCISENESS

In the previous sections, we showed that sometimes you have to add words to make your meaning clear. But adding *needless* words can obscure your meaning.

Writing can suffer from two kinds of wordiness: one kind gives more information than readers need (think of a typical weather report on local television news); the other kind uses too many words to convey information that readers do need ("a great deal of potential for the future" instead of "great potential").

Every word in the document should advance your meaning. We quote an expert on writing:

Writing improves in direct ratio to the number of things we can keep out of it that shouldn't be there. (Zinsser 14)

A concise message conveys most information in fewest words. It is highly informative, but not cluttered.

**Cluttered**    These collar ties weigh 300 pounds apiece, thereby exceeding by a total of 10 percent the load tolerance specified by the building inspector for this project.

A brief but vague message, on the other hand, is useless.

**Brief but vague**    These collar ties are too heavy.

Here is a concise version:

**Brief but informative**    These collar ties weigh 300 pounds apiece, exceeding our specified load tolerance by 10 percent.

First drafts rarely are concise. Get rid of anything that adds no meaning.

**Cluttered**    At this point in time, I would like to say that we are ready to move ahead.

**Concise**    We are ready.

Always use fewer words when fewer will do. But remember the difference between *clear* writing and *compressed* writing that is impossible to decipher.

**Impenetrable**    Give new vehicle air conditioner compression cut-off system specifications to engineering manager advising immediate action.

The following strategies will help you revise for conciseness.

## Avoid Needless Phrases

Don't use a whole phrase when one word will do. Instead of *in this day and age*, write *today*. Each of these wordy phrases can be reduced to one word—without loss in meaning:

| | | |
|---|---|---|
| at a rapid rate | = | rapidly |
| due to the fact that | = | because |
| the majority of | = | most |
| on a personal basis | = | personally |
| give instruction to | = | instruct |
| would be able to | = | could |
| readily apparent | = | obvious |
| a large number | = | many |
| prior to | = | before |
| aware of the fact that | = | know |
| conduct an inspection of | = | inspect |

## Eliminate Redundancy

A redundant expression says the same thing twice, in slightly different form, as in *fellow colleagues*. In these examples, the boldface words needlessly restate the message:

| | |
|---|---|
| a **dead** corpse | **end** result |
| **completely** eliminate | cancel **out** |
| **basic** essentials | consensus **of opinion** |
| enter **into** | **utter** devastation |
| **mental** awareness | **the month of** August |
| **mutual cooperation** | **utmost** perfection |

## Avoid Needless Repetition

Unnecessary repetition of words or phrases can clutter your writing and dilute your meaning.

> **Repetitious**  In trauma victims, breathing is restored by **artificial respiration**. Techniques of **artificial respiration** include mouth-to-mouth **respiration** and mouth-to-nose **respiration**.

Repetition in the passage above can be eliminated when sentences are combined.

> **Concise**  In trauma victims, breathing is restored by artificial respiration, either mouth-to-mouth or mouth-to-nose.

Keep in mind, however, that repetition can be useful. Don't hesitate to repeat, or at least rephrase, material (even whole paragraphs in a longer document) if you feel that readers need reminders in your report. Effective repetition helps avoid cross-references like this: "See page 23" or "Review page 10."

## Avoid *There* Sentence Openers

Save words, strengthen your statement, and improve your emphasis by not using *There is* and *There are* to begin your sentences—whenever your intended meaning allows.

> **Weak**  **There is** sensitive equipment required for this procedure.
>
> **Revised**  This procedure requires sensitive equipment.
>
> **Weak**  **There is** a danger of explosion in Number 2 mineshaft
>
> **Revised**  Number 2 mineshaft is in danger of exploding.

## Avoid Some *It* Sentence Openers

Don't begin a sentence with *It*—unless the *It* clearly refers to something specific in the preceding sentence.

> **Weak**  **It** was his bad attitude that got him fired.
>
> **Revised**  He was fired because of his bad attitude.
>
> **Weak**  **It** is necessary to complete both sides of the form.
>
> **Revised**  Please complete both sides of the form.

## Delete Needless Prefaces

Instead of keeping readers waiting for the new information in your sentence, get to the point right away. Deliver the pitch without too long a windup.

Wordy    **I am writing this letter because** I wish to apply for the position of copy editor.

Concise    I wish to apply for the position of copy editor.

Wordy    **As far as artificial intelligence is concerned,** the technology is only in its infancy.

Concise    Artificial-intelligence technology is only in its infancy.

## Avoid Weak Verbs

Try to choose verbs that express a definite action: *open, close, move, continue, begin*. These strong verbs advance your meaning. Avoid verbs that express no specific action: *is, was, are, has, give, make, come, take*. These weak verbs add words without advancing your meaning. All forms of the verb *to be* are weak. This sentence achieves conciseness because of the strong verb *consider*:

Concise    Please **consider** my offer.

Here is what happens with a weak verb in the same sentence:

Weak and wordy    Please **take into consideration** my offer.

Don't let yourself disappear behind weak verbs—along with their baggage of needless nouns and prepositions.

Weak    My recommendation **is** for a larger budget.

Strong    I **recommend** a larger budget.

Strong verbs, or action verbs, suggest an assertive, positive, and confident writer. Here are some weak verbs converted to strong:

| | | |
|---|---|---|
| is in conflict with | = | conflicts |
| has the ability to | = | can |
| give a summary of | = | summarize |
| make an assumption | = | assume |
| come to the conclusion | = | conclude |
| take action | = | act |
| make a decision | = | decide |

## Delete Needless *To Be* Constructions

In the preceding section we saw that all forms of the verb *to be* (*is, was, are,* and so on) are weak. Sometimes, the *to be* form itself mistakenly appears behind such verbs as *appears, seems,* and *finds*.

Wordy    Your product seems **to be** superior.

I consider this employee **to be** highly competent.

By eliminating *to be* in these examples, we save words without loss in meaning.

## Avoid Using Too Many Prepositions

On page 202, we saw how prepositions can increase clarity by breaking up clusters of stacked modifying nouns. But prepositions combined with forms of the verb *to be* can make wordy sentences.

| | |
|---|---|
| Wordy | The recommendation first appeared **in** the report written **by** the supervisor **in** January **about** that month's productivity. |
| Concise | The recommendation first appeared in the supervisor's productivity report for January. |

Here are some needlessly long prepositional phrases reduced to one or two words—without loss in meaning:

| | | |
|---|---|---|
| with the exception of | = | except for |
| in reference to | = | about (or regarding) |
| in order that | = | so |
| in the near future | = | soon |
| in the event that | = | if |
| at the present time | = | now |
| in the course of | = | during |
| in the process of | = | during (or in) |

## Fight Noun Addiction

Nouns manufactured from verbs (nominalizations) make your sentences weak and wordy. Nominalizations often accompany weak verbs and needless prepositions.

| | |
|---|---|
| Weak and wordy | We ask for the **cooperation** of all employees. |
| Strong and concise | We ask that all employees **cooperate.** |
| Weak and wordy | Give **consideration** to the possibility of a career change. |
| Strong and concise | **Consider** a career change. |

Besides being wordy, nominalizations can be vague—by hiding the agent of an action.

| | |
|---|---|
| Wordy and vague | A **need** for immediate action exists. (Who should take the action? We can't tell.) |
| Precise | We **must act** immediately. |

Here are examples of nominalizations restored to their verb forms:

| | | |
|---|---|---|
| conduct an investigation of | = | investigate |
| provide a description of | = | describe |
| conduct a test of | = | test |
| make a discovery of | = | discover |

Along with weak verbs and needless prepositions, nominalizations take the life out of your style. In cheering for your favorite team, you wouldn't say

<blockquote>
Blocking of that kick is a necessity!<br>
*instead of*<br>
Block that kick!
</blockquote>

Write as you would speak.

## Make Negatives Positive

A positive expression is more easily understood than a negative one. As one expert on writing points out, "To understand the negative, we have to translate it into an affirmative, because the negative only implies what we should do by telling us what we shouldn't do. The affirmative states it directly" (Williams 54).

| | |
|---|---|
| **Indirect and wordy** | Please do not be late in submitting your report. |
| **Direct and concise** | Please submit your report on time. |

Readers have to work even harder to translate sentences with two or more negative expressions:

| | |
|---|---|
| **Confusing and wordy** | Do **not** distribute this memo to employees who have **not** received a security clearance. |
| **Clear and concise** | Distribute this memo only to employees who have received a security clearance. |

Besides the directly negative words (*no, not, never*), some words are indirectly negative (*except, forget, mistake, lose, uncooperative*). When these indirectly negative words combine with directly negative words, readers are forced to translate.

| | |
|---|---|
| **Confusing and wordy** | **Do not neglect** to activate the alarm system. |
| | My diagnosis was **not inaccurate.** |

The second example above shows how multiple negatives can make the writer seem evasive.

| | |
|---|---|
| **Clear and concise** | **Be sure** to activate the alarm system. |
| | My diagnosis was **accurate**. |

Some negative expressions, of course, are perfectly correct, as in expressing disagreement.

| | |
|---|---|
| **Correct negatives** | This is **not** the best plan. |
| | Your offer is **unacceptable.** |
| | This project **never** will succeed. |

Prefer positives to negatives, however, whenever your meaning allows.

Here are some negative expressions translated into their positive versions:

| | | |
|---|---|---|
| did not succeed | = | failed |
| does not have | = | lacks |
| did not prevent | = | allowed |
| not unless | = | only if |
| not until | = | only when |
| not absent | = | present |

### Clean Out Clutter Words

Clutter words stretch a message without adding to its meaning. Here are some of the commonest: *very, definitely, quite, extremely, rather, somewhat, really, actually, currently, situation, aspect, factor.*

> **Cluttered** **Actually**, one **aspect** of a business **situation** that could **definitely** make me **quite** happy would be to have a **somewhat** adventurous partner who **really** shared my **extreme** attraction to risks.

> **Concise** I seek an adventurous business partner who enjoys risks.

If you must use one of these words, be sure it advances your meaning.

### Delete Needless Qualifiers

Qualifiers are expressions that soften the impact of a statement. Here are some common qualifiers: *I feel, It seems, I believe, In my opinion,* and *I think.* Use qualifiers only when you are uncertain about your assertion.

> **Appropriate** Despite Frank's poor grades last year, he will, **I think**, do well in college.
>
> Your product **seems** to be what we need.

But when you are sure of your assertion, eliminate the qualifier so as not to seem evasive.

> **Needless** **It seems** that I've made an error.
>
> **qualifiers** We **appear** to have exceeded our budget.
>
> **In my opinion**, this candidate is outstanding.

## EXERCISES IN REVISING FOR CONCISENESS

**1.** Revise each wordy sentence below to eliminate needless phrases, redundancy, and needless repetition.

> **Needless phrase** I am aware of the fact that Sam is trustworthy.
>
> **Revised** I know that Sam is trustworthy.
>
> **Redundancy** Through mutual cooperation, we can achieve our goals.
>
> **Revised** By cooperation we can achieve our goals.

| **Needless repetition** | This offer is the most attractive offer I've received. |
|---|---|
| **Revised** | This is the most attractive offer I've received. |

a. I have admiration for Professor Jones.
b. Due to the fact that we made the lowest bid, we won the contract.
c. On previous occasions we have worked together.
d. He is a person who works hard.
e. We have completely eliminated the bugs from this program.
f. This report is the most informative report on the project.

2. Revise each sentence below to eliminate *There* and *It* openers and needless prefaces.

| ***There* opener** | There is a coaxial cable connecting the antenna to the receiver. |
|---|---|
| **Revised** | A coaxial cable connects the antenna to the receiver. |
| ***It* opener** | It is necessary for us to complete this report by Friday. |
| **Revised** | We must complete this report by Friday. |
| **Needless preface** | I am writing this memo to spell out evacuation procedures. |
| **Revised** | This memo spells out evacuation procedures. |

a. There was severe fire damage to the reactor.
b. There are several reasons why Jane left the company.
c. It is essential that we act immediately.
d. It has been reported by Bill that several safety violations have occurred.
e. This letter is to inform you that I am pleased to accept your job offer.
f. The purpose of this report is to update our research findings.

3. Revise each wordy and vague sentence below to eliminate weak verbs.

| **Weak verb** | This manual gives instructions to end users. |
|---|---|
| **Revised** | This manual instructs end users. |
| **Weak verb** | I have a preference for Ferraris. |
| **Revised** | I prefer Ferraris. |

a. Our disposal procedure is in conformity with federal standards.
b. Please make a decision today.
c. We need to have a discussion about the problem.
d. I have just come to the realization that I was mistaken.
e. We certainly can make use of this information.
f. Your conclusion is in agreement with mine.

4. Revise each sentence below to eliminate needless prepositions and *to be* constructions, and to cure noun addiction.

| Needless *to be* | I consider George to be an excellent technician. |
| Revised | I consider George an excellent technician. |
| Needless prepositions | Power surges are associated, in a causative way, with malfunctions of computers. |
| Revised | Power surges cause computer malfunctions. |
| Noun addiction | The appearance of this problem was just yesterday. |
| Revised | This problem appeared just yesterday. |

a. Igor seems to be ready for a vacation.
b. Our survey found 46 percent of users to be disappointed.
c. In the event of system failure, your sounding of the alarm is essential.
d. These are the recommendations of the chairperson of the committee.
e. Our acceptance of the offer is a necessity.
f. Please perform an analysis and make an evaluation of our new system.
g. A need for your caution exists.

**5.** Revise each sentence below to eliminate inappropriate negatives, clutter words, and needless qualifiers.

| Confusing negative | Do not accept bids that are not signed. |
| Revised | Accept only signed bids. |
| Clutter words | Our current situation is that we actually cannot offer you employment. |
| Revised | We cannot offer you employment. |
| Needless qualifier | It seems as if I have just wrecked a company car. |
| Revised | I have just wrecked a company car. |

a. Our design must avoid nonconformity with building codes.
b. Never fail to wear protective clothing.
c. Do not accept any bids unless they arrive before May 1.
d. I am not unappreciative of your help.
e. We are currently in the situation of completing our investigation of all aspects of the accident.
f. I appear to have misplaced the contract.

## REVISING FOR FLUENCY

Fluent sentences are easy to read because of clear connections, variety, and emphasis. Their varied length and word order eliminate choppiness and monotony. Fluent sentences enhance *clarity*, allowing readers to see what is most important, with no struggle to sort out relationships. Fluent sentences enhance *conciseness*, often replacing several short, repetitious sentences with one longer, more economical sentence. The strategies discussed below will help you achieve fluent sentences.

## Combine Related Ideas

A series of short, disconnected sentences is not only choppy and wordy, but unclear as well.

**Disconnected**  Jogging can be healthful. You need the right equipment. Most necessary are well-fitting shoes. Without this equipment you take the chance of injuring your legs. Your knees are especially prone to injury. (4 sentences)

**Clear, concise, and fluent**  Jogging can be healthful if you have the right equipment. Shoes that fit well are most necessary because they prevent injury to your legs, especially your knees. (2 sentences)

Never force your readers to figure out connections for themselves.

Most sets of information can be combined in different patterns, depending on what you want to emphasize. Imagine that this group of facts describes an applicant for a junior-management position with your company.

- Roy James graduated from an excellent management school.

- He has no experience.

- He is highly recommended.

Assume that you are a personnel director writing to upper management to convey your impression of this candidate. To give your readers a negative impression, you might combine the facts in this way:

**Strongly negative emphasis**  Although Roy James graduated from an excellent management school and is highly recommended, **he has no experience.**

The *independent* idea (in boldface) receives the emphasis. Earlier ideas are made dependent on (or subordinate to) the independent idea by the subordinating word *although*. When a sentence has two or more ideas of unequal importance, the less important idea is signaled by a subordinating word such as *often, as, because, if, unless, until,* or *while*.

To continue with our Roy James example: If you are undecided, but leaning in a negative direction, you might combine the information in this way:

**Slightly negative emphasis**  Roy James graduated from an excellent management school and is highly recommended, **but** he has no experience.

In the sentence above, both the ideas before and after *but* are independent. These independent ideas are joined by the coordinating word *but*, which suggests that both sides of the issue are equally important (or "coordinate"). Placing the negative idea last, however, gives it a slight emphasis. When a sentence has two or more ideas equal in importance, their equality is signaled by coordinating words such as *and, but, for, nor, or, so,* or *yet*.

Consider once again our Roy James example. To emphasize strong support for the candidate, you could use this combination:

> Although Roy James has no experience, **he graduated from an excellent management school and is highly recommended.**

In the version above, the earlier idea is subordinated by *although*, leaving the two final ideas independent.

Readers interpret our meaning not only by what we say, but by how we put our sentences together.

*Caution:* Combine sentences only to advance your meaning, to ease the reader's task. A sentence with too much information and too many connections can be impossible for readers to sort out.

> **Overstuffed**   Our night supervisor's orders from upper management to repair the overheated circuit were misunderstood by Harvey Kidd, who gave the wrong instructions to the emergency crew, thereby causing the fire.

Readability is affected less by the number of words in a sentence than by the amount of information. Even relatively short sentences can be unreadable if they carry too many details:

> **Overstuffed**   Send three copies of Form 17-e to all six departments, unless Departments A or B or both request Form 16-w.

Although effective combining *enhances* and *streamlines* your meaning, over-combining makes your meaning impenetrable.

## Vary Sentence Construction and Length

Long and short sentences each have their purpose: to express ideas logically or forcefully.[1] We have just seen how related ideas often need to be linked in one sentence, so that readers can grasp the connections:

> **Disconnected**   The nuclear core reached critical temperature. The loss-of-coolant alarm was triggered. The operator shut down the reactor.

> **Connected**   As the nuclear core reached critical temperature, triggering the loss-of-coolant alarm, the operator shut down the reactor.

The ideas above have been combined to show that one action resulted from another. But an idea that should stand alone for emphasis needs a whole sentence of its own:

> **Correct**   Core meltdown seemed inevitable.

Too much of anything loses effect. An unbroken string of long or short

---

[1]My thanks to Professor Edith K. Weinstein, University of Akron, for suggesting this distinction.

sentences can bore and confuse readers; so too can a series with identical openings:

> Dreary  There are some drawbacks about diesel engines. **They** are difficult to start in cold weather. **They** cause vibration. **They** also give off an unpleasant odor. **They** cause sulfur dioxide pollution.
>
> Varied  Diesel engines have some drawbacks. Most obvious are their noisiness, cold-weather starting difficulties, vibration, odor, and sulfur dioxide emission.

Opening sentences repeatedly with *The, This, He, She,* or *I* creates monotony. When you write in the first person, overusing *I* makes you seem self-centered.

Do not, however, avoid personal pronouns if they make the writing more readable (say, by eliminating passive constructions). Instead, to avoid repetitious openings, combine ideas and shift word order.

## Use Short Sentences for Special Emphasis

With all this talk about combining ideas, you might conclude that short sentences have no place in writing. Wrong.

Whereas long sentences show connections and clarify relationships, short sentences (even one-word sentences) isolate an idea for special emphasis. They stick in the reader's mind.

## EXERCISES IN REVISING FOR FLUENCY

**1.** The sentences sets below lack fluency because they are disconnected, have no variety, or have no emphasis. Combine each set into *one* or *two* fluent sentences.

> Choppy  The world's forests are now disappearing. The rate of disappearance is 18 to 20 million hectares a year (an area half the size of California). Most of this loss occurs in humid tropical forests. These forests are in Asia, Africa, and South America.
>
> Revised  The world's forests are now disappearing at the rate of 18 to 20 million hectares a year (an area half the size of California). Most of the loss is occurring in the humid tropical forests of Africa, Asia, and South America.[2]

*a.* The world's population will grow.
It will grow from 4 billion in 1975.
It will reach 6.5 billion in 2000.
This will be an increase of more than 50 percent.

[2]Sample sentences are adapted from *Global Year 2000 Report to the President: Entering the 21st Century* (Washington: GPO, 1980).

*b.* In sheer numbers, population will be growing.
It will be growing faster in 2000 than it is today.
It will add 100 million people each year.
This figure compares with 75 million in 1975.

*c.* Energy prices are expected to increase.
Many less-developed countries will have increasing difficulty.
Their difficulty will be in meeting energy needs.

*d.* One-quarter of humanity depends primarily on wood.
They depend on wood for fuel.
For them, the outlook is bleak.

*e.* The world has finite fuel resources.
These include coal, oil, gas, oil shale, and uranium.
These resources theoretically are sufficient for centuries.
These resources are not evenly distributed.

*f.* Already the populations in parts of Africa and Asia have exceeded the carrying capacity of the immediate area.
This overpopulation has triggered erosion.
This erosion has reduced the land's capacity to support life.

**2.** Combine each set of sentences below into one fluent sentence that provides the requested emphasis.

| | |
|---|---|
| **Sentence set** | John is a loyal employee.<br>John is a motivated employee.<br>John is short-tempered with his colleagues. |
| **Combined for positive emphasis** | Even though John is short-tempered with his colleagues, he is a loyal and motivated employee. |
| **Sentence set** | This word processor has many features.<br>It includes a spelling checker.<br>It includes a thesaurus.<br>It includes a grammar checker. |
| **Combined for positive emphasis** | Among its many features, such as spelling and grammar checkers, this word processor includes a thesaurus. |

*a.* The job offers an attractive salary.
It demands long work hours.
Promotions are rapid.
*(Combine for negative emphasis.)*

*b.* The job offers an attractive salary.
It demands long work hours.
Promotions are rapid.
*(Combine for positive emphasis.)*

*c.* Our office software is integrated.
It has an excellent database management program.
Most impressive is its word processing capability.
It has an excellent spreadsheet program.
*(Combine to emphasize the word processor.)*

*d.* Company X gave us the lowest bid.
Company Y has an excellent reputation.
*(Combine to emphasize Company Y.)*

*e.* Superinsulated homes are energy efficient.
Superinsulated homes create a danger of indoor air pollution.
The toxic substances include radon gas and urea formaldehyde.
*(Combine for a negative emphasis.)*

*f.* Computers cannot *think* for the writer.
Computers eliminate many mechanical writing tasks.
They speed the flow of information.
*(Combine to emphasize the first assertion.)*

## CHOOSING EXACT WORDS

Too often, language in the workplace can be a vehicle for *camouflage* rather than communication. When you write, you might have many reasons to hide behind language, as when you

- speak for your company but not for yourself
- fear the consequences of giving bad news
- are afraid to disagree with company policy
- make a recommendation some readers will resent
- worry about making a bad impression
- worry about being wrong
- pretend to know more than you do
- avoid admitting to a mistake or ignorance

Inflated and unfamiliar words, borrowed expressions, and needlessly technical terms are a few of the poor word choices that camouflage your meaning. Whether intentional or not, poor word choices have only one result: inefficient and often unethical writing that resists interpretation and frustrates the reader.

Following are strategies for choosing words that are *convincing, precise,* and *informative.*

### Use Simple and Familiar Words

If technical vocabulary is essential in your writing, by all means use it. Don't replace technically accurate words with nontechnical words that are less than accurate. Don't write *a part that makes the computer run* when you mean *central processing unit.* Use the technically accurate term, and define its meaning for a nontechnical audience:

> **Correct** Central processing unit: the part of the computer that controls information transfer and carries out arithmetic and logical instructions.

Technical words often may be necessary to your exact meaning. But you usually can simplify the nontechnical words—without significantly altering your meaning. Instead of *answering in the affirmative*, you can *say yes;* or instead of *endeavoring to promulgate* a new policy, you can *try to announce* it.

Don't use three syllables when one will do. Generally, trade for less:

| | | |
|---|---|---|
| aggregate | = | total |
| approximately | = | roughly |
| demonstrate | = | show |
| effectuate | = | cause |
| endeavor | = | effort |
| eventuate | = | result |
| frequently | = | often |
| initiate | = | begin |
| is contingent upon | = | depends on |
| multiplicity of | = | many |
| optimum | = | best |
| subsequent to | = | after |
| utilize | = | use |

Count the syllables. Trim wherever you can. But, most important, choose words that you hear and use in everyday speaking—words that are familiar to all of us.

Don't write *I deem* when you mean *I think*, or *Keep me apprised* instead of *Keep me informed*, or *I concur* instead of *I agree*, or *securing employment* instead of *finding a job*, or *it is cost prohibitive* instead of *we can't afford it.* Experiments have shown that readers have to spend extra time on passages that contain unfamiliar or less familiar words (Bailey 70; Felker et al. 61).

Don't write like the author of a report from the Federal Aviation Administration, who recommended that DC-10 manufacturers reevaluate *the design of the entire pylon assembly to minimize design factors which are resulting in sensitive and/or critical maintenance and inspection procedures* (25 words, 50 syllables). A plain English translation: *Redesign the pylons so they are easier to maintain and inspect* (11 words, 18 syllables).

Besides the annoyance they cause, needlessly big or unfamiliar words can be *ambiguous*.

**Ambiguous**   Make an improvement in the clerical situation.

Should we hire more secretaries or better secretaries? We have no way to tell. Words chosen to impress readers too often confuse them instead.

Of course, now and then the fancier or more impressive word is best—if it expresses your exact meaning. We would not substitute *end* for *terminate* in referring to something with an established time limit.

**Correct**   Our trade agreement terminates this month.

If a fancy word can replace a handful of simpler words—and can sharpen your meaning—use the fancy word.

| | |
|---|---|
| Weak | Six rectangular grooves **around the outside edge** of the steel plate **are needed for** the pressure clamps **to fit into**. |
| Informative and precise | Six rectangular grooves on the steel plate **perimeter accommodate** the pressure clamps. |
| Weak | We need a **one-to-one exchange of ideas and opinions.** |
| Informative and precise | We need a **dialogue**. |
| Weak | Upper management has **taken over for themselves** our authority as decision makers. |
| Informative and precise | Upper management has **usurped** our authority as decision makers. |

Never seek simplicity at the cost of clarity, but always make sure that every word counts. Use the fancy word *only* when your meaning and your audience demand it.

## Avoid Useless Jargon

Every profession has its own "shorthand." Among specialists, technical terms are a precise and economical way to communicate. On page 19, *stat* is medical jargon for *Drop everything and deal with this emergency.* For computer buffs, a *glitch* is a momentary power surge that can erase the contents of internal memory; a *bug* is an error that causes a program to run incorrectly. These are samples of useful jargon that conveys clear meaning to a knowledgeable audience.

Technical language, however, can be used in two ways—appropriately or inappropriately. The latter is useless jargon, meaningless to insiders as well as outsiders.

In the world of useless jargon people don't *cooperate* on a project; instead, they *interface* or *contiguously optimize their efforts*. Rather than *designing a model,* they *formulate a paradigm*. Instead of *observing limits* or *boundaries,* they *function within specific parameters*.

A popular form of useless jargon is adding *-wise* to nouns, as shorthand for *in reference to* or *in terms of*.

| | |
|---|---|
| Useless jargon | **Expensewise** and **schedulewise,** this plan is unacceptable. |
| Revised | In terms of expense and scheduling, this plan is unacceptable. |

Writers create another form of jargon when they invent verbs from nouns or adjectives by adding an *-ize* ending: Don't invent *prioritize* from *priority*; instead use *to rank priorities*. Don't write *deambiguize* when you mean *clarify,* or *finalize* when you mean to *settle on*.

Jargon's worst fault is that it makes the writer seem stuffy and pretentious:

**Pretentious**    Unless all parties interface synchronously within the given parameters, the project will be rendered inoperative.

**Possible**    Unless we coordinate our efforts, the project will fail.
**translation**

Beyond reacting with frustration, readers often conclude that useless jargon is intended as camouflage by a writer who has something to hide.

Before using any jargon, think about your specific readers and ask yourself: "Can I find an easier way to say exactly what I mean?" Use jargon only if it *improves* your communication.

## Use Acronyms Selectively

Acronyms are another form of specialized shorthand, or jargon. They are formed from the first letters of words in a phrase (as in *LOCA* from *loss of coolant accident*) or from a combination of first letter and parts of words (as in *bit* from *b*inary dig*it* or *pixel* from *pic*ture *el*ement).

Computers have given rise to many acronyms, including

DOS  =  disk operating system
RAM  =  random-access memory
VDT  =  video display terminal

Acronyms *can* communicate concisely—but only when your audience already knows their meaning, and only when you use the term often in your document. Always define the acronym (in parentheses) on first use. For unspecialized audiences, try not to use acronyms at all.

## Avoid Triteness

Writers who rely on tired old phrases (clichés) seem either too lazy or careless to find exact ways of saying what they mean. Here are just a few of the countless expressions worn out by overuse:

| | |
|---|---|
| make the grade | the chips are down |
| in the final analysis | not by a long shot |
| close the deal | last but not least |
| in-depth study | welcome aboard |
| water under the bridge | over the hill |
| holding the bag | bite the bullet |
| up the creek | work like a dog |

If it sounds like a "catchy phrase" you've heard often, don't write it.

## Avoid Misleading Euphemisms

Euphemisms are expressions that are aimed at politeness or at making unpleasant subjects seem less offensive. Thus, we *powder our nose* or *use the boys' room* instead of *using the toilet;* we *pass away* or *meet our Maker* instead of *dying.* Euphemisms make the truth less painful.

When euphemisms avoid offending or embarrassing our audience, they are perfectly legitimate. Instead of telling a job applicant that he or she is *unqualified*, we might say, *Your background doesn't meet our needs.*

Euphemisms, however, are unethical if they understate the truth when only the truth will serve. In the sugar-coated world of misleading euphemisms, bad news disappears:

- Instead of being *laid off* or *fired*, workers are *surplused* or *deselected*, or the company is *downsized.*
- Instead of *lying* to the public, governments *engage in disinformation.*
- Instead of *wars* and *nuclear missiles,* we have *conflicts* and *Peacekeepers.*

Language loses all meaning when *criminals* become *offenders*, when *rape* becomes *sexual assault*, and when people who are just plain *lazy* become *underachievers.* Even the prospect of a nuclear holocaust becomes more digestible when one U.S. senator calls it a *nuclear exchange.*

But as benign as *conflicts* and *nuclear exchange* may sound, they kill people just as effectively as do *wars* and a *nuclear holocaust.* Plain talk is always better than deception. If someone offers you a job *with limited opportunity for promotion*, you probably will have a *dead-end job.*

## Avoid Overstatement

Writers lose credibility when they exaggerate to make a point. Be cautious when using words such as *best, biggest, brightest, most,* and *worst.*

| | |
|---|---|
| **Overstated** | **Most** businesses have **no** loyalty toward their employees. |
| **Revised** | **Some** businesses have **little** loyalty toward their employees. |
| **Overstated** | You will find our product to be the **best.** |
| **Revised** | You will **appreciate the high quality** of our product. |

## Avoid Unsupported Generalizations

Be sure your conclusions are based on adequate evidence.

| | |
|---|---|
| **Unsupported generalization** | In 1983, twenty-one murderers were executed, and the murder rate dropped 8.1 percent for that year. These figures prove that the death penalty should be reinstated. |

This isolated piece of evidence does not justify such a sweeping generalization.

| | |
|---|---|
| **Reasonable generalization** | In 1983, twenty-one murderers were executed, and the murder rate dropped 8.1 percent for that year. These figures suggest that the death penalty may deter violent crimes. |

Be aware of the vast differences in meaning among these words:

| | |
|---|---|
| few | never |
| some | rarely |
| many | sometimes |
| most | often |
| all | always |

Unless you specify *few, some, many,* or *most,* readers can interpret your statement as meaning *all.*

> **Misleading**   Assembly-line employees are doing shabby work.

Unless you mean *all* assembly-line employees, be sure to qualify your generalization with *some, most,* or some similar word.


## Avoid Imprecise Words

Poorly chosen words can be offensive, embarrassing, misleading, or ambiguous. Be sure that what you say is what you mean.

Even words listed as synonyms carry a different shade of meaning. Do you mean to say *I'm slender, You're thin, She's lean,* or *He's scrawny*? The wrong choice could be disastrous.

Just one wrong word can offend your readers, as in this job applicant's statement:

> **Offensive**   Another attractive feature of your company is its **adequate** training program.

Word choice here may be accurate, but it is by no means appropriate. Although the program may not have been highly ranked, our writer could have used any of several alternatives (*solid, respectable, growing*—or no modifier at all) without overstating his point or being offensive.

Poor word choice can be embarrassing, as in this example:

> **Imprecise**   **Chaos** is running this project.

Strictly speaking, *chaos* can't run anything!

Be especially aware of similar words with dissimilar meanings, as in these examples:

| | |
|---|---|
| affect/effect | farther/further |
| all ready/already | fewer/less |
| almost dead/dying | healthy/healthful |
| among/between | imply/infer |
| continual/continuous | invariable/inevitably |
| eager/anxious | uninterested/disinterested |
| fearful/fearsome | worse/worst |

Don't write *Skiing is healthy* when you mean that skiing promotes good health. Healthful things keep us healthy.

Be on the lookout for imprecisely phrased (and therefore illogical) comparisons.

| | |
|---|---|
| Imprecise | Your bank's interest rate is higher than Citibank. (Can a rate be higher than a bank?) |
| Precise | Your bank's interest rate is higher than Citibank's. |

Precision is above all essential to the informative value of your writing. Imprecise language can be misleading. Consider how meanings differ:

| | |
|---|---|
| Differing meanings | Include **less** technical detail in your report. |
| | Include **fewer** technical details in your report. |

Imprecise language can result in ambiguous messages as well:

| | |
|---|---|
| Ambiguous | Loan payments are due **bimonthly**. (Every other month or twice monthly?) |
| Clear | Loan payments are due twice monthly. |
| | *or* |
| | Loan payments are due every other month. |

## Be Specific and Concrete

General words name broad classes of things, such as *job, computer,* or *person.* Such words usually need to be clarified by more specific ones.

| | | |
|---|---|---|
| job | = | senior accountant for Softbyte Press |
| computer | = | 512 K Macintosh, with external disk drive |
| person | = | Sarah Jones, production manager |

The more specific your words, the sharper your meaning.

| | |
|---|---|
| General | structure |
| | dwelling |
| | vacation home |
| | log cabin |
| Specific | log cabin in Vermont |
| | a three-room log cabin on the banks of the Battenkill River in Vermont |

Notice how the picture becomes more vivid as we move to lower levels of generality.

Abstract words name qualities, concepts, or feelings *(beauty, luxury, depression)* whose exact meaning has to be nailed down by *concrete* words—words that name things we can know through our five senses.

| | | |
|---|---|---|
| a **beautiful** view | = | snowcapped mountains, a wilderness lake, pink granite ledge, ninety-foot birch trees |
| a **luxurious** condominium | = | imported tiles, glass walls, oriental rugs |
| a **depressed** worker | = | suicidal urge, insomnia, feelings of worthlessness, no hope for improvement |

Informative writing *shows* and *tells*.

> **General**  One of our **workers** was **injured** by a **piece of equipment recently.**

The boldface words only *tell* without showing.

> **Specific**  **Alan Hill** suffered a **broken thumb** while working on a **lathe yesterday.**

Choose high-information words that show exactly what you mean. Don't write *thing* when you mean *lever, switch, micrometer,* or *disk.* Instead of evaluating an employee as *good, great, disappointing,* or *terrible,* use informative words such as *reliable, skillful, dishonest,* or *incompetent*—and give examples.

In some instances, of course, you may wish to generalize for the sake of diplomacy. Instead of writing *Bill, Mary, and Sam have been tying up the office phones with personal calls*, you might prefer to generalize: *Some employees have been tying up* . . . . The second version allows you to get your message across without pointing the finger.

When you can, provide solid numbers and statistics that get your point across:

> **General**  In 1972, thousands of people were killed or injured on America's highways. Many families had at least one relative who was a casualty. After the speed limit was lowered to 55 miles per hour in late 1972, the death toll began to drop.

> **Specific**  In 1972, 56,000 people died on America's highways; 200,000 were injured; 15,000 children were orphaned. In that year, if you were a member of a family of five, chances are that someone related to you by blood or law was killed or injured in an auto accident. After the speed limit was lowered to 55 miles per hour in late 1972, the death toll dropped steadily to 41,000 in 1975.

Concrete and specific information not only is more informative; it is more convincing as well.

## Use Analogies to Sharpen the Image

Ordinary comparison shows similarities between two things *of the same class* (two computer keyboards, two technicians, two methods of cleaning dioxin-contaminated sites). Analogy, on the other hand, shows some essential similarity between two things of *different classes* (report writing and computer programming, computer memory and post office boxes).

Analogies are good for emphasizing a point (*Some rain is now as acidic as vinegar*). But they are especially useful in translating something abstract, complex, or unfamiliar to laypersons, as long as the easier subject is broadly familiar to readers. Analogy therefore calls for particularly careful analyses of audience.

Analogies can save words and convey vivid images. *Collier's Encyclopedia* describes the tail of an eagle in flight as "spread like a fan." On pages 312–314 of this book, the description of a trout feeder mechanism uses this analogy to clarify the positional relationship between two working parts:

> **Analogy**    The metal rod is inserted (and centered, crosslike) between the inner and outer sections of the clip. . . .

Without the analogy *crosslike*, we would need something like this to visualize the relationship:

> **Missing analogy**    The metal rod is inserted, *perpendicular to the long plane and parallel to the flat plane*, between the inner and outer sections of the clip. . . .

This second version is doubly inefficient: more words are needed to communicate, and more work is needed to understand the meaning.[3]

Besides naming things vividly, analogies help *explain* things. The following extended analogy from the *Congressional Research Report* helps us understand something unfamiliar (dangerous levels of a toxic chemical) by comparing it to something more familiar (a human hair).

> **Analogy**    A dioxin concentration of 500 parts per trillion is lethal to guinea pigs. One part per trillion is roughly equal to the thickness of human hair compared to the distance across the United States.

---

## EXERCISES IN REVISING FOR EXACTNESS

**1.** Revise each sentence below for straightforward and familiar language.

> **Big words**    I wish to upgrade my present employment situation.
> **Revised**    I want a better job.
> **Unfamiliar words**    Acoustically attenuate the food-consumption area.
> **Revised**    Soundproof the cafeteria.

*a.* May you find luck and success in all endeavors.
*b.* I suggest you reduce the number of cigarettes you consume.
*c.* Within the copier, a magnetic-reed switch is utilized as a mode of replacement for the conventional microswitches that were in use on previous models.
*d.* A good writer is cognizant of how to utilize grammar in a correct fashion.
*e.* I will endeavor to ascertain the best candidate.
*f.* In view of the fact that the microscope is defective, we expect a refund of our full purchase expenditure.

---

[3]Analogy is, of course, a form of metaphor. For an inspired discussion of metaphor, see John S. Harris, "Metaphor in Technical Writing," *Technical Writing Teacher* 2 (1975): 9–13.

**2.** Revise each sentence below to eliminate useless jargon and triteness.

| | |
|---|---|
| **Useless jargon** | Intercom utilization will be employed to initiate substitute employee operative involvement. |
| **Revised** | Employees who are asked to substitute will be notified on the intercom. |
| **Triteness** | Managers who make the grade are those who can take daily pressures in stride. |
| **Revised** | Successful managers are those who cope with daily pressures. |

    *a.* For the obtaining of the X-33 word processor, our firm will have to accomplish the disbursement of funds to the amount of $6,000.
    *b.* To optimize your financial return, prioritize your investment goals.
    *c.* The use of this product engenders a 50-percent repeat consumer encounter.
    *d.* We'll have to swallow our pride and admit our mistake.
    *e.* We wish to welcome all new managers aboard.
    *f.* Not by a long shot will this plan succeed.

**3.** Revise each sentence below to eliminate euphemism, overstatement, or unsupported generalizations.

| | |
|---|---|
| **Euphemism** | Because of your absence of candor, we can no longer offer you employment. |
| **Revised** | Because of your dishonesty, you're fired. |
| **Overstatement** | Igor is the world's best employee. |
| **Revised** | Igor is the finest employee I've had. |
| **Unsupported generalization** | Television is making students nothing but illiterates. |
| **Revised** | Television seems to negatively affect many students' reading ability. |

    *a.* I finally must admit that I am an abuser of intoxicating beverages.
    *b.* I was less than candid.
    *c.* This employee is poorly motivated.
    *d.* Your faulty machinery traumatically amputated my client's arm.
    *e.* Most entry-level jobs are boring and dehumanizing.
    *f.* Clerical jobs offer no opportunity for advancement.

**4.** Revise each sentence below to make it more precise or informative.

| | |
|---|---|
| **Imprecise** | Push the printer connector into the serial socket. |
| **Revised** | Insert the printer connector into the serial socket. |
| **Too general** | Industrial emissions are causing atmospheric effects on certain bodies of water. Lakes in particular are affected. Many lakes in fact are now dead. |

**Revised**    Sulfur dioxide emissions from coal-burning plants combine with atmospheric water to produce sulfuric acid. The resultant "acid rain" so increases the acidity of lakes that they no longer can support life.

a. Our outlet does more business than Chicago.
b. Anaerobic fermentation is used in this report.
c. Confusion is in control of this office.
d. Your crew damaged a piece of office equipment.
e. His performance was admirable.
f. This thing bothers me.

## ADJUSTING YOUR TONE

Your tone is your personal mark—the personality that appears between the lines. The tone you create in any writing depends on (1) the distance you impose between yourself and the reader, and (2) the attitude you express toward the subject.

Assume that a friend is going to take over a job you've held. You've decided to write instructions for your friend. Here is your first sentence:

> Now that you've arrived in the glamorous world of office work, put on your track shoes; this is no ordinary manager-trainee job.

What kind of tone appears in this sentence? Notice that the writer imposes little distance between herself and the reader (she uses the direct address, "you," and the humorous suggestion to "put on your track shoes"). And the ironic use of "glamorous" suggests just the opposite.

For a different reader (say, in a company training manual written for a stranger), the writer would have chosen some other way to open:

> As a manager trainee at GlobalTech, you will work for many managers. In short, you will spend little of your day at your desk.

The tone has changed; it is no longer intimate, and the writer expresses no distinct attitude toward the job. For yet another audience (say, in an annual report or pamphlet describing the company to clients or investors), the writer might have altered her tone again:

> Manager trainees at GlobalTech are responsible for duties that extend far beyond desk work.

Here the writer imposes even more distance between herself and her audience, especially with the shift from second- to third-person address. The tone is far too impersonal for any document addressed to the trainees themselves.

Your tone changes in response to the situation and audience. When you are introduced to someone, for example, you respond in a tone that defines your relationship:

- Honored to make your acquaintance. [formal tone—greatest distance]

- How do you do? [formal]
- Nice to meet you. [semiformal—medium distance]
- Hello. [semiformal]
- Hi. [informal—least distance]
- What's happening? [informal—slang]

The greeting (and tone) you choose will depend on how much distance you decide is appropriate, and, in turn, the tone will determine how you come across. Each of the above greetings is appropriate in some situations, inappropriate in others. "Hi" might be okay when you are introduced to a new friend, but not the college president.

As a rule,

- Use a formal or semiformal tone with superiors or dignitaries (depending on what you think the reader expects).
- Use a semiformal or informal tone in your professional writing (depending on how close you feel to your readers).
- Use an informal tone when you want your writing to be conversational, or when you want it to sound like a person talking.

Whichever tone you choose, be consistent throughout your document.

| | |
|---|---|
| **Inconsistent tone** | My office isn't fit for a pig [too informal]; it is ungraciously unattractive [too formal]. |
| **Revised** | The shabbiness of my office makes it an unfit place to work. |

In general, lean toward an informal tone without falling into slang. In addition to setting the distance between writer and reader, your tone implies your *attitude* toward the subject. Consider the different attitudes expressed in these examples:

- We dine at seven.
- Dinner is at seven.
- Let's eat at seven.
- Let's chow down at seven.
- Let's strap on the feedbag at seven.
- Let's pig out at seven.

The words we choose tell readers a great deal about where we stand. If your readers expect an impartial report, do not give a biased view by inserting your attitude. But for situations in which your attitude *is* expected, or in which you perceive some danger or ethics violation, speak up and let readers know where you stand.

Say *I enjoyed the fiber optics seminar* instead of *My attitude toward the fiber optics seminar was one of high approval.* Say *Let's liven up our dull rela-*

*tionship* instead of *We should inject some rejuvenation into our lifeless liaison.* Say *Methane levels in No. 3 mineshaft pose the definite risk of explosion* instead of *Rising methane levels in No. 3 mineshaft should be evaluated.* Make sure your attitude is clear and appropriate for the situation.

Besides being clear, your attitude should reflect your relationship with the reader—and what you think the reader expects. In an upcoming meeting about the reader's job evaluation, does your reader expect to *discuss* the evaluation, *talk it over, have a chat,* or *chew the fat?* The words you choose will give the reader a definite notice about where you stand. If the situation calls for a serious tone, don't use language that suggests a casual attitude—or vice versa. Use the following guidelines for making your tone conversational and appropriate.

## Use an Occasional Contraction

Unless you wish to be formal, use contractions to loosen the tone. Balance an *I am* with an *I'm* a *you are* with a *you're*, an *it is* with an *it's* (as we've done throughout this book).

Be careful, though, not to overuse contractions.

| Excessive contractions | It's a shame that Clementine's been crying since she's learned her date'll be late because his tire's flat and his wallet's lost. |
|---|---|

Generally, use contractions only with pronouns, not with nouns or proper nouns (names). Otherwise, the constructions are awkward or ambiguous.

| Awkward contractions | Barbara'll be here soon. |
|---|---|
| | Health's important. |
| Ambiguous contractions | The dog's barking. |
| | Bill's skiing. |

These ambiguous contractions can easily be confused with possessive constructions.

## Address Readers Directly

Use the personal pronouns *you* and *your* generously to create contact with your reader. Otherwise, your writing sounds impersonal.

| Impersonal tone | A writer should use **you** and **your** generously to create contact with his or her readers. |
|---|---|

Notice how distance *and* words increase. Direct address creates a more personal tone.

| Impersonal tone | Students at our college will find the faculty always willing to help. |
|---|---|
| Personal tone | As a student at our college, **you** will find the faculty always willing to help. |

Research shows that readers relate better to something they consider meaningful to them (Felker 33).

*Caution*: Use *you* and *your* only to correspond *directly* with the reader, as in a letter, memo, instructions, or some form of advice, encouragement, or persuasion. By using *you* and *your* when your subject and purpose call for first or third person, you might write something wordy and awkward like this:

| | |
|---|---|
| Wordy and awkward | **When you** are in northern Ontario, **you** can see wilderness lakes everywhere around **you**. |
| Appropriate | Wilderness lakes are everywhere in northern Ontario. |

### Use *I* and *We* When Appropriate

Don't disappear behind your writing. Use *I* or *We* when referring to yourself or your organization.

| | |
|---|---|
| Distant | This writer would like a refund. |
| Revised | I would like a refund. |

A message can become doubly impersonal when both the writer and the reader disappear.

| | |
|---|---|
| Impersonal | The requested report will be sent next week. |
| Personal | **We** will send the report **you** requested next week. |

Avoid opening too many sentences with *I*. Combine ideas and shift word order instead.

### Prefer the Active Voice

Because the active voice is more direct and economical than the passive voice, it creates a more personal and less formal tone.

| | |
|---|---|
| Passive and impersonal | Travel expenses cannot be reimbursed unless receipts are submitted. |
| Active and personal | We cannot reimburse your travel expenses unless you submit receipts. |

### Emphasize the Positive

Readers respond more favorably to encouragement than to criticism. When you give advice, suggestions, or recommendations, try to emphasize benefits rather than flaws.

| | |
|---|---|
| Critical tone | Because of your division's disappointing productivity, a management review may be needed. |
| Encouraging tone | A management review might help stimulate productivity in your division. |

## Remain Unbiased

Biased people make judgments without examining the facts. Your responsibility as a writer is to report accurately and fairly without distorting the evidence or injecting "loaded" words that reflect personal biases. If you *are* asked to include your interpretations and conclusions, base them on the facts. Even subjects that are controversial deserve unbiased treatment.

Imagine you have been sent to investigate the causes of an employee-management confrontation at your company's Omaha branch. Your initial report, written for the New York central office, is intended simply to describe the incident. Here is how an unbiased description might read:

> At 9:00 A.M. on Tuesday, January 21, eighty women employees entered the executive offices of our Omaha branch and remained six hours, bringing business to a virtual halt. The group issued a formal statement of protest, claiming that their working conditions were repressive, their salary scale was unfair, and their promotional opportunities were limited. The women demanded affirmative action, insisting that the company's hiring and promotional policies and wage scales be revised. The demonstration ended when Garvin Tate, vice president in charge of personnel, promised to appoint a committee to investigate the group's claims and to correct any inequities.

Notice the absence of implied judgments; the facts are presented objectively. A less impartial version of the event, from the women's side, might read:

> Last Tuesday, sisters struck another blow against male supremacy when eighty women employees paralyzed the pin-striped world of solidly entrenched sexism for more than six hours. The timely and articulate protest was aimed against degrading working conditions, unfair salary scales, and lack of promotional opportunities. Stunned executives watched helplessly as the group occupied their offices. The women were determined to continue their occupation until their demands for equal rights were met. Embarrassed company officials soon perceived the magnitude of this protest action and agreed to study the group's demands and to revise the company's discriminatory policies. The success of this long overdue confrontation serves as an inspiration to oppressed women employees everywhere.

Judgmental words and qualifiers *(male supremacy, degrading, paralyzed, articulate, stunned, discriminatory)* inject the writer's personal attitude, even though it isn't called for. In contrast, this next version defends the status quo:

> Our Omaha branch was the scene of an amusing battle of the sexes last Tuesday, when a Women's Lib group, eighty strong, staged a six-hour sit-in at the company's executive offices. The protest was lodged against alleged inequities in hiring, wages, working conditions, and promotion for women in our company. The radicals threatened to remain in the building until their demands for "equal rights" were met. Bemused company officials responded to this carnival display with patience and dignity, assuring the militants that their claims and demands—however inaccurate and immoderate—would receive just consideration.

Again, qualifying adjectives and superlatives slant the tone.

Being unbiased, of course, doesn't mean burying your head—and your values—in the sand. Remaining "neutral" on a controversial issue often is insufficient (Kremers 59). You have an ethical responsibility to weigh the facts and make your views known. If, for instance, you conclude that the Omaha protest was clearly justified, don't hesitate to say so.

## Avoid Sexist Language

Another element of bias enters when people use sexist language, stereotyping people on the basis of sex. Your usage is sexist if you refer in general to doctors, managers, lawyers, company presidents, engineers, and other professionals as *he* or *him* while referring to nurses, secretaries, and homemakers as *she* or *her*. Our goal as communicators is to *identify* with our audience, not to exclude half the population.

Follow these guidelines to eliminate sexist expressions from your communications:

- Use neutral expressions:

| | | |
|---|---|---|
| chair, or chairperson | rather than | chairman |
| businessperson | rather than | businessman |
| supervisor | rather than | foreman |
| police officer | rather than | policeman |
| letter carrier | rather than | postman |
| homemaker | rather than | housewife |

- Use plural forms. Instead of using *The manager . . . he*, use *The managers . . . they.*

- When possible (as in direct address), use *you*; for example, *You can begin to eliminate sexual bias by becoming aware of the problem.*

- Drop endings such as *-ess* and *-ette* used to denote females (e.g., *poetess, authoress, bachelorette, majorette*).

- Avoid overuse of pairings (*him or her, she or he, his or hers, he/she*). Too many such pairings are awkward.

- See Chapter 16 for avoiding sexist salutations (e.g., *Dear Sirs,*).

## EXERCISES IN ADJUSTING TONE

**1.** The sentences below have inappropriate tone because of pretentious language, missing contractions, or indirect address. Adjust the tone.

| | |
|---|---|
| Plain English needed | Avoid prolix nebulosity. |
| Revised | Don't be wordy and vague. |
| No contractions | Do not be wordy and vague. |
| Revised | Don't be wordy and vague. |

> **Indirect address** A writer shouldn't be wordy and vague.
>
> **Revised** Don't be wordy and vague.

a. Further interviews are a necessity to our ascertaining the most viable candidate.
b. This project is beginning to exhibit the characteristics of a loser.
c. We are pleased to tell you that you are a finalist.
d. Do not submit the proposal if it is not complete.
e. Employees must submit travel vouchers by May 1.
f. Persons taking this test should use the HELP option whenever they need it.

**2.** These sentences have too few *I* or *We* constructions, too many passive constructions, or express an unclear attitude. Adjust the tone.

> **No *I*** This writer would like to be considered for your opening.
>
> **Revised** Please consider me for your opening.
>
> **Passive** Your request will be given our consideration.
>
> **Revised** We will consider your request.
>
> **Unclear attitude** My disapproval is far more than negligible.
>
> **Revised** I strongly disapprove.

a. Payment will be made as soon as an itemized bill is received.
b. You will be notified.
c. Your help is appreciated.
d. Our reply to your bid will be sent next week.
e. I am not unappreciative of your help.
f. My opinion of this proposal is affirmative.

**3.** The sentences below suffer from negative emphasis, biased expressions, or sexist language. Adjust the tone.

> **Negative emphasis** Aggressive management of this risky project will help you avoid failure.
>
> **Revised** Aggressive management of this risky project will increase the chance for success.
>
> **Biased** Compare our product with its lesser competitors'.
>
> **Revised** Compare our product with its competitors'.
>
> **Sexist** While the girls played football, the men waved pom-poms.
>
> **Revised** While the women played football, the men waved pom-poms.

a. If you want your workers to like you, show sensitivity to their needs.
b. By not hesitating to act, you prevented my death.
c. The union has won its struggle for a decent wage.
d. The group's spokesman demanded salary increases.
e. Each employee should submit his vacation preferences this week.
f. Each applicant must prove that he or she has received his or her state certification.

## AUTOMATED AIDS FOR REVISING

Many of the strategies covered in this chapter could be executed rapidly with good word-processing software. By using the *global search-and-replace function* in some programs, you can command the computer to search for ambiguous pronoun references, overuse of passive voice, *to be* verbs, *There* and *It* sentence openers, negative constructions, clutter words, needless prefaces and qualifiers, overly technical language, jargon, sexist language, and so on. With an on-line dictionary or thesaurus, you can check definitions or see a list of synonyms for a word you have used in your document.

But despite the increasing sophistication of style checkers, diction checkers, and other such editing aids, automation never can eliminate the writer's burden of *choice*. None of the "rules" or advice offered in this chapter applies universally. Our language confronts us with almost infinite choices that cannot be programmed into a computer. Ultimately it is the informed writer's sensitivity to meaning, emphasis, and tone—the human contact—that determines the success and fairness of any document.

# Graphic and
# Design Elements

■

IV

# 12

# Creating
# Visual Aids

Purpose of Visuals

Tables

Graphs

Charts

Illustrations

Computer Graphics

Guidelines for Visuals

A visual aid is any pictorial representation used to simplify or emphasize material for the audience. Visual aids common in the workplace include tables, graphs, charts, and illustrations. These may appear on a printed page or on transparencies and 35-millimeter slides.

Information is useless unless the intended audience interprets it correctly. Overwhelmed by the growing volume of printed information, today's audiences depend on visual aids to get the message straight.

Besides saving space and words, visuals help audiences understand and remember information. They can reveal and display trends, problems, and possibilities that might otherwise remain buried in lists of facts and figures. Because they give audiences a picture, visuals can lead to better-informed and faster decisions. In written reports and oral presentations alike, well-designed visuals help the writer or presenter appear prepared, credible, and persuasive.

## PURPOSE OF VISUALS

Visuals help us connect with our audience. To appreciate the extent of this connection, we can generalize briefly about reading preferences among workplace audiences. What do readers expect in a document?

We know that readers expect more than just raw information; they want the information "processed" for their understanding. Though not especially interested in learning how smart the writer is, readers themselves like to feel smart, to understand the gist of a message at a glance. More receptive to images than words, today's readers resist uninterrupted pages of printed words. Visuals break down a reader's resistance in several ways:

- *Visuals set off and emphasize things.* To show that a microcomputer with 512K of memory has dropped in price 60 percent in four years, a bar graph would be more vivid than a prose statement.

- *Visuals display abstract concepts in concrete, geometric shapes.* ("How do lasers work?" How does the AIDS virus work?")

- *Visuals can compare large amounts of data.* ("How do our profits compare with our expenses?" "How do this month's sales figures compare with last month's?"

- *Visuals depict relationships.* ("How much has our productivity increased since employees received a 15 percent pay hike?" "How has our new incentive program affected employees' absences?")

- *Visuals condense information, displaying it in a meaningful way.* A simple table, for instance, can summarize a long and difficult printed passage, as in the example that follows.

Imagine that you are researching a possible link between chemicals and cancer deaths. From various sources, you might collect these data:

*1.* In 1900, pneumonia and influenza accounted for 11.8 percent of deaths, heart disease for 8.0 percent, cerebrovascular disease for 6.2 percent, all accidents for 4.2 percent, and cancer for 3.7 percent.

*2.* In 1960, pneumonia and influenza caused 3.5 percent of deaths, heart disease 38.7 percent, cerebrovascular disease 11.3 percent, all accidents 5.5 percent, and cancer 15.6 percent.

*3.* In 1970 . . . .

This information is repetitious, tedious, and hard to interpret. When arranged in Table 12.1, however, these statistics are easy to compare.

**TABLE 12-1** Leading Causes of Death in 1900, 1960, and 1970

| | Cause of death (as percentage of all deaths) | | | | |
|---|---|---|---|---|---|
| Year | Pneumonia and influenza | Heart disease | Cerebrovascular disease | All accidents | Cancer |
| 1900 | 11.8 | 8.0 | 6.2 | 4.2 | 3.7 |
| 1960 | 3.5 | 38.7 | 11.3 | 5.5 | 15.6 |
| 1970 | 3.3 | 38.3 | 10.8 | 6.0 | 17.2 |
| Average | 6.2 | 28.3 | 9.4 | 5.2 | 12.1 |

*Source:* Adapted from *Facts of Life and Death* (Washington: GPO, 1974): 31.

Translate your prose into visuals whenever they make your point more clearly than the prose. Use visuals only to *clarify* and to enhance your discussion, not simply to *decorate* it.

## TABLES

Tables are especially useful for displaying exact quantities or for comparing large amounts of information in a small space. Large tables, however, present so much information that readers can find them confusing (Felker 95). An otherwise impressive-looking table (like Table 12.2) can be almost impossible for some readers to interpret because it presents too large a quantity of data in an overly complex format. Unethical writers could too easily use complex tables to bury numbers that are questionable or embarrassing (Williams 12).

Use a table only when you are reasonably certain that it will enlighten—rather than frustrate—your intended audience. For nonspecialized readers, use fewer tables, and keep them simple. Readers of any table need to understand how it is organized, where to find what they need, and how to interpret the information they find (Hartley 90).

To construct a table, use tabulating markers and the tab keys on your typewriter or word processor, and follow these suggestions:

- Number the table in its order of appearance (Table 1, Table 2) and give it a title describing exactly what the table compares or measures.

- Compare your data vertically (in columns) instead of horizontally (in rows). Readers find columns easier to scan than rows.

- Begin each column with a heading naming the category for the items listed *(Heart disease, Cancer)*, and stipulate the units of measurement *(percentage of deaths, miles per gallon)*.

- Convert all fractions into decimals, and align the decimals vertically. For easy comparison, round off insignificant decimals to the nearest whole number.

- Label all parts of the table clearly, so that readers will know what they are looking at.

- Space your listed items so that they are neither cramped not too far apart for easy comparison.

- Whenever possible, include row or column averages, to give readers reference points for comparing individual figures.

- If the table is too wide for the page, turn it 90 degrees and place its top toward the inside of the binding. Or you might divide the data into two tables.

- Try to hold the table to one page. Otherwise write "continued" at the bottom, and begin the second page with the full title, "continued," and the original column headings.

- Try to avoid footnotes, unless you must explain or define entries (as in Table 12.2). If footnotes are necessary, label them with lowercase letters (a, b, c) in the table.

**TABLE 12-2** A Table with Too Much Information

Air Pollutant Emissions, by Pollutant and Source: 1970 and 1983
(In millions of metric tons, except lead in thousands of metric tons. Metric ton = 1.1023 short tons)

| Year and pollutant | Total emissions | Controllable emissions | | | | | | | Percentage of total | | |
| --- | --- | --- | --- | --- | --- | --- | --- | --- | --- | --- | --- |
| | | Transportation | | Fuel combustion[a] | | Industrial processes | Solid waste disposal | Misc. uncontrollable | Transportation | Fuel combustion[a] | Industrial |
| | | Total | Road vehicles | Total | Electric utilities | | | | | | |
| 1970: Carbon monoxide .... | 98.3 | 71.8 | 62.7 | 3.9 | .2 | 9.0 | 6.4 | 7.2 | 73.0 | 4.0 | 9.2 |
| Sulfur oxides .......... | 28.2 | .6 | .3 | 21.3 | 15.8 | 6.2 | (Z) | .1 | 2.1 | 75.5 | 22.0 |
| Volatile organic compounds | 27.0 | 12.3 | 11.1 | .9 | (Z) | 8.7 | 1.8 | 3.3 | 45.6 | 3.3 | 32.2 |
| Particulates[b] .......... | 18.0 | 1.2 | .9 | 4.5 | 2.3 | 10.1 | 1.1 | 1.1 | 6.7 | 25.0 | 56.1 |
| Nitrogen oxides ........ | 18.1 | 7.6 | 6.0 | 9.1 | 4.5 | .7 | .4 | .3 | 42.0 | 50.3 | 3.9 |
| Lead ................ | 203.8 | 163.6 | 156.0 | 9.6 | .3 | 23.9 | 6.7 | (Z) | 80.3 | 4.7 | 11.7 |
| 1983: Carbon monoxide .... | 67.6 | 47.7 | 41.2 | 7.0 | .3 | 4.6 | 2.0 | 6.3 | 70.6 | 10.4 | 6.8 |
| Sulfur oxides .......... | 20.8 | .9 | .5 | 16.8 | 14.0 | 3.1 | (Z) | (Z) | 4.3 | 80.8 | 14.9 |
| Volatile organic compounds | 19.9 | 7.2 | 6.0 | 2.1 | (Z) | 7.5 | .6 | 2.5 | 36.2 | 10.6 | 37.9 |
| Particulates[b] .......... | 6.9 | 1.3 | 1.1 | 2.0 | .5 | 2.3 | .4 | .9 | 18.8 | 29.0 | 33.3 |
| Nitrogen oxides ........ | 19.4 | 8.8 | 7.0 | 9.7 | 6.3 | .6 | .1 | .2 | 45.4 | 50.0 | 3.1 |
| Lead ................ | 46.9 | 40.7 | 38.7 | .6 | .1 | 2.5 | 3.1 | (Z) | 86.8 | 1.3 | 5.3 |

Z Less than 50,000 metric tons.    [a]Stationary.    [b]See footnote a, table 352.

*Source:* U.S. Environmental Protection Agency, *National Air Pollutant Emission Estimates, 1940–1983* Dec. 1984.

- Cite your sources of data beneath any footnotes, even if you make your own table from borrowed data.
- Include a prose explanation of the comparisons (as in Table 12.3).

Tables work well for displaying exact quantities, but for easier interpretation, readers prefer graphs or charts. Also, shapes (bars, curves, circles) are generally easier to remember than lists of numbers (Cochran et al. 25). Always consider which type of visual presentation satisfies your readers' need most efficiently.

Any visual other than a table usually is categorized as a *figure,* and should be so titled *(Figure 1   Aerial View of the Panhandle Site).* Figures covered in this chapter include graphs, charts, and illustrations.

Like all other components in the document, visuals need to be designed with audience and purpose in mind (Journet 3). An accountant doing an audit might need a table listing exact amounts, whereas the average public stockholder reading an annual report would prefer the "big picture" in the easily grasped bar graph or pie chart (Van Pelt 1). The next section will show how the data in Table 12.3 can be presented in a number of displays, depending on the point the writer wants to make.

**TABLE 12-3** Death Rates for Heart Disease and Cancer, 1970–1986

| Year | Number of deaths (per 100,000) population)[a] | | | |
| | Heart disease | | Cancer | |
| | Male | Female | Male | Female |
|---|---|---|---|---|
| 1970 | 419 | 309 | 172 | 135 |
| 1978 | 371 | 300 | 201 | 160 |
| 1986 | 348 | 291 | 219 | 173 |
| Percentage of change, 1970–1986 | −17.6 | −5.8 | +21.4 | +22.0 |

[a]Figures are approximate.

*Source:* Adapted from *Statistical Abstract of the United States* (Washington: GPO, 1986): 205.

As Table 12.3 indicates, both male and female death rates from heart disease decreased from 1970 to 1986, but males showed a sizably larger decrease. Cancer deaths during this period increased at roughly equal rates for both males and females.

## GRAPHS

Graphs translate numbers into pictures. Plotted as a set of points (a *series*) on a coordinate system, a graph shows the relationship between two variables.[1] Graphs are especially useful for displaying comparisons, changes with passing time, or trends. When you decide to use a graph, choose the best type for your purpose: bar graph or line graph.

### Bar Graphs

Easily understood by most readers, bar graphs are effective in showing discrete comparisons, as on a year-by-year or month-by-month basis. Each bar represents a specific quantity. Use bar graphs to compare values that change over equal time intervals (expenses calculated at the end of each month, sales figures totaled at yearly intervals). Use a bar graph only to compare values that are noticeably different. Otherwise, all the bars will appear almost identical.

**Simple Bar Graphs.** The simple bar graph in Figure 12.1 displays one relationship taken from the data in Table 12.3, the rate of male deaths from heart disease. To aid interpretation, you can record exact values above each bar—but only if readers need exact numbers.

[1]Graphs have a horizontal and a vertical axis. The horizontal axis carries categories (the independent variables) to be compared, such as years within a period (1970, 1978, 1986). The vertical axis shows the range of values (the dependent variables) for comparing or measuring the categories, such as the number of people who died from heart failure in a specified year. A dependent variable changes according to activity in the independent variable (say, a decrease in quantity over a set time, as in Figure 12.1). In the equation $y = f(x)$, $x$ is the independent variable and $y$ is the dependent variable.

FIGURE 12.1  A Simple Bar Graph

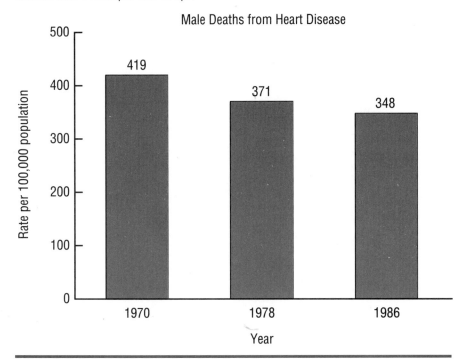

Male Deaths from Heart Disease

**Multiple-Bar Graphs.**   A bar graph can display as many as two or three relationships simultaneously, with each relationship plotted as a separate series. Figure 12.2 expresses two comparisons from Table 12.3, the rate of male deaths from both heart disease and cancer.

The more relationships your bar graph displays, the harder it will be to interpret. As a rule, plot no more than three bars in one graph.

Because Figure 12.2 exhibits more than one relationship (or series), each relationship is represented by a different pattern, and the patterns are identified by a *legend*.

**Horizontal-Bar Graphs.**   To make a horizontal-bar graph, turn a vertical-bar graph (and scales) on its side, or 90 degrees to the right. Horizontal-bar graphs are good for displaying a large quantity of bars arranged in order of increasing or decreasing value, as in Figure 12.3. The horizontal format leaves room for labeling the categories horizontally *(All races, etc.)*. A vertical-bar graph would leave no room for horizontal labeling.

As with all visuals, choose the format that will provide your readers with the clearest and most accurate display.

**Stacked-Bar Graphs.**   Instead of side-by-side clusters of bars, you can display multiple relationships by stacking them on top of one another. Stacked-bar graphs

**FIGURE 12.2** A Multiple-Bar Graph

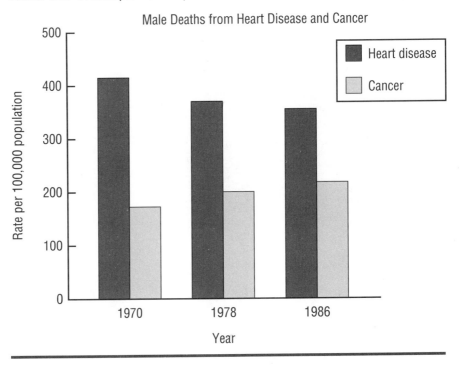

are especially useful for showing how much each item contributes to the whole. Figure 12.4 displays other comparisons from Table 12.3.

Display no more than two or three relationships in a stacked-bar graph. Too many subdivisions and patterns would only confuse your readers.

**100 Percent Bar Graph.** A type of stacked bar graph, the 100 percent bar graph is useful for showing the relative values of the parts that make up the 100 percent value, as in Figure 12.5. Like any bar graph, the 100 percent graph can have either horizontal or vertical bars.

Notice how bar graphs become harder to interpret as bars and patterns increase. For unsophisticated readers, the large quantity of data in Figure 12.5 might be easier to interpret in pie charts (pages 254–256).

**Deviation Bar Graphs.** With the deviation bar graph, you can display both positive and negative values in one graph, as in Figure 12.6. Notice how the vertical axis extends below the zero baseline, following the same incremental division as above the baseline, but in negative values instead.

**Avoiding Distortion in Bar Graphs.** Any one set of data can support contradictory conclusions. Even though your numbers may be accurate, the way in which you display them could distort their meaning.

**FIGURE 12.3** A Horizontal-Bar Graph

Percentage of Adults Who Have Completed Four
Years of High School or More: 1950 to 1980

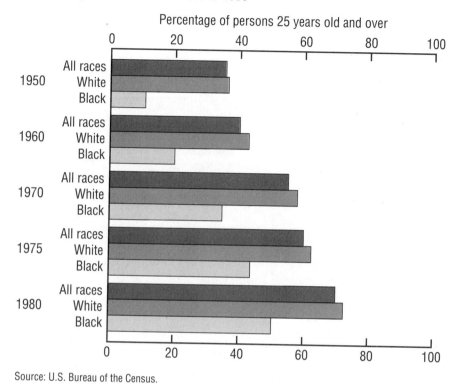

Source: U.S. Bureau of the Census.

---

**FIGURE 12.4** A Stacked-Bar Graph

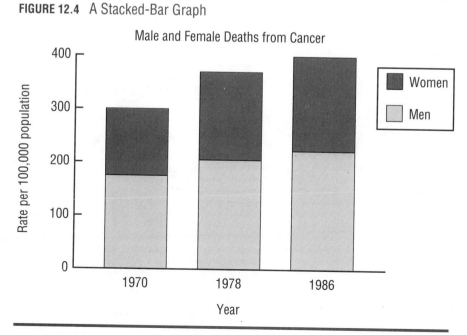

FIGURE 12.5  A 100 Percent Bar Graph

State and Local Government Taxes—
Percentage Distribution by Type, 1960–1980

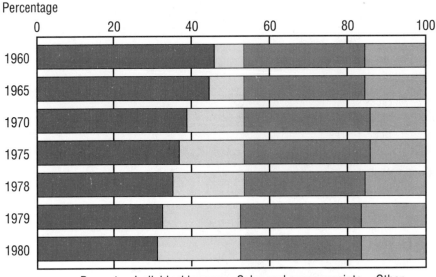

Source: U.S. Bureau of the Census.

FIGURE 12.6  A Deviation Bar Graph

Corporate Profits after Taxes, 1970–1984

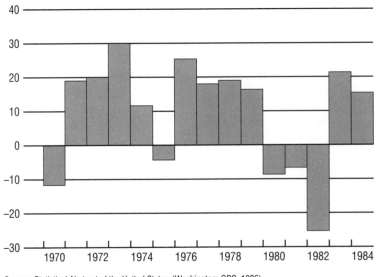

Source: *Statistical Abstract of the United States* (Washington: GPO, 1986).

The visual relationships on your graph should always represent the numerical relationships. Never stretch or compress the scales in trying to reinforce your point. Make your vertical scale at least 75 percent as long as your horizontal scale, and begin the vertical scale at zero. Notice how the visual relationships in Figure 12.7 become distorted when the value scale is compressed or when it fails to begin at zero. In version (a) of Figure 12.7, the bars accurately depict the numerical relationships measured from the value scale. In version (b), Z (400) is depicted as three times X (200). In version (c), the scale is too compressed, causing the shortened bars to understate the differences in quantity. Deliberate distortions are unethical because they imply conclusions contradicted by the actual data.

**FIGURE 12.7**  An Accurate Bar Graph and Two Distorted Versions

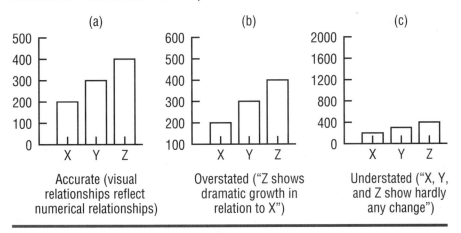

Graphics software usually can plot the scales of your graphs automatically, eliminating distortion. But if you make your graphs by hand, be sure to experiment with scales until you find the most accurate representation.

**Bar Graph Guidelines.**  Once you decide on a type of bar graph, follow these suggestions for presenting the graph to your audience:

- Keep the graph simple and easy to read. Avoid plotting more than three types of bars in each cluster.

- Number your scales in units your audience will find familiar and easy to follow. Units of 1 or multiples of 2, 5, or 10 are best (Lambert 45). Space the numbers equally.

- Label both scales to show what is being measured or compared. If space allows, keep all labels horizontal for easier reading.

- Use *tick marks* to show the points of division on your scale. If the graph has many bars, extend the tick mark into *grid lines* to help readers relate bars to values.

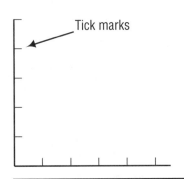 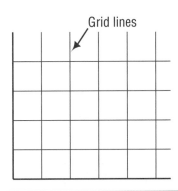

- To avoid confusion, make all bars the same width (unless you are overlapping them, as on page 266). If you must produce your graphs by hand, use graph paper to keep bars and increments evenly spaced.
- In a multiple-bar graph, use a different pattern or color for each bar in a cluster. Be sure to provide a legend identifying each pattern or color.
- If you are trying for emphasis, be aware that darker bars are seen as larger and closer and more important than lighter bars of the same size (Lambert 93).

Many computer graphics programs automatically follow most of these techniques. Anyone producing visuals, however, should know all the conventions.

## Line Graphs

A line graph can accommodate many more data points than a bar graph (say, a twelve-month trend, measured monthly). Line graphs are therefore preferable for evaluating large bodies of information in which exact quantities need not be emphasized. Whereas bar graphs display quantitative differences among items (cities, regions, yearly or monthly intervals), line graphs display data whose value changes repeatedly, as in a trend, forecast, or other change in a specified time (profits, losses, growth). And some line graphs depict cause-and-effect relationships (say, how seasonal patterns affect sales or profits).

**Simple Line Graphs.** A simple line graph, as in Figure 12.8, uses one line to plot time intervals on the horizontal scale and values on the vertical scale. The relationship depicted here would be much harder to express in words alone.

**Multiple-Line Graphs.** A multiple-line graph displays as many as three relationships simultaneously, as in Figure 12.9.

Because readers usually find line graphs harder to interpret than bar graphs, be sure to explain in prose the relationships readers are supposed to see:

As displayed in Figure 12.9, building permits in all three counties increased sharply during summer and early fall. Dade and Monroe counties showed a steady increase as the weather grew warmer, but Shaker County's seasonal increase was more erratic. Shaker's permits actually declined in April and May, but then rose to surpass permits in Dade and Monroe counties from June through September.

**FIGURE 12.8**  A Simple Line Graph

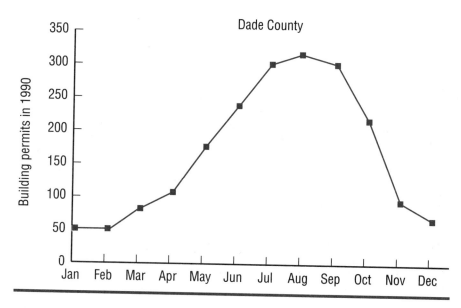

**FIGURE 12.9**  A Multiple-Line Graph

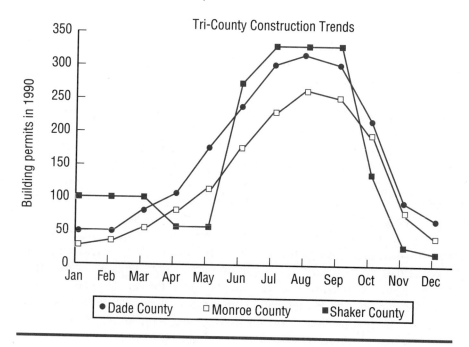

FIGURE 12.10  A Deviation Line Graph

U.S. International Transaction Balances, 1960–1980

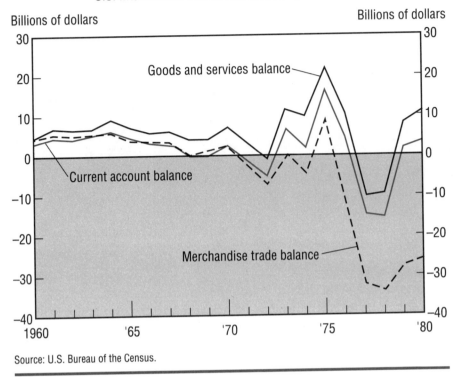

Source: U.S. Bureau of the Census.

**Deviation Line Graphs.**  By extending your vertical scale below the zero baseline, you can display both positive and negative values in one graph, as in Figure 12.10. Be sure to mark the same intervals for values below the baseline.

**Band or Area Graphs.**  A type of line graph, a band or area graph displays changes or trends over a specified period. To attract attention, however, the area beneath each plotted line in a band graph is filled with a pattern. The band graph in Figure 12.11 plots the same data as the line graph in Figure 12.8.

A multiple-band graph can be especially useful for displaying relationships among sums rather than direct comparisons. Figure 12.12 shows how the crisscrossed line graph in Figure 12.9 would appear as a band graph.

Although they might seem more interesting than line graphs, multiple-band graphs are easy to misinterpret unless they are carefully labeled and explained. We recall that every line in a multiple-line graph represents a distance from the zero baseline. In a multiple-band graph, however, the very top line represents the *total* change or trend; each band below the top line represents a part that contributes to the total (like segments in the stacked-bar graph, Figure 12.4). Be sure to explain these relationships to uninitiated readers.

**FIGURE 12.11** A Simple Band Graph

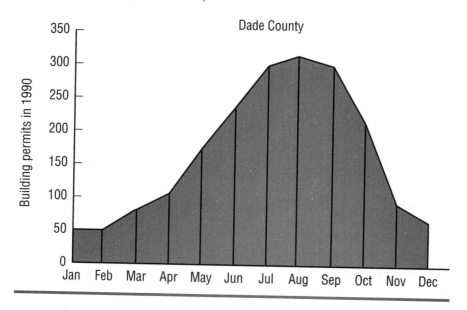

**FIGURE 12.12** A Multiple-Band Graph

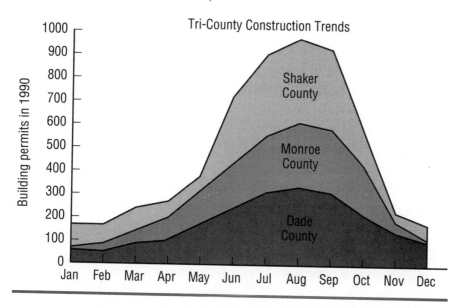

**Avoiding Distortion in Line Graphs.** A poorly constructed line graph can distort visual relationships between numbers that are otherwise accurate. Figure 12.13 shows how one type of distortion can occur, even on a computer, when data that would provide a complete picture are selectively omitted. Version (a) accurately depicts the numerical relationships measured from the value scale. But

FIGURE 12.13  An Accurate Line Graph and a Distorted Version

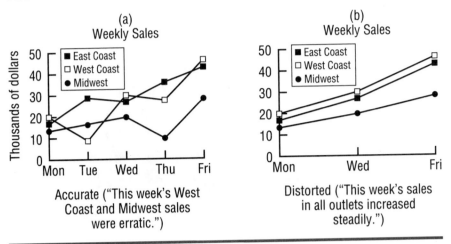

(a)
Weekly Sales

Accurate ("This week's West
Coast and Midwest sales
were erratic.")

(b)
Weekly Sales

Distorted ("This week's sales
in all outlets increased
steadily.")

in version (b), too few points are plotted. If you plot by hand, begin the vertical scale at zero, and avoid stretching or compressing either scale.

Visuals have their own rhetoric and persuasive force, which we can use to advantage—for positive or negative purposes, for the reader's benefit or detriment (Van Pelt 2). Avoiding visual distortion is ultimately a matter of ethics.

**Line Graph Guidelines.**   Line graphs follow the same conventions as bar graphs, with these additions:

- Compare no more than three lines on one graph.
- Mark the individual data points on each line so that readers can see how many points you have used to plot your line.
- In a multiple-line graph, make each line distinct (using colors, dots, dashes, or symbols).
- Label each line clearly so that readers will know what it represents.
- Avoid grid lines; readers might confuse them with the plotted lines.

## CHARTS

The terms *chart* and *graph* are often used interchangeably. But a chart is more precisely a figure that displays relationships (quantitative or cause-and-effect) without being plotted on a coordinate system. Commonly used charts include pie charts, organizational charts, flowcharts, tree charts, column charts, and pictorial charts (pictograms).

### Pie Charts

Considered easy for readers to understand, a pie chart depicts the relative amounts of the parts that make up a whole. In a pie chart, readers can compare the parts to each other as well as to the whole (to show how much was spent on what, how much income comes from which sources, and so on). Figure 12.14 shows a pie chart. Figure 12.15 shows two other versions of the pie chart in Figure 12.14. Version (a) displays dollar amounts, and version (b) the percentage relationships among these dollar amounts.

For constructing pie charts, follow these suggestions:

- Be sure the parts add up to 100 percent.
- If you must produce your charts by hand, use a compass and protractor for precise segments. Each 3.6-degree segment equals 1 percent. Include anywhere from two to eight segments. More than eight segments can be hard to interpret, especially if they are small (Hartley 96).

**FIGURE 12.14**  A Simple Pie Chart

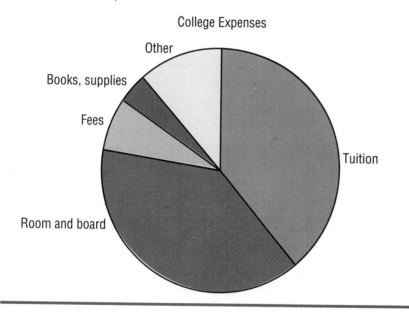

**FIGURE 12.15**  Two Other Versions of Figure 12.14

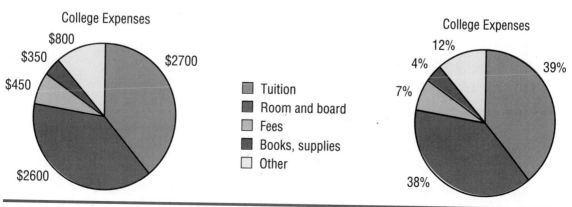

- Combine small segments under the heading "Other."
- Create a sense of spatial logic by locating your first radial line at twelve o'clock and moving clockwise from larger to smaller (except for "Other," which is usually the final segment).
- For easy reading, keep all labels horizontal.

Keep in mind that guidelines for constructing visuals are based on *conventions,* on methods that traditionally have proven effective for visual display. These conventions are by no means inflexible. As your purpose dictates—and accuracy allows—use your own approach.

### Organization Charts

An organization chart divides an organization into its administrative or managerial parts. Each part is ranked according to its authority and responsibility in relation to other parts and to the whole. Figure 12.16 displays the management structure of a typical organization.

An organization chart is a good way to save words. Imagine how much prose would be needed to describe the relationships in Figure 12.16.

**FIGURE 12.16** An Organization Chart for One Corporation

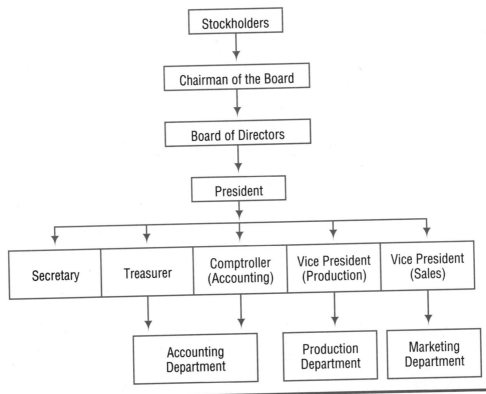

## Flowcharts

A flowchart traces a procedure or process from beginning to end. In displaying the steps in a manufacturing process, the flowchart would begin at the raw materials and proceed to the finished product. Figure 12.17 traces the steps for developing and refining a technical report within one government organization. At a glance, new employees can visualize the collaboration in writing, reviewing, editing, and approval that produces a useful document.

FIGURE 12.17 A Flowchart for Processing a Technical Report

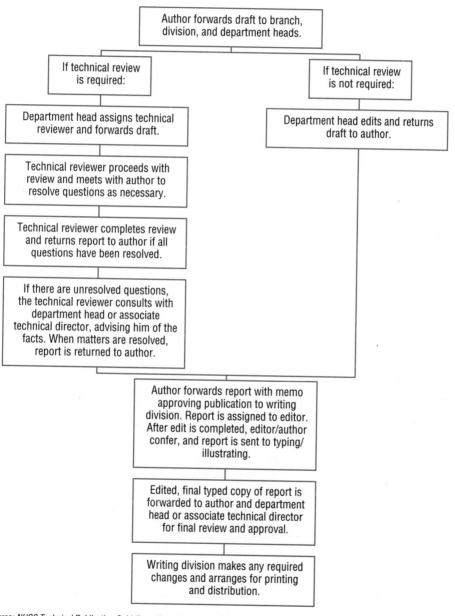

Author forwards draft to branch, division, and department heads.

If technical review is required:

If technical review is not required:

Department head assigns technical reviewer and forwards draft.

Department head edits and returns draft to author.

Technical reviewer proceeds with review and meets with author to resolve questions as necessary.

Technical reviewer completes review and returns report to author if all questions have been resolved.

If there are unresolved questions, the technical reviewer consults with department head or associate technical director, advising him of the facts. When matters are resolved, report is returned to author.

Author forwards report with memo approving publication to writing division. Report is assigned to editor. After edit is completed, editor/author confer, and report is sent to typing/illustrating.

Edited, final typed copy of report is forwarded to author and department head or associate technical director for final review and approval.

Writing division makes any required changes and arranges for printing and distribution.

Source: *NUSC Technical Publication Guidelines,* Naval Underwater Systems Center, Technical Information Dept., 17 Oct. 1977.

### Tree Charts

Whereas flowcharts display the steps in a process, tree charts show how the parts of an idea or concept relate to each other. Figure 12.18 displays the parts of an outline for this chapter so that readers can better visualize the relationships. Notice that the tree version seems clearer and more interesting than the prose listing.

### Pictorial Charts

Pictorial charts, or pictograms, depict numerical relationships with icons or symbols (cars, houses, smokestacks) of the items being measured, instead of using bars. Each symbol represents a stipulated quantity, as in Figure 12.19. Many graphics programs provide an assortment of predrawn symbols.

Use pictograms when you want to make your information more interesting for nontechnical audiences.

## ILLUSTRATIONS

Illustrations consist of diagrams and photographs depicting relationships that are physical rather than numerical. Good illustrations help readers understand

**FIGURE 12.18** An Outline and Its Tree Chart Version

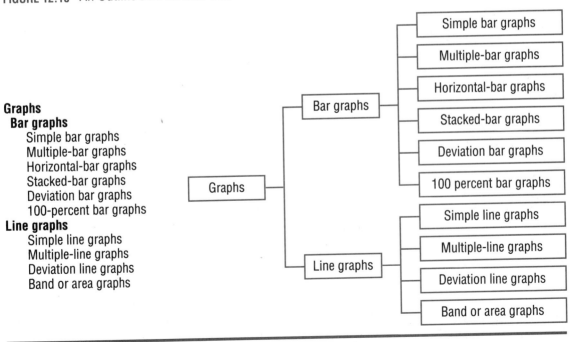

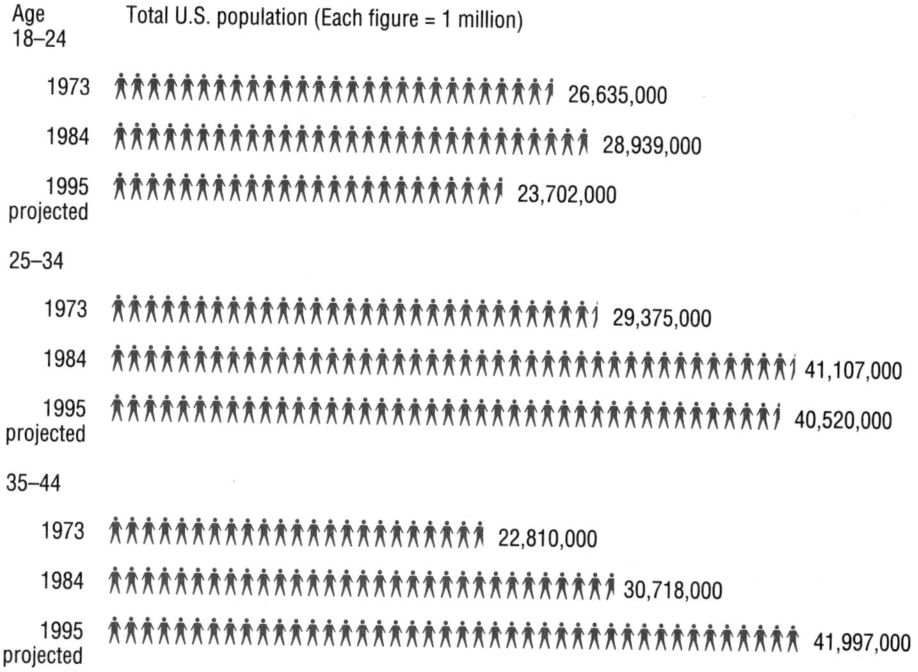

**FIGURE 12.19** A Pictogram

**After increasing between 1973 and 1984, the college-age population is projected to decline over the 1984–1995 period.**

Age 18–24      Total U.S. population (Each figure = 1 million)

| | |
|---|---|
| 1973 | 𝍅𝍅𝍅𝍅𝍅𝍅𝍅 26,635,000 |
| 1984 | 𝍅𝍅𝍅𝍅𝍅𝍅𝍅 28,939,000 |
| 1995 projected | 𝍅𝍅𝍅𝍅𝍅𝍅𝍅 23,702,000 |

25–34

| | |
|---|---|
| 1973 | 𝍅𝍅𝍅𝍅𝍅𝍅𝍅 29,375,000 |
| 1984 | 𝍅𝍅𝍅𝍅𝍅𝍅𝍅 41,107,000 |
| 1995 projected | 𝍅𝍅𝍅𝍅𝍅𝍅𝍅 40,520,000 |

35–44

| | |
|---|---|
| 1973 | 𝍅𝍅𝍅𝍅𝍅𝍅𝍅 22,810,000 |
| 1984 | 𝍅𝍅𝍅𝍅𝍅𝍅𝍅 30,718,000 |
| 1995 projected | 𝍅𝍅𝍅𝍅𝍅𝍅𝍅 41,997,000 |

Source: U.S. Bureau of the Census.

and remember the material (Hartley 82). Consider this information from a government pamphlet, explaining the operating principle of the seat belt:

> The safety-belt apparatus includes a tiny pendulum attached to a lever, or locking mechanism. Upon sudden deceleration, the pendulum swings forward, activating the locking device to keep passengers from pitching into the dashboard.

Without the illustration in Figure 12.20 we have trouble visualizing the mechanism.

Clear and uncluttered, a good diagram eliminates unnecessary details, and is focused only on material useful to the reader. The following pages sample some commonly used diagrams.

## Exploded Diagrams

Exploded diagrams, like that of a faucet in Figure 12.21, show how the parts of an item are assembled; they often appear in repair or maintenance manuals.

**FIGURE 12.20**  A Diagram of a Safety-Belt Locking Mechanism

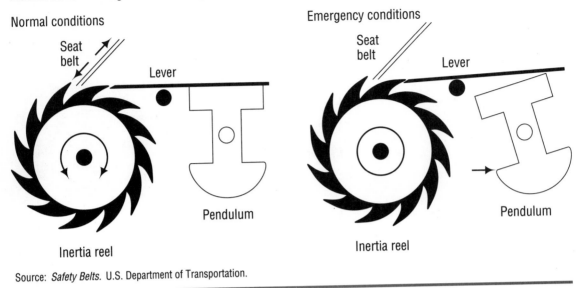

Source: *Safety Belts.* U.S. Department of Transportation.

**FIGURE 12.21**  An Exploded Diagram of a Faucet

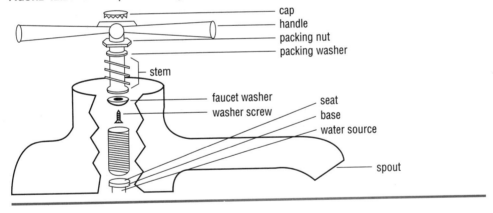

Notice that the faucet is shown in *cutaway view,* where part of the exterior is literally "cut away" to reveal internal parts.

### Diagrams of Procedures

Diagrams can clarify instructions by illustrating steps or actions, as in Figure 12.22.

FIGURE 12.22   A Diagram of a Repair Procedure

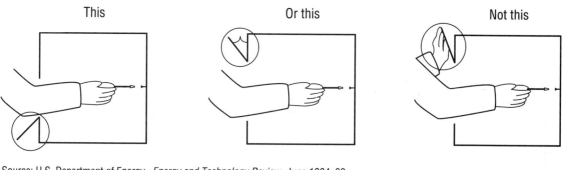

This   Or this   Not this

Source: U.S. Department of Energy, *Energy and Technology Review,* June 1984: 33.

## Block Diagrams

Block diagrams are simplified sketches that represent the relationship between the parts of an item or process. Because block diagrams are designed to illustrate *concepts* (such as current flow in a circuit), the parts are represented as symbols or shapes. The block diagram in Figure 12.23 illustrates how any process can be controlled automatically through a feedback mechanism. Figure 12.24 shows the feedback concept applied as the cruise-control mechanism on a motor vehicle.

FIGURE 12.23   A Block Diagram Illustrating the Basic Feedback Concept

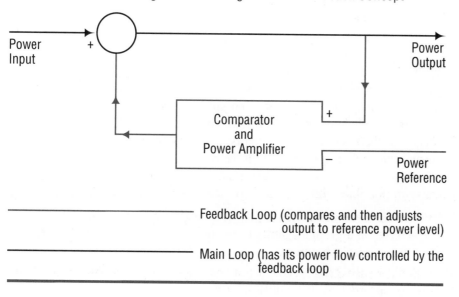

Power Input

+

Power Output

Comparator and Power Amplifier

+

−

Power Reference

Feedback Loop (compares and then adjusts output to reference power level)

Main Loop (has its power flow controlled by the feedback loop

**FIGURE 12.24**  A Block Diagram Showing a Cruise-Control Mechanism

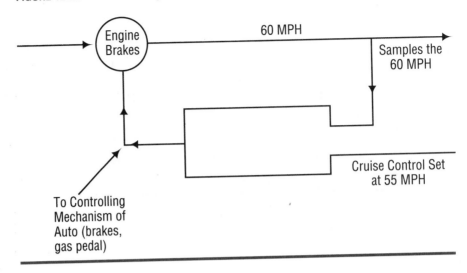

### Photographs

Photographs provide an accurate overall view, but sometimes they can be too "busy." By showing all details as more or less equal, a photograph sometimes fails to emphasize important areas.

When you use photographs, keep them distinct, well focused, and uncluttered. To emphasize features of a complex mechanism, you probably should rely on diagrams instead of photographs, unless you intend simply to show an overall view. Provide a sense of scale by including a person or a familiar object (such as a hand) in your photo.

### COMPUTER GRAPHICS

With one or two commands, today's computer systems can create highly sophisticated, six-color graphic displays. You simply type your numerical data into the program, and choose among types of charts or graphs. Using its own built-in formulas, the system analyzes, plots, and displays the data in the appropriate visual form within seconds. With the right software, all the visual displays in this chapter can be created by computer—as most were. And more complex visuals are possible with computer graphics, as in these few examples:

- With an electronic stylus (a pen with an electronic signal), you can draw pictures on a graphics tablet to be displayed on the monitor, stored, or sent to other computers.

- You can map different concentrations in different colors (say, in a mineral map) in order to distinguish data more clearly.

- You can create three-dimensional effects, showing an object from different angles with shading, shadows, or other techniques.

- You can recreate the visual effect of a mathematical model, as in writing equations to explain what happens when high winds strike a tall building. (As the wind deforms the structure, the equations change. Then you can take those new equations and represent them visually.)

- You can create a design, build a model, simulate the physical environment, and let the computer forecast what will happen with different variables.

- You can integrate computer-assisted design (CAD) with computer-assisted manufacturing (CAM), so that the design will direct the machinery that makes the parts themselves (CAD/CAM).

- You can create animations, to see how bodies move (as in a car crash or in athletics).

- You can practice dealing with toxic chemicals, operating sophisticated machines, or making other rapid decisions in medical or technical environments, without the cost or danger in actual situations.

Once you've created a model, you can do any number of "what-if" projections. Because the computer can generate and evaluate many possibilities rapidly, it enables you to test hypotheses without doing the calculations.

## Microcomputer Graphics

Among all applications for personal computers, graphics production is perhaps the most rapidly expanding (Schmeupe 51). Unlike hand-drawn visuals, computer graphics systems allow you to experiment with scales, formats, colors, perspectives, and patterns. Many systems now offer WYSIWYG (What you see is what you get) packages, in which the on-screen display is almost identical to the hard-copy image that will be printed out. Thus, by testing design options on the screen, you can revise and enhance your visual repeatedly until it achieves your exact purpose.

Here is a sampling of the design options you can expect in many software packages for microcomputer graphics:

- You can update charts and graphs hourly or daily, or whenever the data change. The software will calculate the new data and plot the relationships.

- You can edit your graphics on the screen, adding, deleting, or moving material as needed.

- You can create your image at one scale, and later specify a different scale for the same image.

- You can annotate and label, and create multiple typefaces within one visual.

- You can overlay images in one visual.

- You can adjust bar width and line thickness.

- You can fill a shape with a color or pattern.

Most of these options, and many others, call for no more than a single keystroke.

## Composing and Enhancing a Visual Electronically

Composing for visuals is identical to that for written text: you must decide about the purpose, audience, content, arrangement, and style of your visual message, revising repeatedly until the message conveys your exact meaning. The following short scenario suggests how you might explore options in a typical graphics software package.

### Using Graphics Software

Assume that after a few years with Company X (an international producer of communications hardware and software), you have been appointed personnel recruiter. Your job is to visit college campuses across the nation and recruit the finest talent among graduate as well as undergraduate students. To create interest in your company, you decide to prepare a visual presentation that will complement the lectures you will be giving to student groups during your travels.

Among your most striking data (for this audience) is the dramatic increase in the average salary offered to entry-level people by your company over the past five years; to emphasize the increase, you decide to begin your presentation with a five-year salary comparison. After inserting your graphics software into the computer's main disk drive, you get right to work composing your first visual.

To show change at fixed intervals and to focus on specific numbers, you decide to use a bar graph—a visual whose meaning is apparent, even for those who might have little experience interpreting visual messages. You begin by entering average salaries for entry-level personnel with a B.S. degree:

| | |
|---|---|
| 1986 | 21,500 |
| 1987 | 24,000 |
| 1988 | 26,000 |
| 1989 | 28,500 |
| 1990 | 31,000 |

After receiving a few commands, your computer processes your data to produce the simple graph in Version A.

Most of your audiences will contain both graduate and undergraduate students, and so you decide to include the salary figures for M.S. and Ph.D. entry-level positions. The computer responds by displaying the multiple-bar graph in Version B.

Although this second version is adequate for your own information, it is neither informative nor engaging enough to present to a varied and unfamiliar audience. For greater informative and interest value, your graph will need enhancement. You decide to add a legend to identify the various patterns. Then you emphasize the title with boldface type and a shadowed border. For appeal, you add color. For contrast, you frame the entire graph, as in Version C.

As a further aid to interpretation, you decide to add grid lines, to scale your vertical axis in smaller increments ($5,000 instead of $10,000), and to add major

## VERSION A  A Simple Bar Graph

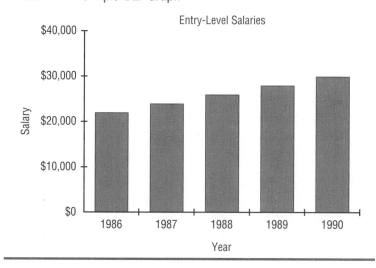

## VERSION B  A Multiple-Bar Graph

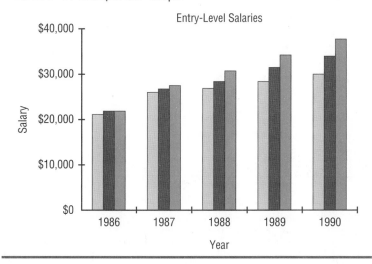

tick marks to the outside of your vertical axis. For easy scanning, you reverse the order of columns in each cluster and overlap them by 50 percent. Your finished product in Version D is a presentation-quality graph that you can project on a large-screen monitor, reproduce as a slide or transparency, or include in a document.

For a different purpose, you might decide to emphasize how salary increases for advanced degrees have outdistanced those for the B.S. One way of achieving this emphasis is to plot the B.S. salary data as a line graph, which would then overlay the original M.S. and Ph.D. bars, as in Version E. This version, of course, is just one among many possibilities for plotting the original data.

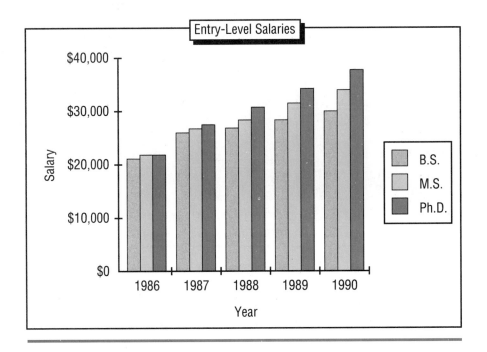

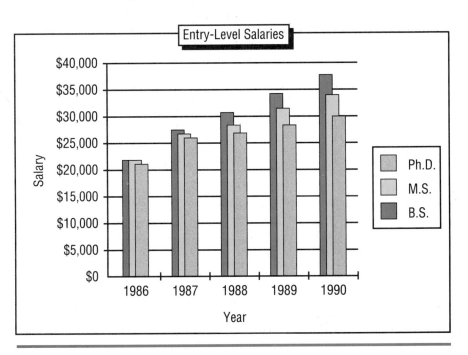

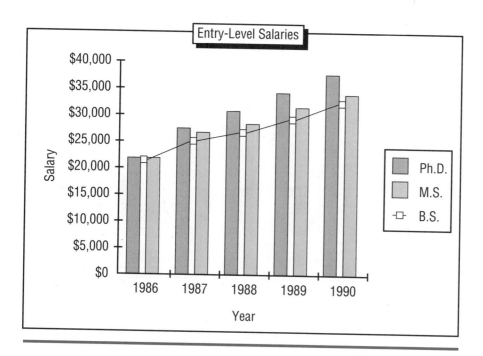

As writers and presenters, we choose our visuals according to the kinds of data, the relationships to be illustrated, the needs of our audience, and the emphasis we desire. With a growing array of computational and artistic software options, we can envision and create the exact visual for almost any purpose.

*Caution:* We have touched on only a few among the array of options (ranging from 3-D to animation to predrawn symbols to freehand drawing tools) available for electronically enhancing and refining your visuals. Faced with such an abundance of possibilities, we can be tempted to overembellish our visuals, as in the unfortunate example in Version F.

Visuals should attract audience attention, but they should *not* bombard the senses with a nauseating excess of textures and patterns. Let restraint and good taste guide you in selecting enhancement options.

## Using Clip Art

*Clip art* is a generic form for collections of ready-to-use images (of computer equipment, maps, machinery, medical equipment, and so on), all stored on computer disks. Various clip-art packages enable you to "import" into your document countless images like the one in Figure 12.25. By running the image through a drawing program such as MacDraw™, you can enlarge, enhance, or customize it, as in Figure 12.26.

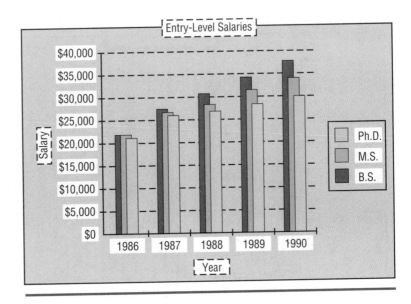

**FIGURE 12.25**  A Clip-art Image     **FIGURE 12.26**  A Customized Image

Source: *Desk Top Art* ®; *Business 1,*
© Dynamic Graphics, Inc.                     Source: Professor R.A. Dumont.

Clip-art software offers images of all kinds from most fields of science, architecture, business, and technology.

One form of clip art especially useful in technical writing is the icon (an image with all nonessential background removed). Icons are used to convey a particular idea visually; for instance, icons appear routinely in computer documentation and in other types of instructions because the image provides readers with an immediate signal of the desired action.

Whenever you use an icon, be sure it is "intuitively recognizable" to your readers ("Using Icons" 3). Otherwise, readers are likely to misinterpret its meaning—in some cases with disastrous results.

A warning:    A direction:

An instruction:    A concept:

(safety)

Source: *Desk Top Art ®*; *Business 1* and *Health Care 1.* © Dynamic Graphics, Inc.

## GUIDELINES FOR VISUALS

An effective visual enables readers to extract the information they need, quickly and easily. To simplify your reader's job, observe these general guidelines:

- *Use a visual only if you need one.* Don't include in visual form information adequately expressed in prose. The visual should clarify or emphasize points or findings that your audience will consider *significant.* Instead of focusing on some minor detail or serving as window dressing, a good visual truly advances the writer's meaning.

- *Number the visual and give it a clear title and labels.* Your title should tell readers what they are seeing. Label all the important material.

- *Choose the visual most appropriate for your purpose.* To depict the operating parts of a mechanism, for instance, an exploded diagram probably would work better than a photograph.

- *Use prose captions to explain important points made by the visual.* Captions help readers interpret a visual (as in Figure 5.2, 12.7, or 14.5). When possible make caption typesize smaller than body typesize, so that captions don't "compete" with body type (The Aldus Guide 35).

- *Match the visual to your audience.* Don't make it too elementary for specialists or too complex for nonspecialists. Be sure your intended audience will be able to interpret the visual correctly.

- *Never include too much information in one visual.* Any visual that contains too many lines, bars, numbers, colors, or patterns will overwhelm readers, causing them to ignore the visual. In place of one complicated visual, use two or more straightforward ones.

- *Use color with caution.* "Color gains impact when it is used selectively. It loses impact when it is overused" (The Aldus Guide 39). Color can bias a reader's interpretation of the relationships. (Green traditionally signifies safety; red, danger; darker colors seem to make a stronger statement than lighter ones.) Also, readers perceive differently the sizes of variously colored objects. Darker items can seem larger and closer than lighter objects of identical size. The brightest colors will attract the reader's attention (Murch 18–20); subdued colors suggest a dignified tone (The Aldus Guide 39). If you do use color, be sure to anticipate accurately its effect on the reader's perception.

- *Never mistake distortion for emphasis.* When you want to emphasize a particular point (a sales increase, a safety record, etc.), be sure your data support the conclusion implied by your visual. For instance, don't use inordinately large visuals to emphasize good news or small ones to downplay bad news (Williams 11). And when using clip-art to dramatize a comparison, be sure the relative size of the images or icons truly reflects the quantities being compared. A visual accurately depicting a 100-percent increase in phone sales at your company might look like this:

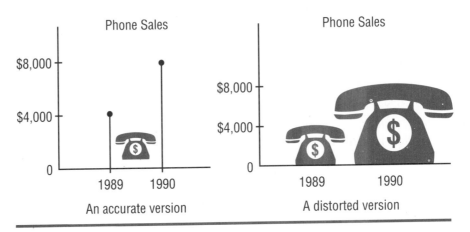

The second version overstates the good news by showing the larger image four times the size instead of twice the size of the smaller (Although the larger image is twice the height, it is also twice the *width,* and so the total area conveys the false impression that sales have *quadrupled.*

Although you are perfectly justified in presenting data in the best light, you are ethically responsible for avoiding misrepresentation.

- *Place the visual where it will best serve your readers.* Avoid clumping visuals in an appendix or at report's end. If the visual clarifies part of your discussion, place it close to that part. But if it will interest only some of your audience, place it in an appendix (discussed on pages 415–418) so that interested readers can refer to it as they wish. Tell readers when to refer to the visual and where to find it.

- *Never refer to a visual that readers cannot easily locate.* In a long document, don't be afraid to repeat a visual if you discuss it again later.
- *Introduce and interpret the visual.* In your introduction, let readers know what to expect.

> Informative   As shown in Table 2, operating costs have increased 7 percent annually since 1980.
>
> Uninformative   See Table 2.

Visuals alone make ambiguous statements (Girill 35); pictures need to be interpreted. Instead of leaving readers with a page of raw data, explain the relationships displayed.

> Informative   This cost increase means that . . . .

Always tell readers what to look for and what it means.

- *Be sure the visual's meaning can stand alone.* Even though it repeats or augments information already in the text, the visual should contain everything readers will need to interpret it correctly.
- *Never crowd a visual into a cramped space.* Set your visual off by framing it with plenty of white space (discussed on page 285), and position it on the page for balance.

The Visual Plan Sheet on page 272 and the Revision Checklist for Visuals on page 273 will help you ensure that your visuals observe all the above guidelines.

## VISUAL PLAN SHEET

### Focusing on Your Purpose

- What is my purpose (to instruct, inform, persuade, create interest)? _____
- What type of information (numbers, shapes, concepts, procedures) will this visual depict? _____
- What kind of relationships will I show (comparison, cause-and-effect, connected-parts, sequence-of-steps)? _____
- What point or conclusion or interpretation do I want to emphasize (that profits have increased, that customers are spending less, that toxic levels are rising, that X is better than Y, that too much time is being spent on a task)? ___
  _____
- Do I need a visual at all? _____

### Focusing on Your Audience

- Is this audience accustomed to interpreting visuals? _____
- Is my audience interested in specific numbers or just an overall view? ___
  _____
- Which type of visual will be most accurate, representative, and attractive, and easiest for my audience to extract the information they need (table, chart, graph, illustration, photograph, diagram)? _____
- In place of one complicated visual, should I use two or more straightforward ones? _____

### Focusing on Your Presentation

- What enhancements, if any, will engage the audience's interest (colors, borders, patterns, legends, labels, varying typefaces, shadowing, enlargement or reduction of some features)? _____
- Which medium—or combination of media—will be most effective for presenting this visual (slides, transparencies, handouts, large-screen monitor, flip chart, report text)? _____
- To achieve the greatest utility and effect, where in my presentation does my visual belong? _____

# REVISION CHECKLIST FOR VISUALS

(Numbers in parentheses refer to the first page of discussion.)

## Content

☐ Does the visual serve a legitimate purpose (clarification, not mere ornamentation) in your document? (241)

☐ Is the visual titled and numbered? (242)

☐ Is the level of complexity appropriate for the audience? (269)

☐ Are all patterns in the visual identified by label or legend? (250)

☐ Are all values or units of measurement specified clearly (grams per ounce, millions of dollars)? (242)

☐ Are the numbers accurate and exact? (242)

☐ Do the visual relationships represent the numerical relationships, without distortion? (249)

☐ Are explanatory notes added only as needed? (242)

☐ Are all data sources cited? (243)

☐ Is the visual introduced, discussed, interpreted, and integrated in the text? (243)

☐ Can the visual itself stand alone in meaning? (271)

## Arrangement

☐ Is the visual easy to locate? (270)

☐ Are all design elements (title, line thickness, legends, notes, borders, white space) positioned for balance? (269)

☐ Is the visual positioned on the page for balance? (271)

☐ Is the visual set off by adequate white space or borders? (271)

☐ Does the top of a wide visual face the inside binding? (242)

☐ Is the visual in the best report location? (270)

## Style

☐ Is this the best type of visual for your purpose? (269)

☐ Are all decimal points in each column vertically aligned? (242)

☐ Is the visual uncrowded and uncluttered? (269)

☐ Is it sufficiently engaging (patterns, colors, shapes), without being too busy? (269)

☐ Is it tasteful? (267)

## EXERCISES

**1.** The following statistics are based on data from three colleges in a large western city. They give the number of applicants to each college over six years.

- In 1981, X College received 2,341 applications for admission, Y College received 3,116, and Z College received 1,807.
- In 1982, X College received 2,410 applications for admission, Y College received 3,224, and Z College received 1,784.
- In 1983, X College received 2,689 applications for admission, Y College received 2,976, and Z College received 1,929.
- In 1984, X College received 2,714 applications for admission, Y College received 2,840, and Z College received 1,992.
- In 1985, X College received 2,872 applications for admission, Y College received 2,615, and Z College received 2,112.
- In 1986, X College received 2,868 applications for admission, Y College received 2,421 applications for admission, and Z College received 2,267.

Illustrate these data in a line graph, a bar graph, and a table. Which version seems most effective for a reader who (a) wants exact figures, (b) wonders how overall enrollments are changing, or (c) wants to compare enrollments at each college in a certain year? Write a paragraph interpreting one of these versions.

**2.** Devise a flowchart for a process in your field or area of interest. Include a title and a brief discussion.

**3.** Devise an organization chart showing the lines of responsibility and authority in an organization where you hold a part-time or summer job.

**4.** Devise a pie chart to depict your yearly expenses. Title the chart and discuss it.

**5.** Obtain enrollment figures at your college for the past five years by sex, age, race, or any other pertinent category. Construct a stacked-bar graph to illustrate one of these relationships over the five years.

**6.** Keep track of your pulse and respiration at thirty-minute intervals over a four-hour period of changing activities. Record your findings in a line graph, noting the times and specific activities below your horizontal coordinate. Write a prose interpretation of your graph and give it a title.

**7.** In textbooks or professional journal articles, locate each of these visuals: a table, a multiple-bar graph, a multiple-line graph, a diagram, and a photograph. Evaluate each according to the revision checklist, and discuss the most effective visual in class.

**8.** We have discussed the importance of choosing an appropriate scale for your graph, and the most effective form for presenting your data. Study this presentation carefully:

Strong evidence now indicates that not only nicotine and tar in cigarette smoke can be lethal, but also carbon monoxide. Much of cigarette smoke is carbon monoxide. The bar graph in Figure 12.33 (page 275) lists the ten leading U.S. cigarette brands according to the carbon monoxide given off per pack of inhaled cigarettes.

Is the scale effective? If not, why not? Can these data be presented in a bar graph? Present the same data in some other form that seems most effective.

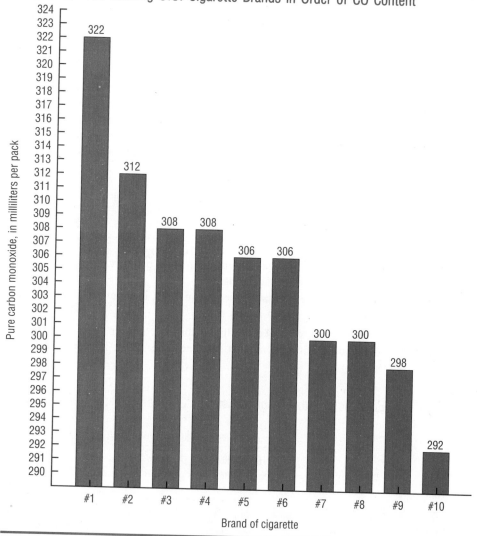

**FIGURE 12.27** Ten Leading U.S. Cigarette Brands in Order of CO Content

Pure carbon monoxide, in milliliters per pack

#1 322
#2 312
#3 308
#4 308
#5 306
#6 306
#7 300
#8 300
#9 298
#10 292

Brand of cigarette

**9.** Choose the most appropriate visual aid for illustrating each of these relationships. Justify each choice in a short paragraph.

   *a.* A comparison of three top brands of fiberglass skis, according to cost, weight, durability, and edge control.

   *b.* A breakdown of your monthly budget.

   *c.* The changing cost of an average cup of coffee, as opposed to that of an average cup of tea, over the past three years.

   *d.* The percentage of college graduates finding desirable jobs within three months after graduation, over the last ten years.

   *e.* The percentage of college graduates finding desirable jobs within three months after graduation, over the last ten years—by gender.

   *f.* An illustration of automobile damage for an insurance claim.

   *g.* A breakdown of the process of radio wave transmission.

   *h.* A comparison of five cereals on the basis of cost and nutritive value.

*i.* A comparison of the average age of students enrolled at your college in summer, day, and evening programs, over the last five years.

*j.* Comparative sales figures for three items made by your company.

**10.** *Computer graphics:* Using the scenario on pages 264–266 as a model, compose and enhance one or more visuals electronically. You might begin by looking through a recent volume of the *Statistical Abstract of the United States* (in the government documents section of your library). From the *Abstract,* or from a source you prefer, select a body of numerical data that will interest your classmates. After completing the Visual Plan Sheet, compose one or more visuals to convey a message about your data, to make a point, as in these examples:

- Consumer buying power has increased or decreased since 1980.
- Defense spending, as a percentage of the federal budget, has increased or decreased since 1980.
- Average yearly temperatures across America are rising or falling.

Experiment with formats and design options, and enhance your visual(s) as appropriate. Add any necessary prose explanations.

Be prepared to present your visual message in class, using either an overhead or opaque projector or a large-screen monitor.

**11.** Revise the layout of Table 12.4 according to the directions on pages 242–243, and explain to readers the significant comparisons in the table. (*Hint:* The unit of measurement is percentage.)

**TABLE 12.4** Educational Attainment of Persons 25 Years Old and Over

| | Highest level completed | |
|---|---|---|
| Year | High school (4 years or more) | College (4 years or more) |
| 1970 | 52.3164 | 10.7431 |
| 1980 | 66.5432 | 16.2982 |
| 1982 | 71.0178 | 17.7341 |
| 1984 | 73.3124 | 19.1628 |

*Source:* Adapted from *Statistical Abstract of the United States* (Washington: GPO, 1986): 164.

**12.** Display each of these sets of information in the visual format most appropriate for the stipulated audience. Complete the Visual Plan Sheet for each visual. Explain why you selected the type of visual as most effective for that audience. Include with each visual a brief prose passage interpreting and explaining the data.

*a.* (For general readers.) The 1986 *Statistical Abstract of the United States* breaks down energy sources for electrical energy production into these categories, by percentage: In 1970, coal, 46.2; natural gas, 24.3; hydro, 16.2; nuclear, 1.4; oil, 11.9. In 1980, coal, 51.1; natural gas, 15.1; hydro, 12.1; nuclear, 11.0; oil, 10.7.

*b.* (For experienced investors in rental property.) As an aid in estimating annual heating and air-conditioning costs, here are annual maximum and minimum temperature averages from 1951 to 1980 for five Sunbelt cities (in Fahrenheit degrees): In Jacksonville, the average maximum was 78.7; the minimum was 57.2. In Miami, the maximum was 82.6; the minimum was 67.8. In Atlanta, the maximum was 71.3; the minimum was 51.1. In Dallas, the maximum was 76.9; the minimum was 55. In Houston, the maximum was 77.5; the minimum was 57.4. (From U.S. National Oceanic and Atmospheric Administration.)

*c.* (For the student senate.) Among the students who entered our university four years ago, here are the percentages of those who graduated, withdrew, or are still enrolled: In Nursing, 71 percent graduated; 27.9 percent withdrew; 1.1 percent are still enrolled. In Engineering, 62 percent graduated; 29.2 percent withdrew; 8.8 percent are still enrolled. In Business, 53.6 percent graduated; 43 percent withdrew; 3.4 percent are still enrolled. In Arts and Sciences, 27.5 percent graduated; 68 percent withdrew; 4.5 percent are still enrolled.

*d.* (For the student senate.) Here are the enrollment trends from 1978 to 1987 for two colleges in our university. In Engineering: 1978, 455 students enrolled; 1979, 610; 1980, 654; 1981, 758; 1982, 803; 1983, 827; 1984, 1,046; 1985, 12,00; 1986, 1,115; 1987, 1075. In Business: 1978, 922; 1979, 1,006; 1980, 1,041; 1981, 1,198; 1982, 1,188; 1983, 1,227; 1984, 1,115; 1985, 1,220; 1986, 1,241; 1987, 1,366.

**13.** Anywhere on campus or at work, locate at least one visual that needs revision for accuracy, clarity, appearance, or appropriateness. Look in computer manuals, lab manuals, newsletters, financial aid or admissions or placement brochures, student or faculty handbooks, or newspapers—or in your textbooks. Use the Visual Plan Sheet and the Revision Checklist as guides to revise and enhance the visual. Submit to your instructor a copy of the original, along with a memo explaining your improvements. Be prepared to discuss your revision in class.

## TEAM PROJECT

Assume that your technical writing instructor is planning to purchase five copies of a graphics software package for students to use in designing their documents. The instructor has not yet decided which general-purpose package would be most useful. Your group's task is to test one package and to make a recommendation.

In small groups, visit your school's microcomputer lab and ask for a listing of the graphics packages that are available to students and faculty. Here are some of the most popular:

| *For the Macintosh* | *For the IBM PC* |
|---|---|
| Cricket Graph | Graph Station |
| Easy 3D | GraphWriter |
| MacChart | Harvard Graphics |
| MacDraw | PC Draw |
| Microsoft Chart | Microsoft Chart |
| Thunderscan | |

Select one graphics package and learn how to use it. Design at least four representative visuals. In a memo or presentation to your instructor and classmates, describe the package briefly and tell what it can do. Would you recommend purchasing five copies of this package for general-purpose use by writing students? Explain. Submit your report, along with the sample graphics you have composed. Appoint one member to present your group's recommendation in class.

Do the same assignment, comparing various clip-art packages. Which package offers the best selection of images for technical writers?

# Designing
# Effective Formats

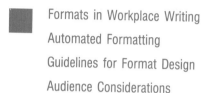

Formats in Workplace Writing

Automated Formatting

Guidelines for Format Design

Audience Considerations

**B**esides having worthwhile content, sensible organization, and readable style, a useful document looks inviting and accessible. Format is the *look* of a page, the layout of words and graphics. A well-formatted document invites readers in, motivates them to pay attention, and helps them understand your message.

Readers like things that look good. Their *first* impression of a document is a purely visual, aesthetic judgment. Readers are drawn to a document that seems attractive, carefully crafted, tastefully designed, and easy to follow. Readers are repelled by a document that seems unappealing, carelessly produced, tasteless, or hard to follow.

## FORMATS IN WORKPLACE WRITING

Format decisions become especially significant when we consider these realities about writing and reading in the workplace:

1. *Technical information generally is designed differently from material in most novels, news stories, and other forms of writing.* To be accessible, a technical document requires more than just an unbroken sequence of paragraphs. To find their way through complex material, readers may need the help of charts, diagrams, lists, various type sizes and typefaces, different headings, and other page-design elements.

2. *Technical documents rarely get the readers' undivided attention.* Readers may be skimming the document while they jot down ideas, talk on the phone, or take a coffee break. Or they may refer to sections of the document while they sit in a meeting or listen to a presentation. In a world of frequent distractions, readers must be able to leave the document and return easily.

3. *Many people don't like to read work-related documents.* Novels, newspapers, and magazines are read for relaxation, but work-related documents mean *work*. The more complex the document, the harder your readers will have to work. If these readers had other ways of acquiring your information, they would prefer not to read at all.

4. *As computers generate more and more paper, any document is forced to compete for readers' attention.* Even the most brilliantly written document is useless unless it is read by its intended audience. Suffering from information overload, today's readers will tune out any document that appears overwhelming. They increasingly want formats that will help them find the information they really need. A "user-friendly" document has an accessible format: at a glance, readers can see how the document is organized, where they are in the document, which items matter most, and how the items relate.

Having decided at a glance whether your document is inviting and accessible, readers will draw conclusions about the value of your information, the quality of your work, the extent of your interest in your audience, and your overall credibility. For an otherwise useful document, a carefully crafted format becomes your initial and final stamp of professionalism.

No matter how vital your information, an ineffective format surely will alienate readers. Even though the information in Figure 13.1 is worthwhile and accurate, its format resists our attention. Without visual cues, we have no way of grouping this information into organized units of meaning. Figure 13.2 shows the same information after a format overhaul. Figure 13.3 shows one version possible on a computer. As readers, we take good format for granted—that is, we hardly notice format *unless* it is offensive.

## AUTOMATED FORMATTING

For anyone who writes in the workplace, computer technology has large implications. Planning, drafting, and revising at their workstations, writers themselves are becoming increasingly responsible for all stages in document preparation.

*Desktop publishing* means less reliance on secretaries, print shops, and graphic artists. Using microcomputers, page-design software, and laser printers, writers at their desks can control the entire production task: designing, illustrating, laying out, and printing the final document.

**FIGURE 13.1** An Ineffective Format

```
                         Sunspaces
        Either as an addition to a home or as an integral part of a
        new home, sunspaces have gained considerable popularity.
             A sunspace should face within 30 degrees of true south. In
        the winter, sunlight passes through the windows and warms the
        darkened surface of a concrete floor, brick wall, water-filled
        drums, or other storage mass. The concrete, brick, or water
        absorbs and stores some of the heat until after sunset, when
        the indoor temperature begins to cool. The heat not absorbed
        by the storage elements can raise the daytime air temperature
        inside the sunspace to as high as 100 degrees Fahrenheit. As
        long as the sun shines, this heat can be circulated into the
        house by natural air currents or drawn in by a low-horsepower
        fan.
             In order to be considered a passive solar heating system,
        any sunspace must consist of these parts: a collector, such as
        a double layer of glass or plastic; an absorber, usually the
        darkened surface of the wall, floor, or water-filled contain-
        ers inside the sunspace; a storage mass, normally concrete,
        brick, or water, which retains heat after it has been ab-
        sorbed; a distribution system, the means of getting the heat
        into and around the house by fans or natural air currents; and
        a control system, or heat-regulating device, such as movable
        insulation, to prevent heat loss from the sunspace at night.
        Other controls include roof overhangs that block the summer
        sun, and thermostats that activate fans.
```

Source: *Sunspaces and Solar Greenhouses*. U.S. Department of Energy, 1984.

As automated design technology continues to improve the look of workplace writing, audiences raise their standards. Now more than ever, readers expect documents to be inviting and accessible. Of course, an electronically produced document—no matter how attractive—is useless if it carries the same old mistakes found in typewriter-produced documents, or if it says nothing worthwhile.

## GUIDELINES FOR FORMAT DESIGN

Whether you write with a typewriter or a computer, approach your formatting decisions from the top down: first, consider the overall look of your pages; next, the shape of each paragraph; and finally, the size and style of individual letters and words (Kirsh 112). Figure 13.4 shows how format design follows a top-down sequence, moving from large matters to small.[1]

[1]For advice on formatting specific documents, see the chapters on memos and letters.

**FIGURE 13.2** An Effective Format (Typewritten)

SUNSPACES

Either as an addition to a home or as an integral part of a new home, sunspaces have gained considerable popularity.

How Sunspaces Work

A sunspace should face within 30 degrees of true south. In the winter, sunlight passes through the windows and warms the darkened surface of a concrete floor, brick wall, water-filled drums, or other storage mass. The concrete, brick, or water absorbs and stores some of the heat until after sunset, when the indoor temperature begins to cool.

The heat not absorbed by the storage elements can raise the daytime air temperature inside the sunspace to as high as 100 degrees Fahrenheit. As long as the sun shines, this heat can be circulated into the house by natural air currents or drawn in by a low-horsepower fan.

The Parts of a Sunspace

In order to be considered a passive solar heating system, any sunspace must consist of these parts:

1. A collector, such as a double layer of glass or plastic.
2. An absorber, usually the darkened surface of the wall, floor, or water-filled containers inside the sunspace.
3. A storage mass, normally concrete, brick, or water, which retains heat after it has been absorbed.
4. A distribution system, the means of getting the heat into and around the house (by fans or natural air currents).
5. A control system (or heat-regulating device) such as movable insulation, to prevent heat loss from the sunspace at night. Other controls include roof overhangs that block the summer sun, and thermostats that activate fans.

**FIGURE 13.3**  A Version Prepared on a Computer

# Sunspaces

Either as an addition to a home or as an integral part of a new home, sunspaces have gained considerable popularity.

## How Sunspaces Work

A sunspace should face within 30 degrees of true south. In the winter, sunlight passes through the windows and warms the darkened surface of a concrete floor, brick wall, water-filled drums, or other storage mass. The concrete, brick, or water absorbs and stores some of the heat until after sunset, when the indoor temperature begins to cool.

The heat *not* absorbed by the storage elements can raise the daytime air temperature inside the sunspace to as high as 100 degrees Fahrenheit. As long as the sun shines, this heat can be circulated into the house by natural air currents or drawn in by a low-horsepower fan.

## The Parts of a Sunspace

In order to be considered a passive solar heating system, any sunspace must consist of these parts:

1. A *collector,* such as a double layer of glass or plastic.
2. An *absorber*, usually the darkened surface of the wall, floor, or water-filled containers inside the sunspace.
3. A *storage mass*, normally concrete, brick, or water, which retains heat after it has been absorbed.
4. A *distribution system,* the means of getting the heat into and around the house (by fans or natural air currents).
5. A *control system* (or heat-regulating device) such as movable insulation, to prevent heat loss from the sunspace at night. Other controls include roof overhangs that block the summer sun, and thermostats that activate fans.

**FIGURE 13.4** A Flowchart for Decision Making in Format Design

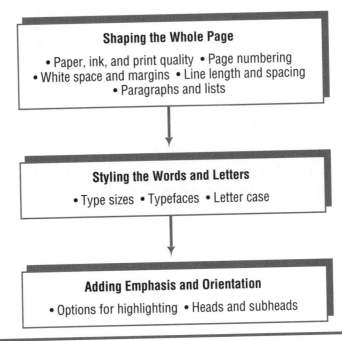

Different audiences may require different formats, but any format must be inviting and accessible. The following guidelines illustrate basic design principles that will enable your formats to satisfy any reader's expectations.

### Use the Right Paper and Ink

Type or print your finished document in black ink, on 9½-by-11-inch plain white paper. Use rag-bond paper (2 pounds or heavier) with a high fiber content (25 percent minimum). Shiny paper produces glare and tires the eyes. Use low-gloss paper instead.

### Type or Print Neatly and Legibly

Keep erasures to a minimum, and retype all smudged pages. Typewriter erasures with opaquing fluid or opaquing film (correction tape) are neatest. Use a fresh ribbon and keep your typewriter keys clean. If you write on a computer, print your hard copy on a letter-quality printer, a laser printer, or a dot-matrix printer in the near-letter-quality mode. Many older dot-matrix printers produce copy which is hard to read or which looks unprofessional.

### Number Pages Consistently

For a long document, count your title page as page i, without numbering it, and number all front-matter pages, including your table of contents and abstract, with lowercase roman numerals (ii, iii, iv). Number the first page of your report and subsequent pages with arabic numerals (1, 2, 3). Place all page numbers in the upper right-hand corner, two spaces from the top and five spaces from the right edge. (Some word processors may limit page-number placement; check the manual.)

## Use Adequate White Space

White space on a page is all the space not filled by text. White space separates sections in a document, headings and visuals from text, paragraphs on a page, sentences in a paragraph, words in a sentence, letters in a word. Instead of being random, white space can be deliberately designed to enhance a document's appearance, clarity, and emphasis.

Well-designed white space imparts a shape to the whole document, a shape that orients readers and gives form to the printed matter by

- keeping related things together
- isolating and emphasizing important things
- creating a path for the reader's eyes
- providing breathing room between blocks of information.

Good use of white space makes a page look uncluttered, inviting, easy to follow. It conveys an immediate sense of user-friendliness.

## Leave Ample Margins

Small margins make a page look crowded and difficult. On your 8½-by-11-inch page, leave margins no smaller than these:

|                |   |            |
|---------------:|---|------------|
| top margin     | = | 1 ¼ inches |
| bottom margin  | = | 1 ½ inches |
| right margin   | = | 1 ¼ inches |
| left margin    | = | 2 inches   |

The larger left margin allows space for binding the document.

## Keep Line Length Reasonable

Long lines tire the eye. The longer the line, the harder it is for readers to return to the left margin and locate the next line (White 25).

Notice how your eye labors to follow the message that here seems to stretch in lines that continue long after your eye was prepared to move down to the next line. After reading more than a few of these lines, you begin to feel tired and bored and annoyed.

Short lines force the eye to jump back and forth (Felker 79). "Too-short lines disrupt the normal horizontal rhythm of reading" (White 25).

Lines that are too
short cause your eye
to stumble from one
fragment to another
at a pace that too
soon becomes
annoying, if not
nauseating.

A reasonable line length is 50-60 characters (or 9 to 12 words) per line for an 8½-by-11-inch single-column page. The number of characters will depend on print size on your typewriter or word processor.

Line length, of course, is affected by the number of columns (vertical blocks of print) on your page. Two-column pages often appear in newsletters and brochures, but research indicates that single-column pages work best for complex, specialized information (Hartley 148).

If you write using a computer, you can usually choose between *unjustified* text (uneven or "ragged" right margins) and *justified* text (even right margins).

To make the right margin even in justified text, the spaces vary between words and letters on a line, sometimes creating "channels" or rivers of white space. The reader's eyes are then forced to adjust continually to these space variations within a line or paragraph. And because each line ends at an identical vertical space, the eyes must work harder to differentiate one line from another (Felker 85). Moreover, to preserve the even margin, words at line's end must often be hyphenated. And too many hyphenated line endings can be distracting.

Unjustified text, on the other hand, uses equal spacing between letters and words on a line, and an uneven right margin (as traditionally produced by a typewriter). For some readers, a ragged right margin makes reading easier. These differing line lengths can prompt the eye to move from one line to another (Pinelli 77). In contrast to justified text, an unjustified page seems to look less formal, less distant, and less "official."

Justified text seems preferable for books, annual reports, brochures, newsletters, and other materials published for a broad readership. Unjustified text seems preferable for more personal forms of communication such as letters, memos, and in-house reports.

### Keep Line Spacing Consistent

For any document that your audience is likely to read completely (letters, memos, instructions), single-space within paragraphs and double-space between. Instead of indenting single-spaced paragraphs, use an extra line of space to separate them.

For longer documents that your audience is likely to read selectively (proposals, formal reports), double-space within paragraphs. Indent your double-spaced paragraphs or separate them with an additional line of space. This open spacing enables readers to scan a long document, and to quickly locate what they need.

## Tailor Each Paragraph

Many readers merely skim a long document to find what they want. Most paragraphs should therefore begin with a topic sentence forecasting the content of the paragraph. Think of each paragraph as a self-contained block of information.

Make the paragraph long enough to cover one concept only. Use a long paragraph (no more than fifteen lines) for clustering material that is closely related (such as history and background, or any body of information that is best understood in one block).

Use short paragraphs for making complex material more digestible, for giving step-by-step instructions, or for emphasizing vital information.

Instead of indenting a series of short paragraphs, separate them by inserting an extra line of space (as here).

For any paragraph, avoid "orphan" lines: the paragraph's opening line on the bottom of a page, or the paragraph's closing line on the top line of a page.

## Make Lists for Easy Reading

Workplace readers prefer documents that have information in list form rather than in continuous prose paragraphs (Hartley 51). Whenever you have a paragraph or long sentence containing a series of distinct but related items, consider displaying these items one by one.

Here are typical items you might list: advice or examples, conclusions and recommendations, criteria for evaluation, errors to avoid, materials and equipment for a procedure, parts of a mechanism, or steps or events in a sequence. Notice how the information in the preceding sentence becomes easier to grasp and remember when displayed in the list below.

Here are typical items you might list:

- advice or examples
- conclusions and recommendations
- criteria for evaluation
- errors to avoid
- materials and equipment for a procedure
- parts of a mechanism
- steps or events in a sequence

A list of brief items usually needs no punctuation at the end of each line. But a list of full sentences (page 293) or questions (page 18) would call for appropriate punctuation at the end of each item.

Depending on the list's contents, set off each item with a distinct visual or verbal signal. If your items require a strict sequence (as in a series of steps, or parts of a mechanism), use arabic numbers (1, 2, 3) or the words *First, Second, Third,* and so on. If the items require no strict sequence (as in some examples or advice), use dashes, asterisks, or bullets (heavy dots).

Always introduce your list with an explanatory statement. And phrase all listed items in parallel grammatical form (pages 504–505). If the items suggest no strict sequence, try to impose some logical ranking (most important to least important, alphabetical, or some such). Set off the list with extra white space above and below.

Keep in mind that a document with too many lists will appear busy, disconnected, and splintered (Felker 55). And long lists could be used by unethical writers to camouflage this or that piece of bad or embarrassing news (Williams 12). As with all format options, your restraint and good judgment are essential.

### Use Standard Type Sizes

Many word processing programs allow you to choose among a variety of type sizes:

9 point
10 point
12 point
14 point
18 point
## 24 point

The standard type size for most documents is 10 to 12 point. Use larger or smaller sizes only for certain headings, titles, captions (brief explanation of a picture or other visual), or special emphasis.

### Select an Appropriate Typeface

Typeface is the style of individual letters and characters. Each typeface can be said to have its own *personality:* "The typefaces you select for . . . [heads], subheads, body copy, and captions affect the way readers experience your ideas" (*Aldus Guide* 24).

Word-processing programs generally offer a variety of typefaces (or fonts) like the examples on page 289, listed by name.
Except for special emphasis, stick to the more conservative typefaces and avoid ornate ones altogether.

All typefaces divide into two broad categories: *serif* or *sans serif.* Serifs are fine lines that extend horizontally from the main strokes of a letter:

Geneva typeface
Monaco typeface
New York typeface
**Chicago typeface**
*Venice typeface*
**Athens typeface**
𝕷𝖔𝖓𝖉𝖔𝖓 𝖙𝖞𝖕𝖊𝖋𝖆𝖈𝖊

---

Serif type makes body copy more readable because the horizontal lines "bind the individual letters" and thereby guide the reader's eyes from letter to letter—as in the type you are now reading (White 14).

In contrast, sans serif type is purely vertical (like this). Clean looking and "businesslike," sans serif is considered ideal for technical material (numbers, equations, etc.), marginal comments, headings, examples, tables, and captions to pictures and visuals, and any other material "set off" from the body copy (White 16).

### Avoid Sentences in Full Caps

Sentences or long passages in full capitals (uppercase letters) are hard to read because all uppercase letters and words have the same visual outline (Felker 87).

MY DOG HAS MANY FLEAS.

My dog has many fleas.

The longer the passage, the harder readers must work to sort things out and grasp your intended emphasis.

| Hard | ACCORDING TO THE NATIONAL COUNCIL ON RADIATION PROTECTION, YOUR MAXIMUM AL-LOWABLE DOSE OF LOW-LEVEL RADIATION IS 500 MILLIREMS PER YEAR. |
| --- | --- |
| Easier | According to the National Council on Radiation Protection, your MAXIMUM allowable dose of low-level radiation is 500 millirems per year. |

Lowercase letters take up less space, and the distinctive shapes make each word easier for readers to recognize and remember (Benson 37).

Use full caps as section headings (INTRODUCTION) or to highlight a word or phrase (WARNING: NEVER TEASE THE ALLIGATOR). As with other highlighting options discussed in the next section, use full caps sparingly in your document.

## Highlight for Emphasis and Appeal

Effective highlighting helps readers distinguish the important from the less important elements. Highlighting options include typefaces, type sizes, white space, and other graphic devices that

- emphasize key points
- make headings prominent
- separate sections of a long document
- set off examples, warnings, and notes

On a typewriter you can highlight with underlining, FULL CAPS, dashes, parentheses, and asterisks.

You can indent to set off examples, explanations, or any material that should be separate from other elements in your document.

Using ruled (or typed) horizontal lines, you can separate sections in a long document:

---

Using ruled lines, broken lines, or ruled boxes, you can set off crucial information such as a warning or a caution:

— — — — — — — — — — — — — — — — — — — — — — — — — — — —

*Caution:* A document with too many highlights can appear confusing, disorienting, and tasteless.

— — — — — — — — — — — — — — — — — — — — — — — — — — — —

Most software now offers an array of highlighting options that might include **boldface,** *italics,* SMALL CAPS, varying type sizes and typefaces, and color. For specific highlighted items, some options are better than others:

**Boldface works well for emphasizing a single sentence or statement, and is perceived by readers as being "authoritative"** (*Aldus Guide* 42).

*Italics suggest a more subtle or "refined" emphasis than boldface (Aldus Guide 42). Italics can highlight words, phrases, book titles, or anything else you would have underlined on a typewriter. But multiple lines (like these) of italic type are hard to read.*

SMALL CAPS WORK FOR HEADINGS AND SHORT PHRASES. BUT ANY LONG STATEMENT ALL IN CAPS IS HARD TO READ.

SMALL TYPE SIZES (USUALLY SANS SERIF) WORK WELL FOR CAPTIONS AND AS LABELS FOR VISUALS OR TO SET OFF OTHER MATERIAL FROM THE BODY COPY.

*Large type sizes and dramatic typefaces are both hard to miss and hard to digest. Be conservative—unless you really need to convey a sense of forcefulness.*

Color is appropriate only in some documents, and only when used sparingly. (Page 269 discusses how color can influence readers' perception and interpretation of a message.)

Whichever highlights you select for a document, be consistent. Make sure that all headings at one level are highlighted identically, that all warnings and cautions are set off identically, and so on. And *never* combine too many highlights.

## Use Headings That Provide Access

Most long documents are read not like novels but like textbooks. Readers look back or jump ahead to sections that interest them most. Without headings, readers don't know how the document is organized, and they can't find what they need. Headings make your writing task easier as well: like guideposts or points on a road map, they keep you on course and they provide transitions.

Headings signal readers that something is ending and something else is beginning. They make a document more attractive and less intimidating by dividing it into accessible chunks. An informative heading can help a reader decide whether a section is worth reading (Felker 17). Besides cutting down on reading and retrieval time, headings help readers remember information (Hartley 15).

To be effective, headings should conform to the guidelines that follow.

**Make Headings Informative.** A heading should be long enough to be informative, without being wordy. Informative headings orient readers, showing them what to expect. Vague or general headings can be more misleading than no headings at all (Redish 144). Whether your heading takes the form of a phrase, a statement, or a question, be sure it advances thought.

**Uninformative heading** Document Formatting

What should we expect here: specific instructions, an illustration, a discussion of formatting policy in general? We can't tell.

**Informative versions**
How to Format Your Document

Format Your Document in This Way

How Do I Format My Document?

When you use questions as headings, phrase the questions in the same way as readers might ask them.

**Make Headings Specific as Well as Comprehensive.**   Focus the heading on a specific topic. Do not preface a discussion of the effects of acid rain on lake trout with a broad heading such as "Acid Rain." Use instead "The Effects of Acid Raid on Lake Trout."

Also, provide enough headings to contain each discussion section. If, say, chemical, bacterial, and nuclear wastes are three *separate* discussion items, provide a heading for each. Do not simply lump them under the sweeping heading "Hazardous Wastes." Adapt major and minor headings from your outline.

**Make Headings Grammatically Consistent.**   All major topics or minor topics in a document share equal rank; to emphasize this equality, express topics at the same level in identical — or parallel — grammatical form.

**Nonparallel headings**
How to Avoid Damaging Your Disks:

*1.* Clean Disk-Drive Heads
*2.* Keep Disks Away from Magnets
*3.* Writing on Disk Labels with a Felt-Tip Pen
*4.* It Is Crucial That Disks Be Kept Away from Heat
*5.* Disks Should Be Kept Out of Direct Sunlight
*6.* Keep Disks in Their Protective Jackets

In items 3, 4, and 5, the lack of parallelism (no verbs in the imperative mood) obscures the relationship between individual steps, and confuses readers. This next version emphasizes the equal rank of these items.

**Parallel headings**
*3.* Write on Disk Labels with a Felt-Tip Pen
*4.* Keep Disks Away from Heat
*5.* Keep Disks Out of Direct Sunlight

Parallelism helps make a document readable and accessible.

**Make Headings Visually Consistent.**   For an understanding of how your document is organized, readers consider not only what the headings *say* but also how they *look:* "Wherever heads are of equal importance, they should be given similar visual expression, because that regularity itself becomes an understandable symbol" (White 104). Use identical type size and typeface for all headings at a given rank.

**Lay Out Headings by Rank.**   Like a road map, your headings should identify clearly the large and small segments in your document. A long document ordinarily has section headings to mark its introduction, body, and conclusion. The body section in particular can have several lower ranks of headings: major and minor topic headings, and perhaps subtopic headings as well. (Use the logical divisions from your outline as a model for heading layout.) Think of each heading at a particular rank as an "event in a sequence" (White 95). Your job is to ensure that readers can follow the sequence.

Figure 13.5 shows how headings (typewritten) vary in their positioning and highlighting, depending on their rank. The layout in the figure embodies these guidelines:

- *Ordinarily, use no more than four levels of heading (section, major topic, minor topic, subtopic).* Too many heads and subheads can make a document seem cluttered.

- *To divide logically, be sure each higher-level heading yields at least two lower-level headings.*

- *Always insert one additional line of space above your heading.* For double-spaced text, triple-space before the heading, and double-space after; for single-spaced text, double-space before the heading, and single space after. Without the additional space, readers can't tell whether the heading belongs with the text which precedes it or that which follows it. The spacing provides a visual clue.

- *Never begin the sentence right after the heading with "this," "it," or some other pronoun referring to the heading.* Instead, make the sentence's meaning independent of the heading.

- *Never leave a heading floating on the final line of a page.* Instead, carry the heading over to start the next page.

- *If possible, use different type sizes to highlight and reflect the ranks of heads.* Readers tend to equate large type size with importance (White 95). On a typewriter, set major heads in full caps. On a word processor set major heads in a larger type size, and perhaps in boldface (as shown in Figure 13.6).

When headings show the relationships among all the parts, readers can grasp at a glance how your document is organized.

FIGURE 13.5  Recommended Format for Typewritten Headings

```
                        SECTION HEADING

     Write section headings in full caps, centered horizontally, one
extra line space below any preceding text. Do not underline or
further highlight these heads.


Major Topic Heading
     Begin each word (except articles and prepositions) in major
topic heads with an uppercase letter. Abut the left margin (flush
left), one extra line space below preceding text. Underline.


     Minor Topic Heading
     Begin each important word in minor topic heads with an upper-
case letter. Indent the heads five spaces from the left margin, one
extra line space below preceding text. Underline these heads.


Subtopic Heading. Begin each important word in subtopic heads with
an uppercase letter. Instead of indenting, place the heads flush
left, one extra line space below preceding text, and on the same
line as following text (separated from the text by a period).
```

## AUDIENCE CONSIDERATIONS

Like any decisions in writing, formatting choices are by no means random. An effective writer designs a document for specific use by a specific audience.

How do you decide on a format? Your best bet is to work from a detailed audience-and-use profile (Wight 11). Know who your readers are and how they will use your information. Design a format to meet their particular needs, as in these examples:

- If readers will use your document for reference only (as in a repair manual), make sure you have plenty of headings.
- If readers will follow a sequence of steps, show that sequence in a numbered list.
- If readers will need to evaluate something, give them a checklist of criteria (as in this book at the end of most chapters).

**FIGURE 13.6** Recommended Format for Word-Processed Headings

# Section Heading

On a word processor, section headings in boldface and enlarged type are more appealing and readable than heads in full caps. Use a type size roughly 4 points larger than body copy (say, 16-point section heads for 12-point body copy). Avoid *overly large* heads, and use no other highlights. Set these and *all* lower heads one extra line space below any preceding text.

## Major Topic Heading

Major topic heads abut the left margin (flush left), and each important word begins with an uppercase letter. Use boldface and a type size roughly 2 points larger than body copy, with no other highlights.

### *Minor Topic Heading*

Minor topic heads also are set flush left. Use boldface, italics (optional), and the same type size as in the body copy, with no other highlights.

**Subtopic Heading.** Instead of indenting the first line of body copy, place subtopic heads flush left on the same line as following text, and set off by a period. Use boldface and the same type size as in the body copy, with no other highlights.

- If readers need a warning, highlight the warning so that it cannot possibly be overlooked.
- If readers have asked for your one-page résumé, save space by using the 10-point type size.
- If readers will be facing complex information or difficult steps, widen the margins, increase all white space, and shorten the paragraphs.

How effectively you combine your format options will depend mostly on how carefully you have analyzed your audience. But regardless of your audience, never make the document look "too intellectually intimidating" (White 4).

Finally, keep in mind that even the most brilliant formatting cannot redeem a document the content of which is worthless, organization chaotic, or style unreadable. The value of a document ultimately depends on elements that are more than merely skin deep.

## REVISION CHECKLIST FOR DOCUMENT FORMATS

Use this checklist to evaluate and revise a document's format. (Numbers in parentheses refer to the first page of discussion.)

☐ Is the paper white, low-gloss, rag bond, with black ink? (284)

☐ Is all type or print neat and legible? (284)

☐ Does the white space guide the reader's eyes? (285)

☐ Are the margins ample? (285)

☐ Is line length reasonable? (285)

☐ Is the right margin unjustified? (286)

☐ Is line spacing appropriate and consistent? (286)

☐ Does each paragraph begin with a topic sentence? (287)

☐ Does the length of each paragraph suit its subject and purpose? (287)

☐ Are all paragraphs free of "orphan" lines? (287)

☐ Do parallel items in strict sequence appear in a numbered list? (287)

☐ Do parallel items of any kind appear in a list whenever a list is appropriate? (287)

☐ Are pages numbered consistently? (284)

☐ Is the body type size 10 or 12 point? (288)

☐ Do full caps highlight only words or short phrases? (289)

☐ Is the highlighting consistent and subdued? (290)

☐ Are all format patterns distinct enough so that readers will find what they need? (290)

□ Are there enough headings so that readers always know where they are in the document? (291)

□ Are headings informative, comprehensive, specific, parallel, and visually consistent in relation to headings of equal rank? (291)

□ Are all headings clearly differentiated according to rank? (293)

□ Is the overall format sufficiently inviting without being overwhelming? (284)

## EXERCISES

**1.** Find an example of effective formatting, in a textbook or elsewhere. Photocopy a representative selection (two to three pages), and attach a memo explaining to your instructor and classmates why this format is effective. Be specific in your evaluation. Now do the same for an example of poor formatting. Bring your examples and explanations to class, and be prepared to discuss why you chose them.

**2.** These are headings from a set of instructions for listening. Rewrite the headings to make them parallel.

- You Must Focus on the Message
- Paying Attention to Nonverbal Communication
- Your Biases Should Be Suppressed
- Listen Critically
- Listening for Main Ideas
- Distractions Should Be Avoided
- Provide Verbal and Nonverbal Feedback
- Making Use of Silent Periods
- Are You Allowing the Speaker Time to Make His or Her Point?
- Keeping an Open Mind Is Important

**3.** Using the format checklist on page 296, redesign an earlier assignment or a document you've prepared on the job. Submit to your instructor the revision and the original, along with a memo explaining your improvements. Be prepared to discuss your format design in class.

**4.** Anywhere on campus or at work, locate a document with a format that needs revision. Candidates include career counseling handbooks, financial aid handbooks, student or faculty handbooks, software or computer manuals, medical information, newsletters, or registration procedures. Redesign the document or a two- to five-page selection from it. Submit to your instructor a copy of the original, along with a memo explaining your improvements. Be prepared to discuss your revision in class.

## TEAM PROJECT

Working in small groups, redesign the format of a document your instructor provides. Prepare a detailed explanation of your group's revision. Appoint a group member to present your revision to the class, using an opaque or overhead projector, a large-screen computer monitor, or photocopies.

# Specific
# Documents
# and
# Applications

# 14

# Descriptions and Specifications

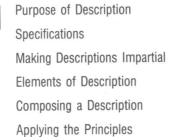

Purpose of Description

Specifications

Making Descriptions Impartial

Elements of Description

Composing a Description

Applying the Principles

To describe is to create a picture with words (and visuals). Descriptions convey information about a product or mechanism to someone who will use it, buy it, operate it, assemble it, manufacture it, or to someone who has to know more about it. Any item can be described from many different angles. Therefore, *how* you describe — your angle — depends on your purpose and on the needs of your audience.

## PURPOSE OF DESCRIPTION

Description is part of all writing. Manufacturers use descriptions to sell products; banks require detailed descriptions of any business or construction venture before approving a loan; medical personnel maintain daily or hourly descriptions of a patient's condition and treatment.

No matter what the subject of description, readers expect answers to as many of these questions as are applicable: *What is it? What does it do? What does it look like? What is it made of? How does it work? How was it put together?* The description in Figure 14.1, part of an installation and operation manual, answers applicable questions for do-it-yourself homeowners.

## SPECIFICATIONS

Airplanes, bridges, smoke detectors, and countless other items are produced according to certain *specifications*. A particularly exacting type of description, specifications (or "specs") prescribe standards for performance, safety, and quality. For almost any product, specifications spell out

- the methods for manufacturing, building, or installing the product
- the materials and equipment to be used
- the size, shape, and weight of the product

These requirements define an acceptable level of quality, and so specifications have ethical and legal implications. Any product "below" specifications provides grounds for a lawsuit. And when injury or death results (as in a bridge collapse caused by inferior reinforcement) the contractor or supplier who "cut corners" is criminally liable.

Federal and state regulatory agencies routinely issue specifications to insure safety. The Consumer Product Safety Commission specifies that power lawn mowers be equipped with a "kill switch" on the handle, a blade guard to prevent foot injuries, and a grass thrower that aims downward to prevent eye and facial injury. This same agency issues specifications for baby products, as in governing the fire retardency of pajama material. Passenger airline specifications for aisle width, seat-belt configurations, and emergency equipment are issued by the Federal Aviation Administration. State and local agencies issue specifications in the form of building codes, fire codes, and other standards for safety and reliability.

Government departments (Defense, Interior, etc.) issue specifications for all types of military hardware and other equipment. A set of NASA specifications for spacecraft parts can be hundreds of pages long, prescribing the standards for even the smallest nuts and bolts, down to screw-thread depth and width—in millimeters.

The private sector issues specifications for countless products or projects, to help ensure that customers get *exactly* what they want. Figure 14.2 illustrates partial specifications drawn up by an architect for a building that will house a small medical clinic. This section of the specs covers only the structure's "shell." Other sections detail the requirements for plumbing, wiring, and interior finish work.

Specifications like those in Figure 14.2 (page 304) must be clear enough for *identical* interpretation by the widest possible range of readers (Glidden 258–259):

- *the customer,* who has the "big picture" of what is needed and who wants the best product at the best price
- *the designer* (architect, engineer, computer scientist, etc.), who must translate the customer's wishes into the specifications themselves
- *the contractor or manufacturer,* who won the job by making the lowest bid, and so must preserve profit by doing *only* what is prescribed
- *the supplier,* who must provide the exact materials and equipment

**FIGURE 14.1** A Mechanism Description

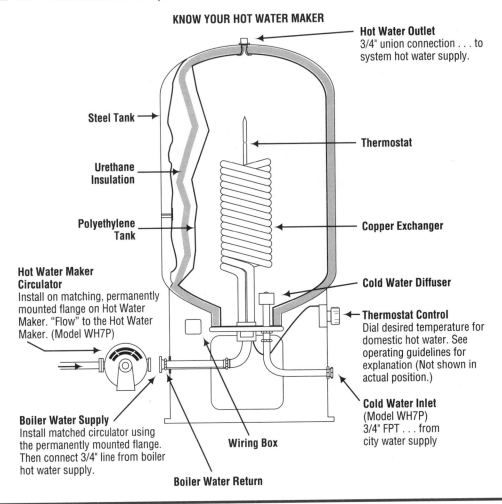

**KNOW YOUR HOT WATER MAKER**

**Hot Water Outlet**
3/4" union connection . . . to system hot water supply.

**Steel Tank**

**Thermostat**

**Urethane Insulation**

**Polyethylene Tank**

**Copper Exchanger**

**Hot Water Maker Circulator**
Install on matching, permanently mounted flange on Hot Water Maker. "Flow" to the Hot Water Maker. (Model WH7P)

**Cold Water Diffuser**

**Thermostat Control**
Dial desired temperature for domestic hot water. See operating guidelines for explanation (Not shown in actual position.)

**Boiler Water Supply**
Install matched circulator using the permanently mounted flange. Then connect 3/4" line from boiler hot water supply.

**Wiring Box**

**Cold Water Inlet**
(Model WH7P)
3/4" FPT . . . from city water supply

**Boiler Water Return**

- *the workforce,* who will do the actual assembly or construction or installation (managers, supervisors, subcontractors, and workers—some who will work on only one part of the product, such as plumbing or electrical)

- *the inspectors,* who evaluate how well the product conforms to the specifications (such as building or plumbing or electrical inspectors)

Each of these parties has to understand and agree on exactly *what* is to be done and *how* it is to be done. And in the case of a lawsuit over failure to meet specifications, the readership broadens to include judges, lawyers, and jury. Figure 14.3 shows how a clear set of specifications binds all readers (their various viewpoints, motives, and levels of expertise) in a community of shared understanding.

**FIGURE 14.1** A Mechanism Description *Continued*

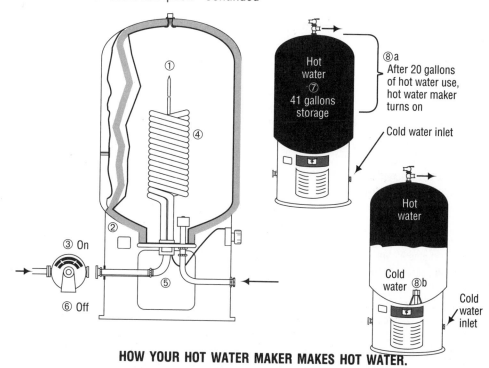

① ④ ② ③ On ⑥ Off ⑤

⑧a
After 20 gallons of hot water use, hot water maker turns on

Hot water
⑦
41 gallons storage

Cold water inlet

Hot water

Cold water ⑧b

Cold water inlet

## HOW YOUR HOT WATER MAKER MAKES HOT WATER.

1. The thermostat calls for energy to make hot water in your Hot Water Maker.

2. The built-in relay signals your boiler/burner to generate energy by heating boiler water.

3. The Hot Water Maker circulator comes on and circulates hot boiler water through the inside of the Hot Water Maker heat exchanger.

4. Heat energy is transferred, or "exchanged" from the boiler water inside the exchanger to the water surrounding it in the Hot Water Maker.

5. The boiler water after the maximum of heat energy is taken out of it, is returned to the boiler so it can be reheated.

6. When enough heat has been exchanged to raise the temperature in your Hot Water Maker to the desired temperature, the thermostat will de-energize the relay and turn off the Hot Water Maker circulator and your boiler/burner. This will take approximately 23 minutes—when you first start up

the Hot Water Maker. During use, reheating will be approximately 9–12 minutes.

7. You now have 41 gallons of hot water in storage . . . ready for use in washing machines, showers, sinks, etc. This 41 gallons of hot water will stay hot up to 10 hours, if you don't use it, without causing your boiler/burner to come on. (Unless, of course, you need it for heating your home in the winter.)

8. When you do use hot water, you will be able to use approximately 20 gallons, before the Hot Water Maker turns on. Then you will still have 21 gallons of hot water left for use, as your Hot Water Maker "recoups" 20 gallons of cold water. This means, during normal use (3½ GPM Flow), you will never run out of hot water.

9. You can expect substantial energy savings with your Hot Water Maker, as its ability to store hot water and efficiently transfer energy to make more hot water will keep your boiler off for longer periods of time.

Source: Courtesy of AMTROL, Inc.

FIGURE 14.2 Partial Specifications for a Building Project

# Ruger, Filstone, and Grant
# Architects

## SPECIFICATIONS FOR THE POWNAL CLINIC BUILDING

*Foundation*
    footings: 8" x 16" concrete (load-bearing capacity 3000 lbs. per sq. in.)
    frost walls: 8" x 4' @ 3000 psi
    slab: 4" @ 3000 psi, reinforced with wire mesh over vapor barrier

*Exterior walls*
    frame: eastern pine #2 timber frame with exterior partitions set inside posts
    exterior partitions: 2" x 4" kiln-dried spruce set at 16" on center
    sheathing: 1/4" exterior-grade plywood
    siding: #1 red cedar with a 1/2" x 6" bevel
    trim: finished-pine boards ranging from 1" x 4" to 1" x 10"
    painting: 2 coats of Clear Wood Finish on siding; trim primed and finished with one coat of bone-white oil base paint

*Roof system*
    framing: 2" x 12" kiln-dried spruce set at 24" on center
    sheathing: 5/8" exterior-grade plywood
    finish: 240 Celotex 20-year fiberglass shingles over #15 impregnated felt roofing paper
    flashing: copper

*Windows*
    Andersen casement and fixed-over-awning models, with white exterior cladding, insulating glass and screens, and wood interior frames

*Landscape*
    driveway: gravel base, with 3" traprock surface
    walks: timber defined, with traprock surface
    cleared areas: to be rough graded and covered with wood chips
    plantings: 10 assorted lawn plants along the road side of the building

**FIGURE 14.3** Readers and Potential Readers of Specifications

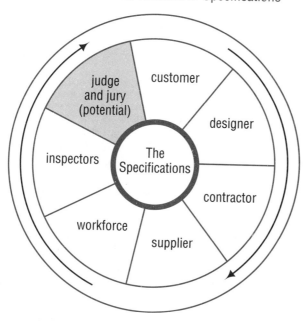

In addition to guiding a product's design and construction, specifications can facilitate the product's use and maintenance. For instance, specifications in a computer manual include the product's performance limits, or *ratings:* its power requirements, work or processing or storage capacity, environment requirements, the makeup of key parts, etc. Product support literature for appliances, power tools, and other items routinely contains ratings to help readers select a good operating environment or replace worn or defective parts (Riney 186). The ratings in Figure 14.4 are taken from the owner's manual for a dot-matrix printer.

## MAKING DESCRIPTIONS IMPARTIAL

Descriptions are mainly *subjective* or *objective:* based on feelings or fact. Subjective description emphasizes the writer's attitude toward the thing, whereas objective description emphasizes the thing itself. Because they show an *impartial* view, objective descriptions are more precise.

### Subjective Description

Essays describing "An Unforgettable Person" or "A Beautiful Moment" express opinions, a personal point of view. Subjective description aims at expressing feelings, attitudes, and moods. You create an *impression* of your subject ("The weather was miserable"), more than communicating factual information about it ("All day, we had freezing rain and gale-force winds").

FIGURE 14.4  Partial Specifications for the ImageWriter™ II Printer

**TABLE C-1**  Continued. Printer Specifications

| | |
|---|---|
| Paper Types: | Single sheets<br>Pin-feed paper<br>(hole centers 4.0–9.5 inches) |
| Ribbon: | Cassette containing black inked fabric ribbon<br>13 mm wide by 13000 mm long, continuous<br>Four color ribbon optional 21 mm wide by 18000 mm long, continuous |
| Power Options: | American — 120 volts AC ± 10%, 60 hertz<br>Universal — 100 volts AC ± 10%, 50/60 hertz<br>120 volts AC ± 10%, 50/60 hertz<br>140 volts AC ± 10%, 50/60 hertz<br>200 volts AC ± 10%, 50/60 hertz<br>220 volts AC ± 10%, 50/60 hertz<br>240 volts AC ± 10%, 50/60 hertz |
| Power Consumption: | Operating: 180 watts maximum<br>Standby: 20 watts maximum |
| Data Interface: | 8-bit serial |
| Weight: | 11.36 kilograms (25 pounds) |
| Dimensions: | Width — 431.8 / 17.0  Depth — 304.8 / 12.0  Height — 127.0 millimeters / 5.0 inches |
| Ambient Temperature:<br>Operating<br>Storage | 10 to 40 degrees Celsius (50 to 104 degrees F.)<br>−40 to +47 degrees Celsius (−40 to +116 degrees F.) |
| Humidity:<br>Operating<br>Storage | 20% to 95% relative humidity, noncondensing<br>10% to 95% relative humidity, noncondensing |

Source: Partial specifications for the ImageWriter™ II Printer. Reprinted by permission of Apple Computer, Inc.

## Objective Description

Strictly speaking, no writing can be purely objective in that an individual writer makes subjective choices about what to focus on and what to ignore. So-called objective description, however, filters out—as much as possible—personal impressions, and focuses instead on observable details.

Except for promotional writing, most descriptions on the job are impartial. The six questions on page 300 require objective answers. Notice that "What is your opinion of the item?" is not included. If you *are* asked for an opinion, give it only *after* the details. Here are guidelines for remaining impartial.

**Record Observable Details Faithfully.** To identify observable details, ask yourself these questions: What could any observer recognize? What would a camera record?

> Subjective His office has an *awful* view, *terrible* furniture, and a *depressing* atmosphere.

The italicized words only *tell;* they do not *show.*

> Objective His office has broken windows looking out on a brick wall, a rug with a six-inch hole in the center, chairs with bottoms falling out, missing floorboards, and a ceiling with plaster missing in three or four places.

**Use Precise and Informative Language.** Use high-information words that create pictures. Name specific parts without calling them "things," "gadgets," "watchamacallits," or "doohickeys." Avoid judgmental words ("good," "super," "impressive," "poor"), unless your judgment is requested and can be supported by facts. Instead of "large," "small," "long," and "near," give exact measurements, weights, dimensions, and ingredients.

Use words that specify location and spatial relationships: "above," "oblique," "behind," "tangential," "adjacent," "interlocking," "abutting," and "overlapping." Use position words: "horizontal," "vertical," "lateral," "longitudinal," "in crosssection," "parallel."

| Indefinite | Precise |
|---|---|
| an adequate memory | a 512K memory |
| a late-model car | a 1989 Ford Taurus sedan |
| an inside view | a cross-sectional, cutaway, or exploded view |
| right next to the foundation | adjacent to the right side |
| a small red thing | a red activator button with a 1-inch diameter and a concave surface |

The following description of a stethoscope is indefinite:

> The standard stethoscope is an *ordinary-looking* item that is *small* and *light* in weight. This *well-made gadget* consists of *several* parts that work together to perform an *important* function.

If you knew nothing about a stethoscope, this description would not help. In fact, the item could be just about anything with parts that do something. The italicized words tell us nothing.

Here is a version with precise details about appearance and function.

> The stethoscope is roughly 24 inches long and weighs about 5 ounces. The instrument consists of a sensitive sound-detecting and amplifying device whose flat surface is pressed against a bodily area. This amplifying device is attached to rubber

and metal tubing that transmits the body sound to a listening device inserted in the ear.

Seven interlocking pieces contribute to the stethoscope's Y-shaped appearance: (1) diaphragm contact piece, (2) lower tubing, (3) Y-shaped metal piece, (4) upper tubing, (5) U-shaped metal strip, (6) curved metal tubing, and (7) hollow ear plugs. These parts form a continuous unit.

Do not confuse precise language, however, with overly complicated technical terms. Don't say "phlebotomy specimen" instead of "blood" or "thermal attenuation" instead of "insulation." The clearest writing uses precise but plain English language. General readers prefer nontechnical language, as long as the simpler words do the job. Always think about your specific readers' needs.

## ELEMENTS OF DESCRIPTION

### Clear and Limiting Title

Promise exactly what you will deliver—no more and no less. "A Description of a Velo Ten-Speed Racing Bicycle" promises a complete description, down to the smallest part. If your intention is to describe the braking mechanism only, be sure your title so indicates: "A Description of the Velo's Center-Pull Caliper Braking Mechanism."

### Overall Appearance and Component Parts

Allow readers to see the whole item before you describe each part. Begin with the big picture (as in the previous stethoscope description).

### Visuals

Use drawings, diagrams, or photographs generously (see Chapter 12). Our overall description of the stethoscope is greatly clarified by Figure 14.5.

### Function of Each Part

Explain what each part does. In the stethoscope description, the function of the diaphragm contact piece can be explained in this way:

> The diaphragm contact piece is caused to vibrate by body sounds. This part is the "heart" of the stethoscope that receives, amplifies, and transmits the sound impulse.

### Appropriate Details

Your choice of details and level of technicality is crucial. Provide enough details for a clear picture. But do not burden readers needlessly. Identify your readers and their reasons for reading your description.

Assume you set out to describe a brand and model of bicycle. The picture you create will depend on the details you select. How will your reader use this

**FIGURE 14.5** Stethoscope with diaphragm contact piece (front view)

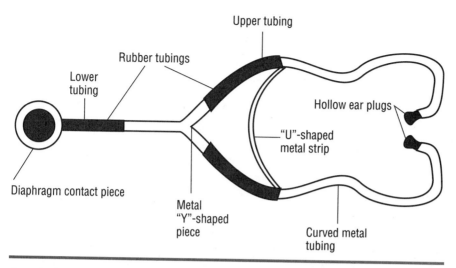

description? What is his or her level of technical understanding? Is this a customer interested in the bike's appearance—its flashy looks and racy style? Is it a repair technician who needs to know the order in which the parts operate? Or is your reader a helper in your bicycle shop who needs to know how to assemble this kind of bicycle? If you had designed the bike, you would give the manufacturer detailed specifications.

The description of the hot-water maker on pages 302–303 focuses on what this particular model looks like, what it's made of, and how it works. Its intended audience of do-it-yourselfers will know already what a hot-water maker is and what it does. And so this audience needs no background on hot-water makers in general. A description of how it was put together appears with the installation and maintenance instructions later in the manual.

Specifications for readers who will manufacture the hot-water maker would describe each part in exacting detail (say, the steel tank's required thickness and pressure rating as well as the required percentages of iron, carbon, and other constituents in the steel alloy). Choice of details in any description depends on its readers' needs and interests.

## Clearest Descriptive Sequence

Almost any item has its own logic of organization, based on (1) the way it appears as a static object, (2) the way its parts operate in order, or (3) the way its parts are assembled. We describe these relationships, respectively, in a spatial, functional, or chronological sequence.

**Spatial Sequence.**  Part of all physical descriptions, a spatial sequence answers these questions: *What is it? What does it do? What does it look like? What parts and material is it made of?* Use this sequence when you want your reader to visualize the item as a static object or mechanism at rest (a house interior, a document, the Statue of Liberty, a plot of land, a chainsaw, or a computer keyboard). Will your reader best understand the item from outside to inside, front to rear, left to right, top to bottom? (What logical path do the parts create?) A retractable pen would logically be viewed from outside to inside. The specifications in Figure 14.2 proceed from the ground upward.

**Functional Sequence.**   The functional sequence answers: *How does it work?* It is best used in describing a mechanism in action, such as a 35-millimeter camera, a nuclear warhead, a smoke detector, or a car's cruise-control system. The logic of the item is reflected by the order in which its parts function. Like the hot-water maker on page 303, a mechanism usually has only one functional sequence. The stethoscope description on page 308 follows the sequence of parts through which sound travels.

In describing a solar home-heating system, you would begin with the heat collectors on the roof, moving on through the pipes, pumping system, and tanks for the heated water, to the heating vents in the floors and walls—from the source to the outlet. After describing the heating system according to the functional sequence of operating parts, you could describe each part according to a spatial sequence.

**Chronological Sequence.**   A chronological sequence answers: *How has it been put together?* The chronology follows the sequence in which the parts are assembled.

Use the chronological sequence for an item that is best visualized by its assembly (such as a piece of furniture, an umbrella tent, or a prehung window or door unit). Architects might find a spatial sequence best for describing a proposed beach house to clients; however, they would use a chronological sequence (of blueprints) for specifying to the builder the prescribed dimensions, materials, and construction methods at each stage of construction.

**Combined Sequences.**   The description of a trout-activated feeder mechanism in Figure 14.6 employs all three sequences; first, a spatial sequence (top-to-bottom) for describing the overall mechanism at rest; next, a chronological sequence for explaining the order in which the parts are assembled; finally, a functional sequence for describing the order in which the parts operate.

---

**Audience-and-Use Profile**
The audience for the description in Figure 14.6 (owners/operators of trout hatcheries) are technically informed. They already know about other types of feeder mechanisms. Therefore, they will need no background explanation about the function of feeder mechanisms in general. Nor will they need definitions of technical terms such as *surface feeders* and *gravity feed*. Many readers may want to make their own feeders based on this design, and so they are interested in *exact* dimen-

sions and materials (specifications), the order in which parts are assembled, and the way this mechanism works. Each section of the description is clarified by visuals, including an enlarged drawing of the paper-clip attachment (the most crucial part of the mechanism at rest).

## COMPOSING A DESCRIPTION

Description of a complex mechanism almost invariably calls for an outline. This model is adaptable to any description.

I. INTRODUCTION: GENERAL DESCRIPTION[1]
  A. Definition, Function, and Background of the Item
  B. Purpose (and Audience—for classroom only)
  C. Overall Description (with general visuals, if applicable)
  D. Principle of Operation (if applicable)
  E. List of Major Parts

II. DESCRIPTION AND FUNCTION OF PARTS
  A. Part One in Your Descriptive Sequence
    1. Definition
    2. Shape, dimensions, material (with specific visuals)
    3. Subparts (if applicable)
    4. Function
    5. Relation to adjoining parts
    6. Manner of attachment (if applicable)
  B. Part Two in Your Descriptive Sequence (and so on)

III. SUMMARY AND OPERATING DESCRIPTION
  A. Summary (used only in a long, complex description)
  B. Interrelation of Parts
  C. One Complete Operating Cycle

This outline is tentative, because you might modify, delete, or combine certain parts to suit your subject, purpose, and reader.

### Introduction: General Description

Give your readers only as much background as they need to get the picture.

### A DESCRIPTION OF THE STANDARD STETHOSCOPE

**Introduction**

The stethoscope is a listening device that amplifies and transmits body sounds, to aid in detecting physical abnormalities.   *Definition and function*

This instrument has evolved from the original wooden, funnel-shaped instrument invented by a French physician, R. T. Lennaec, in 1819. Because of his female patients' modesty, he found it necessary to develop a device, other than his ear, for auscultation (listening to body sounds).   *History and background*

[1]In most descriptions, the subdivisions in the introduction can be combined and need not appear as individual headings in the document.

FIGURE 14.6  A Descripton Using Combined Sequences

# AN INEXPENSIVE TROUT-ACTIVATED FEEDER MECHANISM: DESCRIPTION AND SPECIFICATIONS

## The Overall Mechanism

Our trout-activated feeder mechanism is designed so that trout can obtain food at will. This description explains to trout hatchery owners and operators the appearance, assembly, and operation of a simple and inexpensive feeder mechanism.

The mechanism consists of a 1-pint plastic funnel, a 5 x 1-1/4-inch wood strip, a 12-inch galvanized metal rod (1/4-inch diameter), a wood disk (1/8 inch thick) with a 1/4-inch diameter hole in the center, and a 2-inch galvanized paper clip (Figure 1). A rod passes through its two openings. The cone's large opening is on top. Centered on the rod under the smaller opening is a wood disk of slightly larger diameter than the opening. This disk fastens to the rod by a paper clip.

**FIGURE 1**  A Side View of the Feeder Mechanism

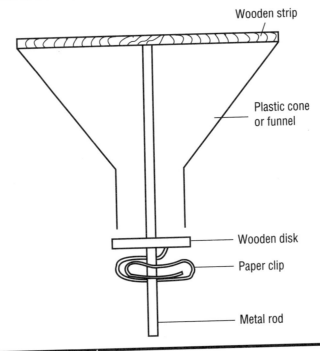

Wooden strip

Plastic cone or funnel

Wooden disk

Paper clip

Metal rod

FIGURE 14.6   A Descripton Using Combined Sequences *Continued*

2

### Order of Assembly

The plastic funnel is large enough to hold more food than fifteen two-pound trout need in one day. Placed across the funnel's top opening, the wood strip holds the metal rod that runs down through the cone. This rod then passes through the smaller bottom opening (and the disk) until it touches the water's surface. The wood disk is attached to the rod just under the lower opening by a paper clip (Figure 2).

**FIGURE 2**   A Side View of the Feeder Mechanism's Assembly and Operation

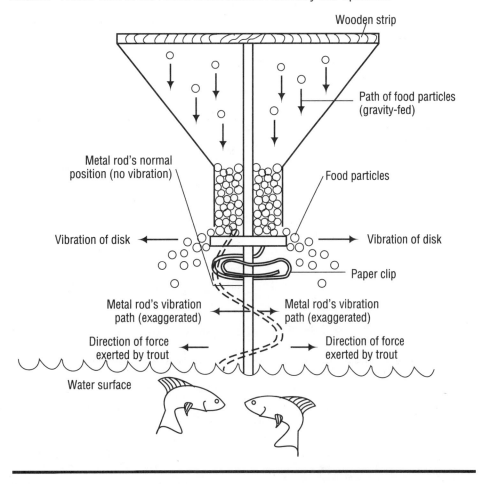

**FIGURE 14.6**  A Descripton Using Combined Sequences  *Continued*

3

Bent twice, the paper clip forms a notch where the metal rod is inserted. One bend is made in the inner section of the clip, slightly behind the 180-degree arc. The metal rod is inserted (and centered, crosslike) between the inner and outer sections of the clip, and the 180-degree arc of the inner section is bent around the rod at a 90-degree angle to the flat plane of the clip. At the open end of the clip is a tip, bent 60 degrees straight upward from the clip's edge, and inserted in the wood disk (Figure 3).

**Mechanism in Operation**

When the trout surface, they strike the tip of the metal rod, which is touching the water. The vibration travels up the rod and shakes the disk, which is covered with food from the funnel's gravity feed. As the disk vibrates, food resting on it falls to the water surface (Figure 2).

**FIGURE 3**  A Side View of the Paper Clip Attachment

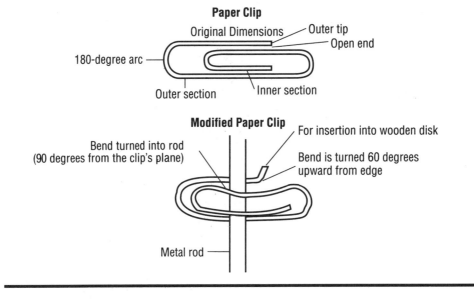

**Paper Clip**

Original Dimensions — Outer tip
— Open end

180-degree arc —

Outer section          Inner section

**Modified Paper Clip**

For insertion into wooden disk

Bend turned into rod
(90 degrees from the clip's plane)

Bend is turned 60 degrees
upward from edge

Metal rod —

This report explains to the beginning paramedical or nursing student the structure, assembly, and operating principle of the stethoscope. [Omit this section if you submit to your instructor an audience-and-use profile or if you write for a working-world audience.]

Purpose and audience

Finally, give a brief, overall description of the item, discuss its principle of operation, and list its major parts, as in the overall stethoscope description on pages 307–308.

## Description and Function of Parts

The body of your text describes each major part. After arranging the parts in sequence, follow the logic of each part. Provide only as much detail as your readers need.

Readers of this description will use a stethoscope daily, and so need to know how it works, how to take it apart for cleaning, and how to replace worn or broken parts.[2]

### Diaphragm Contact Piece

The diaphragm contact piece is a shallow metal bowl, about the size of a silver dollar (and twice its thickness), which is caused to vibrate by various body sounds.

Definition, Size, Shape, and Material

Three separate parts make up the piece: hollow steel bowl, plastic diaphragm, and metal frame, as shown in Figure 1.

Subparts

The stainless steel metal bowl has a concave inner surface, with concentric ridges that funnel sound toward an opening in the tapered base, then out through a hollow appendage. Lateral threads ring the outer circumference of the bowl to accommodate the interlocking metal frame. A fitted diaphragm covers the bowl's upper opening.

EXHIBIT 1   Front and Top View of the Three Parts of a Diaphragm Contact Piece

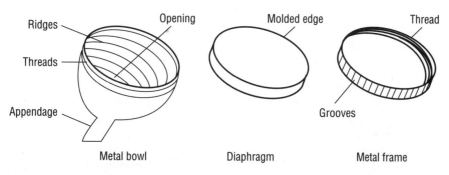

[2]Specifications for the manufacturer would require many more technical details (dimensions, alloys, curvatures, and so on).

The diaphragm is a plastic disk, 2 millimeters thick, 4 inches in circumference, with a molded lip around the edge. It fits flush over the metal bowl, and vibrates sound toward the ridges. A metal frame that screws onto the bowl holds the diaphragm in place.

The stainless steel frame fits over the disk and metal bowl. A ¼-inch ridge between the inner and outer edge accommodates threads for screwing the frame to the bowl. The frame's outside circumference is notched with equally spaced, perpendicular grooves—like those on the edge of a dime—to provide a gripping surface.

**Function and Relation to Adjoining Parts**

The diaphragm contact piece is the "heart" of the stethoscope that receives, amplifies, and transmits sound through the system of attached tubing.

**Manner of Attachment**

The diaphragm contact piece is attached to the lower tubing by an appendage on its apex (narrow end), which fits inside the tubing.

Each part of the stethoscope, in turn, is described according to its own logic of organization.

## Summary and Operating Description

Conclude by explaining how the parts work together to make the whole item function.

### Conclusion

The seven major parts of the stethoscope provide support for the instrument, flexibility of movement for the operator, and ease in auscultation.

**How parts interrelate**

**One Complete Operating Cycle**

In an operating cycle, the diaphragm contact piece, placed against the skin, picks up sound impulses from the body surface. These impulses cause the plastic diaphragm to vibrate. The amplified vibrations, in turn, are carried through a tube to a dividing point. From here, the amplified sound is carried through two separate but identical series of tubes to hollow ear plugs.

## APPLYING THE PRINCIPLES

This description of an automobile jack, aimed toward a general audience, follows our outline model.

### Audience-and-Use Profile

Some readers of this description (written for an owner's manual) will have no mechanical background. Before they can follow instructions for *using* the jack safely, they will have to learn what it is, what it looks like, what its parts are, and how, generally, it works. They will *not* need precise dimensions (e.g., "The rectangular base is 8 inches long and 6½ inches wide, sloping upward 1½ inches from the front outer edge to form a secondary platform 1 inch high and 3 inches square"). The engineer who designed the jack might include such data in specifications for the manufacturer. Laypersons, however, need only the dimensions that will help them recognize specific parts, for safe use and assembly.

Also, this audience will need only the broadest explanation of how the leverage mechanism operates. Although the physical principles *(torque, fulcrum)* would interest design engineers, they would be of little use to readers who simply need to operate the jack safely.

## DESCRIPTION OF A STANDARD BUMPER JACK

### Introduction — General Description

The standard bumper jack is a portable mechanism for raising the front or rear of a car through force applied with a lever. This jack enables even a frail person to lift one corner of a 2-ton automobile.

The jack consists of a molded steel base supporting a free-standing, perpendicular, notched shaft (Figure 1). Attached to the shaft are a leverage mechanism, a bumper catch, and a cylinder for insertion of the jack handle. Except for the main shaft and leverage mechanism, the jack is made to be dismantled and to fit neatly in the car's trunk.

The jack operates on a leverage principle, with the human hand traveling 18 inches and the car only ⅜ of an inch during a normal jacking stroke. Such a device requires many strokes to raise the car off the ground, but may prove a lifesaver to a motorist on some deserted road.

Five main parts make up the jack: base, notched shaft, leverage mechanism, bumper catch, and handle.

### Description of Parts and Their Function

#### The Base

The rectangular base is a molded steel plate that provides support and a point of insertion for the shaft (Figure 2). The base slopes upward to form a platform containing a 1-inch depression that provides a stabilizing well for the shaft. Stability is increased by a 1-inch cuff around the well. As the base rests on its flat surface, the bottom end of the shaft is inserted into its stabilizing well.

#### The Shaft

The notched shaft is a steel bar (32 inches long) that provides a vertical track for the leverage mechanism. The notches, which hold the mechanism in its position on the shaft, face the operator.

The shaft vertically supports the raised automobile, and attached to it is the leverage mechanism, which rests on individual notches.

#### The Leverage Mechanism

The leverage mechanism provides the mechanical advantage needed for its operator to raise the car. It is made to slide up and down the notched shaft. The main body of this pressed-steel mechanism contains two units: one for transferring the leverage and one for holding the bumper catch.

The leverage unit has four major parts: the cylinder, connecting the handle and a pivot point; a lower pawl (a device that fits into the notches to allow forward and prevent backward motion), connected directly to the cylinder; an upper pawl, connected at the pivot point; and an "up-down" lever, which applies or releases pressure on the upper pawl by means of a

FIGURE 1 A Side View of the Standard Bumper Jack

Bumper
catch
holder

Pivot point

Cylinder

Handle

Bumper
catch

Upper pawl

Lower pawl

Lip

"Up, down" lever

Bolt

Leverage mechanism

Notched shaft

Base

FIGURE 2 A Top View of the Jack Base

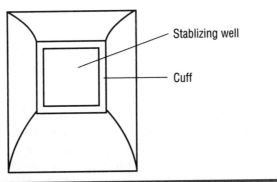

Stablizing well

Cuff

spring (Figure 1). Moving the cylinder up and down with the handle causes the alternate release of the pawls, and thus movement up or down the shaft—depending on the setting of the "up-down" lever. The movement is transferred by the metal body of the unit to the bumper-catch holder.

The holder consists of a downsloping groove, partially blocked by a wire spring (Figure 1). The spring is mounted in such a way as to keep the bumper catch in place during operation.

### The Bumper Catch

The bumper catch is a steel device that attaches the leverage mechanism to the bumper. This 9-inch molded plate is bent to fit the shape of the bumper. Its outer ½ inch is bent up into a lip (Figure 1), which hooks behind the bumper to hold the catch in place. The two sides of the plate are bent back 90 degrees to leave a 2-inch bumper contact surface, and a bolt is riveted between them. This bolt slips into the groove in the leverage mechanism and provides the attachment between the leverage unit and the car.

### The Jack Handle

The jack handle is a steel bar that serves both as lever and lug-bolt remover. This round bar is 22 inches long, ⅝ inch in diameter, and is bent 135 degrees roughly 5 inches from its outer end. Its outer end is a wrench made to fit the wheel's lug bolts. Its inner end is flattened for removing the wheel covers and for insertion into the cylinder on the leverage mechanism.

### Conclusion and Operating Description

The jack's five main parts (base, notched shaft, leverage mechanism, bumper catch, and handle) make up an efficient, lightweight device for raising a car to change a tire.

The jack is quickly assembled by inserting the bottom of the notched shaft into the stabilizing well in the base, the bumper catch into the groove on the leverage mechanism, and the flat end of the jack handle into the cylinder.

The bumper catch is then attached to the bumper, with the lever set in the "up" position. As the operator exerts an up-down pumping motion on the jack handle, the leverage mechanism gradually climbs the vertical notched shaft until the car's wheel is raised above the ground. When the lever is in the "down" position, the same pumping motion causes the leverage mechanism to descend the shaft.

## REVISION CHECKLIST FOR DESCRIPTIONS

Use this list to refine the content, arrangement, and style of your description. (Numbers in parentheses refer to first page of discussion.)

### Content

☐ Does the title promise exactly what the description delivers? (308)

☐ Have you described the item's overall features, as well as each part? (308)

□ Have you defined each part before discussing it? (311)

□ Have you explained the function of each part? (308)

□ Have you used visuals whenever they provide clarification? (308)

□ Will readers be able to visualize the item? (307)

□ Any details missing, needless, or confusing for this audience? (308)

### Arrangement

□ Is there an introduction-body-conclusion structure? (311)

□ Does the description follow the clearest possible sequence(s)? (309)

### Style

□ Is the description sufficiently impartial? (306)

□ Is the language informative and precise? (307)

□ Will the level of technicality connect with the audience? (308)

□ Is the description in plain English? (308)

□ Is each sentence clear, concise, and fluent? (199)

□ Is the description in correct English? (Appendix A)

---

### EXERCISES

1. Rewrite these subjective statements, making them more impartial:
   a. This classroom is (attractive, unattractive).
   b. The weather today is (beautiful, awful, mediocre).
   c. (He, she) is the best-dressed person in this class.
   d. My textbook is in (good, poor) condition.
   e. This room is (too large, too small, just the right size) for our class.

2. Select three possessions (choose simple objects easily described). Write a description of each so that police could identify the items if they were lost or stolen.

3. These statements are indefinite. Choose one statement and expand it into a precise description in one paragraph:
   a. Come to my place for a dinner you will never forget.
   b. Writing is done more efficiently on a word processor.
   c. The job opportunities in my major are numerous.
   d. I plan to be highly successful in my career.
   e. Attending a small college has (advantages, disadvantages).
   f. Regular access to a computer makes my work much easier.

4. a. Choose an item from this list, or select a device used in your major field. Using the general outline as a model, develop an objective description. Write for a specific use by a specified audience. Attach your written audience profile (based on the worksheet, page 33) to your report.

| | |
|---|---|
| a soda-acid fire extinguisher | a 100-amp electrical panel |
| a breathalyzer | a computer printer |
| a sphygmomanometer | a Skinner box |
| a ditto machine | a simple radio |
| a microprocessor | a flat-plate solar collector |
| a transit | a distilling apparatus |
| a saber saw | a bodily organ |
| a drafting table | a Wilson cloud chamber |
| a hazardous waste site | a programming flowchart |
| a blowtorch | a brand of woodstove |
| a photovoltaic panel | a catalytic converter |

Remember, you are simply describing the item, its parts, and its function: *do not* provide directions for its assembly or operation.

*b.* As an optional assignment, describe a place you know well. You are trying to convey a visual image, not a mood; therefore, your description should be impartial, discussing only the observable details.

**5.** The bumper-jack description in this chapter is aimed toward a general reading audience. Evaluate it by using the revision checklist. In one or two paragraphs, discuss your evaluation, and suggest revisions.

**6.** Figure 14.7 shows a description from an owner's manual for a telephone answering machine. Study the description and prepare *detailed* answers to these questions:

- Is this description impartial? Explain why.
- Compare Figure 14.7 with Figure 14.1. One description focuses on the workings of the *internal* mechanism itself, while the other describes the function of *external* parts. Which is which, and why?
- Are the details in Figures 14.1 and 14.7 adequate and appropriate for the respective audiences? Why?
- What is the descriptive sequence in Figure 14.7? How does it differ from the sequence in 14.1? Why is each sequence appropriate?

## TEAM PROJECTS

**1.** Divide into small groups. Assume that your group works in the product-development division of a large and diversified manufacturing company.

Your division has just invented an idea for an inexpensive consumer item with a potentially vast market. (Choose one from the following list.)

- an air pump for bicycle tires
- a flashlight
- nail clippers
- a retractable ball-point pen
- a safety razor
- scissors
- a stapler
- vise grips
- any simple mechanism—as long as it has moving parts

**FIGURE 14.7** A Description of an Answering Machine

## Guide To Model 2200 Controls

In the installation and operating instructions that follow, you'll be directed to switches, keys, indicator lights and other controls on your Model 2200. For orientation and easy reference, they're all mapped out here.

**1. Tape Counter**
Three-digit counter keeps track of tape used for incoming messages. Button resets tape counter to zero.

**2. Microphone**
Use it to record your outgoing messages or dictate memos.

**3. POWER key**
Turns the unit on or off. Red light above key indicates "power on."

**4. REW (rewind) key**
Rewinds your message cassette. Green light above key indicates use.

**5. PLAY key**
Plays your incoming or outgoing message cassette. Green light above key indicates use.

**6. FFW (fast forward) key**
Scans the message cassette quickly during playback. Green light above key indicates use.

**7. STOP key**
Stops record, play, rewind or fast forward functions on incoming message cassette.

**8. RECORD key**
Records your announcements on the outgoing cassette. Records memos on the incoming cassette. Red light above key indicates use.

**9. Function switch**
Set to AUTO to have your calls answered automatically. Set to MESSAGE to play back your incoming messages. Set to ANNOUNCE to record and test your outgoing announcements.

**10. AUTO indicator**
Red indicator glows steadily when your Model 2200 is set to answer calls automatically. It flashes to signal that messages have been received.

**11. VOLUME control**
Controls speaker volume during message playback.

(At the back of the unit:)

**12. AC cord**

**13. MESSAGE SELECT switch**
Incoming message length can be set to a 30-second or 270-second limit. The ANS ONLY (answer only) position can be used when you want your callers to hear your outgoing announcement, but you do not want to record any incoming messages.

**14. RING SELECT switch**
Program the unit to answer calls on the second or fourth ring.

**15. Modular telephone line cord**
Connects your Model 2200 to a modular wall jack.

**16. Modular telephone jack**
Accepts modular plug of a telephone or related equipment.

(Not visible in diagrams)

Cassette compartment
Under the unit's cover. The incoming message cassette goes on the left. The outgoing message cassette goes on the right.

Speaker
On the bottom of the unit behind circular grillwork. Plays back messages.

Identification plate
On the bottom of the unit. Includes the Model 2200 FCC registration number, the ringer equivalence number, and the remote control Touch-Tone codes.

Source: Courtesy of AT&T

**FIGURE 14.7** Description of an Answering Machine *Continued*

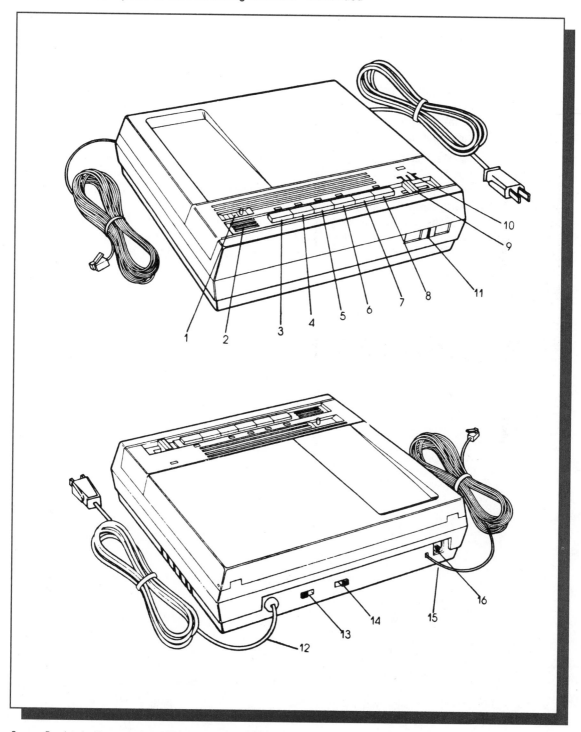

Your group's assignment is to prepare two descriptions of this invention:

**a.** for company executives who will decide whether to produce and market the item

**b.** for the engineers, machinists, etc., who will design and manufacture the item.

Before writing for each audience, be sure to collectively complete audience-and-use profiles (page 33).

Appoint a group manager, who will assign tasks (visuals, typing, etc.) to members. When the descriptions are fully prepared, the group manager will appoint one member to present them in class. The presentation should include explanations of how the two descriptions differ for the different audiences.

**2.** Assume that your group is an architectural firm designing buildings at your college. Develop a set of specifications for duplicating the interior of the classroom in which this course is held. Focus only on materials, dimensions, and equipment (chalkboard, desk, etc.) and use visuals as appropriate. Your audience includes the firm that will construct the classroom, teachers, and school administrators. Use the same format as in Figure 14.2 or design a better one. Appoint one member to present the completed specifications in class. Compare versions from each group for accuracy and clarity.

# Procedures
# and Processes

A process is a series of actions or changes leading to a product or result. A procedure is a way of carrying out a process.

## PURPOSE OF PROCESS-RELATED EXPLANATION

Writers explain procedures and processes as a way of answering these two questions for readers: (1) *How do I do it?* (2) *How does it happen?* In the workplace, you might instruct a colleague or customer in a specific procedure: how to test a soil sample for chlordane contamination, how to access a database, how to ship radioactive waste. Besides procedural instructions, you might have to explain how something happens: how the budget for various departments is determined, or how the electronic mail system in your company works. In this chapter we discuss both types of process-related explanation: instructions and process analysis.

## ELEMENTS OF INSTRUCTIONS

Instructions tell how to carry out a procedure. Almost anyone with a responsible job writes instructions, and almost everyone has to read instructions. The new employee will use instructions for operating office equipment or industrial machinery; the employee going on vacation writes instructions for the person filling in. The person who buys a computer reads the manuals (or *documentation*) for instructions on connecting a printer or running a program.

Because they tell readers what to do, instructions have legal implications. As many as 10 percent of workers are injured yearly on the job (Clement 14). And countless other injuries result from consumer misuse of power tools, car jacks, and other products—misuse often resulting from defective instructions.

Any reader injured because of inaccurate or incomplete or unclear instructions can sue the writer. The court has ruled that a writing defect in product support literature carries liability, as would a design or manufacturing defect in the product itself (Girill 37). Some legal experts argue that defects in the instructions carry a greater liability than defects in the product because they are more easily demonstrated to an unsophisticated jury (Bedford and Stearns 28).

To insure that your own instructions meet professional and legal requirements for accuracy, completeness, and clarity, observe the following guidelines.

## Clear and Limiting Title

Make your title promise exactly what your instructions deliver—no more and no less. The title "Instructions for Cleaning the Carriage and Key Faces of an Electronic Typewriter" is effective because it tells readers what to expect: instructions for a specific procedure on selected parts. But a title like "The Electronic Typewriter" gives no forecast; for all we can tell, the document might be a history of the typewriter, typing instructions, or a description of each part. On the other hand, a title like "Instructions for Cleaning an Electronic Typewriter" would be misleading if the instructions were limited to selected parts.

## Informed Content

Make sure you know *exactly* what you're talking about. Ignorance on your part makes you no less liable:

> If the author of [a car repair] manual had no experience with cars, yet provided faulty instructions on the repair of the car's brakes, the home mechanic who was injured when the brakes failed may recover from the author (Walter and Marsteller 165).

Do not write instructions unless you know the procedure down to its smallest detail and unless you have actually performed it.

## Logically Ordered Steps

Good instructions not only divide the process into steps; they also guide users through the steps *in order,* to reduce the likelihood of mistakes.

> You can't splice two wires to make an electrical connection until you have removed the insulation. To remove the insulation, you will need . . .

The logic of a procedure requires that you explain the steps in chronological order, as they are performed. Try to keep all information for one step close together so that readers won't have to flip pages.

## Visuals

Today's readers are image-oriented. Well-chosen visuals help keep words to a minimum and allow readers to picture what to do.

## Extension Cord Retainer

1. Look into the end of the Switch Handle and you will see 2 slots. The WIDER end of the Retainer goes into the TOP slot (Figure 8).

2. Plug extension cord into Switch Handle and weave cord into Retainer, leaving a little slack (Figure 9).

    NOTE: The retainer may be removed from the tool for replacement if damaged, or if improperly assembled. Two "disengage slots" are located in the top and bottom of the handle (Figure 8). To disengage the retainer, pull on each end of the retainer in turn while pushing a screwdriver gently into the appropriate slot.

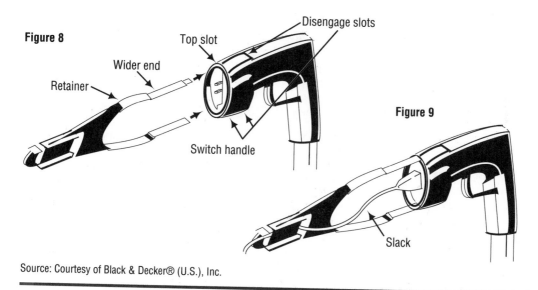

Source: Courtesy of Black & Decker® (U.S.), Inc.

Whenever possible, accompany a specific step with a specific illustration. Diagrams are often better than photographs because a diagram can selectively omit details unessential to the reader.

Always show the same angle of vision the reader will have in doing the activity or in using the equipment illustrated—and name the angle (*side view, top view*) if you think readers will have trouble figuring it out for themselves.

The less specialized your audience, the more visuals they are likely to need. Do not, however, illustrate a step such as "Press the RETURN key," or any action simple enough for readers to visualize on their own.

## Appropriate Level of Technicality

Unless you know that your readers have the relevant background and skills, write for general readers, and do three things:

1. Give them enough background to understand *why* they should follow your instructions.

2. Give appropriately detailed explanations so that readers can understand *what* the instructions say to do.

3. Give examples that show *how* to follow the procedure correctly.

Assume you are preparing instructions titled "How to Copy a Disk Using a Two-Drive Macintosh System." Here is how you might adapt your instructions for a general audience.

**Background Information.** Begin by explaining the purpose of the procedure.

> You might easily lose information stored on a computer disk if
>
> 1. the disk is damaged by direct sunlight, extreme temperature, or moisture
> 2. the disk is erased by a faulty disk drive, a power surge, or a user error
> 3. the stored information is scrambled by a nearby magnet (telephone, computer terminal, or the like)
>
> Always make at least one backup copy of any disk that contains important material.

Also, state your assumptions about your reader's level of technical understanding.

> To follow these instructions, you should be able to identify these parts of a Macintosh system: terminal, keyboard, mouse, external disk drive, and 3.5-inch disk.

Define any specialized terms that appear in your instructions.

> **Working definition**
> *Initialize:* Before you can store or retrieve information on a disk, you must initialize the blank disk. Initializing creates a format the computer can understand—a directory of specific memory spaces (like post office boxes) on the disk, where you can store and retrieve information as needed.

When your reader understands *what* and *why,* you are ready to explain *how* the reader can carry out the procedure.

**Detailed Explanations.** Explain the procedure in detail, so that your reader knows exactly what to do. Vague instructions result from the writer's failure to consider the intended readers' needs, as in these unclear instructions for giving first aid to an electrical shock victim:

1. Check vital signs.
2. Establish an airway.
3. Administer external cardiac massage as needed.
4. Ventilate, if cyanosed.
5. Treat for shock.

These instructions might be clear to medical experts, but not to average readers. Not only are the details inadequate, but terms such as "vital signs" and "cyanosed" are too technical for the layperson. Such instructions posted for workers in a high-voltage industrial area would be useless. The instructions need to be rewritten with illustrations and explanations, as in a Red Cross manual.

It's easy to assume that others know more about a procedure than they really do, especially when the procedure is almost automatic for you. Think about the days when a relative or friend was teaching you to drive a car, or when you have tried to teach someone else. Whenever you write instructions, remember that the reader knows less than you. A colleague will know at least a little less; a layperson will know a good deal less—maybe nothing—about this procedure.

Exactly how much information is enough? This is a hard question for any writer. Here are suggestions to help you find an answer:

- Give readers everything they need so that the instructions can stand alone.

- Give readers no more than they need to complete a step. Don't tell them how to build a computer when they need to know only how to insert a disk in the disk drive.

- Omit steps *(Seat yourself at the computer)* obvious to readers.

- Adjust the information rate ("the amount of information presented in a given page," Meyer 17) to the reader's background and the difficulty of the step. When explaining difficult or sensitive steps, slow the rate at which you present your information accordingly.

- Whenever possible, provide visuals that reinforce the prose. And don't be afraid to repeat information at various stages, if it will save readers from flipping pages.

- When writing instructions for consumer products, assume "a barely literate reader" (Clement 151). Simplify.


**Examples.**  Procedures require specific examples (how to load a program, how to order a part), to help readers follow the steps correctly.

To load your program, type this command:

| Load "Style Editor" and press RETURN |

Like visuals, examples *show* readers what to do. Examples in fact often appear as visuals.

In the sample procedure that follows, careful use of background, detailed explanation, and visual examples create a "user-friendly" level of technicality for readers who are just learning to use a computer.

---

**First Step: How to Initialize Your Blank Disk**
Before you can copy or store information on a blank disk, you must initialize the disk. Follow this procedure:

1. Switch the computer on.
2. Insert your application disk in the main disk drive.
3. Insert your blank disk in the external disk drive. Unable to recognize this new disk, the computer will respond with a message asking whether you wish to initialize the disk (Figure 1).

FIGURE 1   The "Initialize" Message

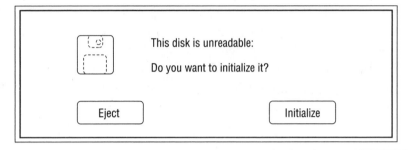

4. Using your mouse, place the tip of the on-screen pointer (a small arrow) inside the "Initialize" box.
5. Press and quickly release the mouse button. Within 15–20 seconds the initializing will be completed, and a message will appear, asking you to name your disk (Figure 2).

FIGURE 2   The "Disk-naming" Message

**Second Step: How to Name Your Initialized Disk** (and so on)

In the previous sample, notice that instructions and illustrations repeat the same information—they are redundant (Weiss 100). Granted, this prose-visual redundancy does make a longer document, but such repetition helps prevent misinterpretation.[1] When you can't be sure how much is enough, you are safer overexplaining than underexplaining.

[1]You may recall how Chapter 11 shows how writers avoid *style redundancy* (extra words that give no extra information—pages 209–210). For instructions, however, a writer deliberately seeks *content redundancy*, giving the same information in prose and then in visuals.

## Warnings, Cautions, and Notes

Here are the only items that should interrupt the steps in a set of instructions (Van Pelt 3):

- A *note* clarifies a point, emphasizes vital information, or describes options or alternatives.

    NOTE: The computer will not initialize a disk that is scratched or imperfect. If your blank disk is rejected, try a new disk.

- A *caution* prevents possible mistakes that could result in injury or damage:

    CAUTION: A momentary electrical surge or power failure will erase the contents of internal memory. To avoid losing your work, every few minutes save on disk what you have just typed into the computer.

- A *warning* alerts users against potential hazards to life or limb:

    WARNING: To prevent electrical shock, always disconnect your printer from its power source before cleaning internal parts.

- A *danger* notice identifies an immediate hazard to life or limb:

    DANGER: The red canister contains **DEADLY** radioactive material. **Do not break the safety seal** under any circumstances.

In addition to prose warnings, attract readers' attention and help them identify hazards by using symbols, or icons (Bedford and Stearns 128):

Do not enter     Radioactivity     Fire danger

Preview the warnings, cautions, and notes in your introduction, and place them, *clearly highlighted,* immediately before the respective steps.

Use notes, warnings, and cautions only when needed; overuse will dull their effect, and readers may overlook their importance.

## Appropriate Words, Sentences, and Paragraphs

Of all communications, instructions have the strictest requirements for clarity, because the words on the page lead to *immediate action.* Readers are impatient, and often won't read the entire instructions before plunging into the first step. Poorly phrased and misleading instructions can be disastrous.

Like descriptions, instructions name parts, use location and position words, and state exact measurements, weights, and dimensions. Instructions additionally require your closest attention to phrasing, sentence structure, and paragraph structure.

**Active Voice and Imperative Mood.** Use the active voice and imperative mood ("Insert the disk") to address the reader directly. Otherwise, your instructions can lose authority ("You should insert the disk"), or can become ambiguous ("The disk is inserted"). In the ambiguous example, we can't tell if the disk should be inserted or if it has already been inserted.

> **Indirect or confusing**
> The user types in his or her access code.
> You should type in your access code.
> It is important to type in the access code.
> The access code is typed in.
>
> **Direct and clear**
> Type in your access code.

The imperative makes instructions more definite because the action verb—the crucial word that tells what the next action will be—comes first. Instead of burying your verb in midsentence, *begin* with an action verb ("raise," "connect," "wash," "insert," "open") to give readers an immediate signal.

**Transitions to Mark Time and Sequence.** Transitional words provide a bridge between related ideas (Appendix A). Some transitional words ("in addition," "next," "meanwhile," "finally," "ten minutes later," "the next day," "before") mark time and sequence. They help readers understand the step-by-step process, as in these instructions:

> **Preparing the Ground for a Tent**
> Begin by clearing and smoothing the area that will be under the tent. This step will prevent damage to the tent floor and eliminate the discomfort of sleeping on uneven ground. *First,* remove all large stones, branches, or other debris within a level 10 × 13-foot area. Use your camping shovel to remove half-buried rocks that cannot easily be moved by hand. *Next,* fill in any large holes with soil or leaves. *Finally,* make several light surface passes with the shovel or a large, leafy branch to smooth the area.

**Parallel Phrasing.** Like any items in a series, steps in a set of instructions should be in identical grammatical form. Parallelism (Appendix A) is important in all writing but in instructions especially, because repeating grammatical forms emphasizes the step-by-step organization of the instructions.

> **Incorrect**
> To log on to the DEC 20, follow these steps:
> 1. Switch the terminal to "on."
> 2. The CONTROL key and C key are pressed simultaneously.
> 3. Typing LOGON, and pressing the ESCAPE key.
> 4. Type your user number, and press the ESCAPE key.

**Correct**   To log on to the DEC 20, follow these steps:

1. Switch the terminal to "on."
2. Press the CONTROL key and the C key simultaneously.
3. Type LOGON, and press the ESCAPE key.
4. Type your user number, and press the ESCAPE key.

Parallelism makes steps more readable and lends continuity to the instructions.

**Carefully Shaped Paragraphs and Sentences.**   Much of the introductory and explanatory material in a set of instructions takes the form of standard prose paragraphs, with enough sentence variety (page 218) to keep readers interested. But the steps themselves have unique paragraph and sentence requirements. Unless the procedure consists of simple steps (as in "Preparing the Ground for a Tent" on page 332), be sure to separate each step by using a numbered list—one step for one activity. If the activity is especially complicated, use a new line (not indented) to begin each sentence in the step.

Instructions ordinarily use short sentences. But brief is not always best, especially when readers have to fill in the information gaps. Never telegraph your message by omitting articles *(a, an, the),* as shown on page 200.

Unlike many types of documents, instructions call for very little sentence variety. Use sentences with similar structure ("Do this. Then do that.") to avoid confusing the reader trying to follow the procedure.

On page 332 you have seen how the steps usually begin with the action verb. But if a single step covers two related actions, follow the sequence of the actions that are required:

**Confusing sequence**   Insert the disk in the drive before switching on the computer.

The second action required is mistakenly given before the first.

**Logical sequence**   Before switching on the computer, insert the disk in the drive.

Make your explanatory material easier to follow by using a familiar-to-unfamiliar sequence (page 202):

**Hard**   You must initialize a blank disk before you can store information on it.

We can better understand this sentence if the familiar material comes first:

**Easier**   Before you can store information on a blank disk, you must initialize the disk.

Shape every sentence and every paragraph for the reader's access.

## Accessible Format

Instructions rarely get undivided attention. The reader, in fact, is doing two things more or less at once: interpreting the instructions and performing the task. *Accessibility,* then, becomes crucial. Readers of instructions have to find immediately what they need. Here are suggestions for designing an accessible instruction format:

- Use informative headings that tell readers what to expect. A heading such as "How to Initialize Your Blank Disk" is more informative than "Disk Initializing."

- Arrange all steps in a numbered list.

- Single-space within steps, and double-space between them.

- Double-space to signal a new paragraph, instead of indenting.

  Use white space and highlighting to separate the discussion from the step.

    Set off your discussion on a separate line (like this), indented or highlighted or both. On a typewriter, you can highlight with underlining, capitals, dashes, parentheses, and asterisks. On some computers, you can also use **boldface,** *italics,* varying type sizes, and varying typefaces.

- Set off warnings, cautions, and notes in ruled boxes or use highlighting and plenty of white space.

- Keep the visual and the step as close as you can. If you have room, place the visual right beside the step; if not, right after the step. Set off the visual with plenty of white space.

- Strive for format variety that is appealing but not overwhelming or inconsistent. Readers can be overwhelmed by a page with excessive or inconsistent graphic patterns.

The more accessible you make your format, the more likely it is that your readers will follow the instructions successfully. Don't be afraid to experiment with formats until you come up with an appealing and accessible design.

## COMPOSING INSTRUCTIONS

You can adapt the introduction-body-conclusion structure to any instructions. Here are the possible sections to include:

I. INTRODUCTION
  A. Definition, Benefits, and Purpose of the Procedure
  B. Intended Audience (usually omitted for workplace audiences)
  C. Prior Knowledge and Skills Needed by the Audience
  D. Brief Overall Description of the Procedure
  E. Principle of Operation
  F. Materials, Equipment (in order of use), and Special Conditions

**G.** Working Definitions (always in your introduction)

**H.** Warnings, Cautions, and Notes (mentioned here, and spelled out in the body)

**I.** List of Major Steps

**II.** INSTRUCTIONS FOR PERFORMANCE
  - **A.** First Major Step
    1. Definition and purpose
    2. Materials, equipment, and special conditions for this step
    3. Substeps (if applicable)
       - **a.**
       - **b.**
       - **c.**
  - **B.** Second Major Step (and so on)

**III.** CONCLUSION
  - **A.** Review of Major Steps (for a complex procedure only)
  - **B.** Interrelation of Steps
  - **C.** Troubleshooting or Follow-up Advice (as needed)

This outline is only tentative: because you might modify, delete, or combine some elements, depending on your subject, purpose, and audience.

## Introduction

Begin by giving readers any background they will need to carry out your instructions safely and effectively. Discuss the operating principle briefly; most readers are interested primarily in "how to use it or fix it," and need only a general sense of "how it works." Try neither to bury your readers in a long introduction, nor to set them loose on the procedure itself without adequate preparation. Know your audience—what they need and don't need.

We reproduce here an introduction from instructions written for any user of the college library. Some of these users will be computer experts; some will be novices—but all will require a detailed introduction to computerized literature searches before they conduct a search on their own.

---

### HOW TO USE THE OCLC TERMINAL TO SEARCH FOR A BOOK

Clear and limiting title

#### Introduction

The Southeastern Massachusetts University library's OCLC (On-line Computer Library Center) terminal offers an efficient way of searching for books, journals, government publications, and other printed materials. This terminal is connected to the OCLC database in Columbus, Ohio.

Definition of the procedure

The Ohio database contains more than 8 million records of books and other published materials in libraries nationwide. Each rec-

Definition *(continued)*

ord lists the information found on a catalog card, and identifies the libraries holding the work.

These instructions will enable you to discover whether a book you seek can be found in our library or in other libraries throughout the country. **Purpose**

Any library patron can use the OCLC terminal to search for a book. To operate the terminal, you need only be familiar with the keyboard (Figure 1) and to have an operator's manual handy for general reference use. **Audience and required skills**

After logging on to the system, you can search for a book by TITLE, AUTHOR, or AUTHOR and TITLE. Once you have viewed the catalog entry for your book you seek, you can then use the terminal to determine the libraries from which you might borrow the book. These instructions show a search by TITLE only. **General description of the procedure**

FIGURE 1  The OCLC Terminal Keyboard

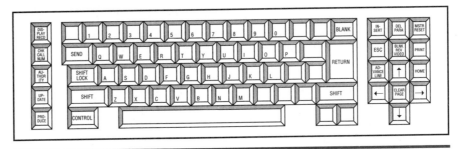

The only additional equipment you need is a copy of the *Manual of OCLC Participating Institutions,* to find a listing of library names according to their OCLC symbol displayed on your terminal screen. **Materials and equipment**

Specialized terms that appear in these instructions are here defined: **Working definitions**

*Cursor:* a small, blinking rectangle that indicates the screen position of the next character you will type.
*HOME Position:* the cursor location at the extreme top left of your screen. HOME position is where you will type most of your messages to the computer, and where the computer will display its messages to you.

Pay close attention to the NOTES that accompany the logging-on step and to the CAUTION preceding the logging-off step. **Cautions and notes**

The major steps in using OCLC to search for a book by its TITLE are (1) logging on to the system, (2) initiating the search, (3) viewing the information on your title, (4) locating the book, and (5) logging off. **List of major steps**

## Body

In your body section (labeled *Required Steps*), give each step and substep in order. Insert warnings, cautions, and notes as needed. Begin each step with its definition or purpose or both. Readers who understand the reasons for a step will do a better job. A numbered list (like the following) is an excellent way to segment the steps.

---

### Required Steps

1. *Logging On to the OCLC System*
   To activate the system, you must first log on.

   a. Flick on the red power switch at the keyboard's right edge.

   NOTE: Always press HOME before typing, to make sure your cursor is at HOME position.

   b. Type the authorization number (07-34-6991).

   NOTE: If you make a typing error, place the cursor over the incorrect character, and type the correct character. To move the cursor one space, press and release one of the four keys marked with an arrow (lower right of keyboard).

   c. Press SEND, and wait for the | HELLO | response.

2. *Initiating the Search*
   To initiate a computer search, you must enter your title's exact code name.

   a. Type in the code for your desired title, excluding articles *(a, an, the)* when they appear as the first word of the title. Always type your entry exactly as in the following example:

   Suppose your title is *Presentation Graphics on the Apple Macintosh;* you type

   pre,gr,on,t

   b. Press DISPLAY/RECD and then SEND.
   The computer will display a numbered list of titles that match the code you have sent (Figure 2).

**FIGURE 2**  The computer's list of titles that match the code you typed

```
1 Presentation Graphics on the Apple Macintosh: how to use
  Microsoft Chart to create dazzling graphics for profes-
  sional and corporate applications / Lambert, Steve.
  Microsoft Press: Distributed in the U.S. and Canada by
  Simon and Schuster, c 1984.

2 Presentation Graphics on the IBM PC and compatibles: how
  to use Microsoft Chart to create dazzling graphics for
  professional and corporate applications / Lambert,
  Steve. Microsoft Press: Distributed in the U.S. and
  Canada by Simon and Schuster, c 1986.
```

*Marginal notes:*

Section heading

Heading previews the step

Capitals and spacing set off note

Explanation begins new line

Example shows what to do

Single-spacing within steps; double-spacing between

Visual repeats the prose information

3. *Viewing the Information on Your Title*

Now that you have found your title, check to see if our library has the book. To view the full information on your title, follow these steps:

All steps have parallel phrasing

a. Type in the number of the title you seek.

b. Press DISPLAY/RECD and then SEND.
   The computer will display a complete catalog entry for your title (Figure 3).

Result of step begins new line

**FIGURE 3** The partial catalog entry for your title

Visual (near the step it illustrates, and set off by white space and ruled box)

```
NO HOLDINGS IN SMU-FOR HOLDINGS ENTER dh DEPRESS DISPLAY
RECD. SEND OCLC 10753933
  3 020 091484511X (pbk. : cover) : $18.95
  5 050 T385b.L34 1984
  9 100 Lambert, Steve, 1945-
 10 245 Presentation graphics on the Apple Macintosh: how
         to use Microsoft Chart to create dazzling graphics
         for professional and corporate applications / c
         Steve Lambert.
 11 260 Bellevue, Wash.: Microsoft Press; distributed in
         the U.S. and Canada by Simon and Schuster, New York;
         c 1984.
```

Because SMU does not have this title, you will have to determine where it can be found.

4. *Locating the Book*

To determine which OCLC libraries hold the title you seek, follow these steps:

a. Type dh and then press DISPLAY/RECD and then SEND. The computer will list symbols of holding libraries by state (Figure 4).

"Then" signals transition between substeps

**FIGURE 4** Display of symbols for regional libraries holding your title

```
REGIONAL LOCATIONS-FOR OTHER HOLDINGS ENTER dhs, dha, OR
dh. DISPLAY RECD. SEND; FOR BIBLIOGRAPHIC RECORD ENTER
bib. DISPLAY RECD. SEND
 STATE       LOCATIONS
  CT           ETN
  MA           LDV LIN MRK
  NY           NYP VYC VYM YQR ZBM ZNC ZQP ZSA
  VT           VTU
```

b. Identify the closest library by looking up symbols in the *Manual of OCLC Participating Institutions.*

Steps begin with action verbs

Your title, *Presentation Graphics,* can be found in these Massachusetts libraries: (LDV) Lotus Development Corporation, (LIN) Massachusetts Institute of Technology, and so on.

Example shows results of the step

c. For reference, obtain a printout of the library listing by pressing PRINT.

5. *Logging Off*

Once you have your printout, you can log off the system.

CAUTION: Before logging off, press CLEAR PAGE simultaneously with CONTROL. This step will clear any images that might damage the screen's phosphorus coating.

To log off, type *end* and then press SEND. The computer will respond with GOODBYE.

---

## CONCLUSION

The conclusion of a set of instructions has several possible functions:

- You might summarize the major steps in a long and complex procedure, to help readers review their performance.
- You might describe the results of the procedure.
- You might offer follow-up advice, about what could be done next or refer the reader to further sources of documentation.
- You might give troubleshooting advice, about what to do if anything goes wrong.

Depending on your subject and your audience, you might do all these things — or none of them. If your procedural instructions have given readers all they need, omit the conclusion altogether.

Our instructions for OCLC searching do require some concluding remarks, as follows.

---

**Conclusion**

Once you have located the titles you seek, ask the reference librarian for help in requesting the book through Interlibrary Loan. (Books usually arrive within two days.)

Follow-up advice

In case the first library is unable to supply the book requested, you initially should choose several libraries from your printout. The Interlibrary Loan system will automatically forward your request to each lender you have specified, until one lender indicates it can supply the book.

Troubleshooting advice

---

## APPLYING THE PRINCIPLES

The instructions for *doing something* (felling a tree) in Figure 15.1 are patterned after our general outline, shown earlier. They will be printed in one of a series of brochures for forestry students about to begin summer jobs with the Idaho Forestry Service.

### Audience-and-Use Profile

I'm writing these instructions for partially informed readers who know how to use chainsaws, axes, and wedges but who are approaching this dangerous procedure for the first time. Therefore, I'll include no visuals of cutting equipment (chainsaws and so on) because the audience already knows what these items look like. I can omit basic information (such as what happens when a tree binds a chainsaw) because the audience already has this knowledge also. Likewise, these readers need no definition of general forestry terms such as *culling* and *thinning*, but they *do* need definitions of terms that relate specifically to tree felling (*undercut, holding wood,* and so on).

To ensure clarity, I will illustrate the final three steps with visuals. The conclusion, for these readers, will be short and to the point—a simple summary of major steps with emphasis on safety. I'll experiment with formats until I create a design that is most accessible.

## COMPOSING A PROCESS ANALYSIS

An explanation of how things work or happen divides the process into its parts or principles. Colleagues and clients need to know how stock and bond prices are governed, how your bank reviews a mortgage application, how transmitters propagate radio waves, how an optical fiber conducts an impulse, and so on. Process analysis emphasizes *the process itself* so that readers can follow it as "observers."

Like instructions, a process analysis must be detailed enough to enable readers to follow the process step by step. Because it emphasizes the process itself, rather than the reader's role, process analysis is written in the third person. Instead of leading to immediate action, process analysis is designed to help readers gain information. Format and paragraph structure, therefore, are traditional.

Much of your writing in college term papers and essay exams explains how things happen. Your audience is the professor, who will evaluate what you have learned. Because this informed reader knows *more* than you do about the subject, you often discuss only the main points, omitting details.

But your real challenge comes in analyzing a process for readers who know *less* than you. Here you translate specialized information for readers who are neither willing nor able to fill in the blank spots; you then become the teacher, and your readers become the students.

**FIGURE 15.1** A Set of Instructions

# INSTRUCTIONS FOR FELLING A TREE

## INTRODUCTION

Felling is the cutting down of trees. Forestry Service personnel fell trees for various purposes: to cull or thin a forested plot, to eliminate the hazard of dead trees standing near power lines, to clear an area for road or building construction, and the like.

These instructions describe how to remove sizable trees for personnel who know how to use a chainsaw, axe, and wedge safely.

When you set out to fell a tree, expect to spend most of your time planning the operation and preparing the area around the tree. To fell the tree, you will make two chainsaw cuts, severing the stem from the stump. Depending on the direction of the cuts, the weather, and the terrain, the severed tree will fall into a predetermined clearing. You can then cut the tree into smaller sections for removal.

> **WARNING:** Although these instructions cover the basic procedure, felling is very dangerous. Trees, felling equipment, and terrain vary greatly. Even professional fellers are sometimes killed because of errors in judgment or misuse of tools.
>
> Your main consideration is safety. *Be sure* to have a skilled feller demonstrate this procedure before you try it on your own. Also, pay close attention to the warnings in steps 1 and 4.

To fell sizeable trees, you will need this equipment:
  —a 3- to 5-horsepower chainsaw with a 20-inch blade
  —a single-blade splitting axe
  —two or more 12-inch steel wedges

The major steps in felling a tree are (1) choosing the lay, (2) providing an escape path, (3) making the undercut, and (4) making the backcut.

## REQUIRED STEPS

1. *How to Choose the Lay*
   To "choose the lay" is to decide where you want the tree to fall. If the tree stands on level ground in an open field, which way you direct the fall makes little difference. But such ideal conditions are rare.

   Consider ground obstacles and topography, surrounding trees, and the condition of the tree to be felled. Plan your escape path and the location of your cuts depending on surrounding houses, electrical wires, and trees. Then follow these steps:

   a. Make sure the tree still is alive.

**FIGURE 15.1** A Set of Instructions *Continued*

2

b. Determine the direction and amount of lean.

**WARNING:** If the tree is dead and leaning substantially, do not try to fell it without professional help. Many dead, leaning trees have a tendency to split along their length, causing a massive slab to fall spontaneously.

c. Find an opening into which the tree can fall in the direction of lean or as close to the direction of lean as possible.

Because most trees lean downhill, try to direct the fall downhill. If the tree leans slightly away from your desired direction, use wedges to direct the fall.

2. *How to Provide an Escape Path*
Falling trees are unpredictable. Avoid injury by planning a definite escape path. Follow these steps:

a. Locate the path in the direction opposite that in which you plan to direct the fall (Figure1).

**FIGURE 1** Escape-path Location

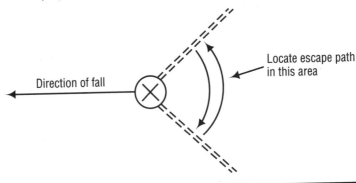

Direction of fall

Locate escape path in this area

b. Clear a path about 2 feet wide for a distance beyond where the top of the tree could land.

3. *How to Make the Undercut*
Making the undercut is the cutting of a triangular slab of wood from the trunk on the side toward which you want the tree to fall. The undercut directs and controls that fall. Follow these steps:

a. Start the chainsaw.

b. Holding the saw with the blade parallel to the ground, make the first cut 2 to 3 feet above the ground. Cut horizontally into the tree, to no more than 1/3 its diameter (Figure 2).

**FIGURE 15.1** A Set of Instructions *Continued*

**FIGURE 2**  Making the Undercut

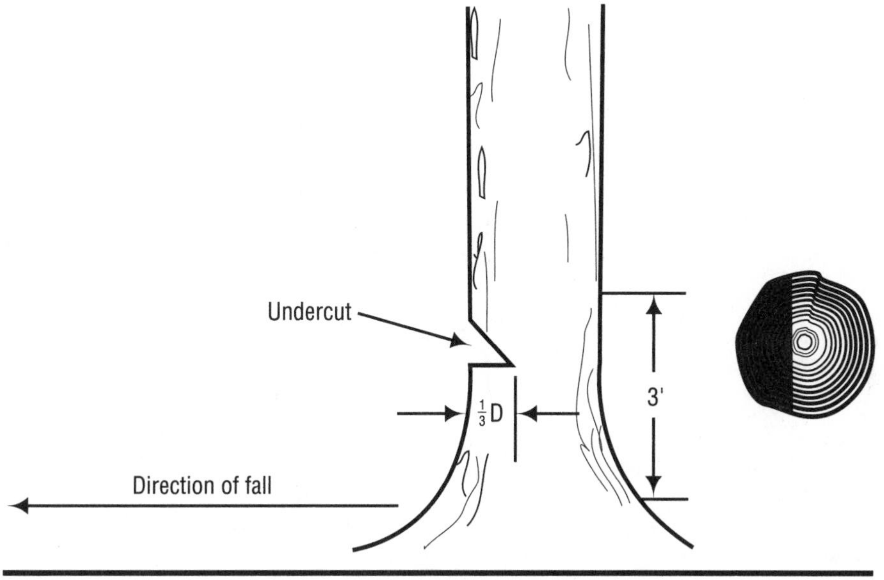

Undercut

$\frac{1}{3}$D

3'

Direction of fall

c. Make a downward-sloping cut, starting 4 to 6 inches above the first so that the cuts intersect at 1/3 the diameter (Figure 2).

4. *How to Make the Backcut*
After completing the undercut, you make the backcut to sever the stem from the stump. This step requires good reflexes and absolute concentration.

> **WARNING:** Observe tree movement closely during the backcut. If the tree shows any sign of falling in your direction, drop everything and move out of its way.
>
> Also, do not cut completely through to the undercut; but instead, leave a narrow strip of "holding wood" as a hinge, to help prevent the butt end of the falling tree from jumping back at you.

To make your backcut, follow these steps:

a. Holding the saw with the blade parallel to the ground, start your cut about 3 inches above the undercut, on the opposite side of the trunk. Leave a narrow strip of "holding wood" (Figure 3).

**FIGURE 15.1**  A Set of Instructions  *Continued*

**4**

**EXHIBIT 3**  Making the Backcut

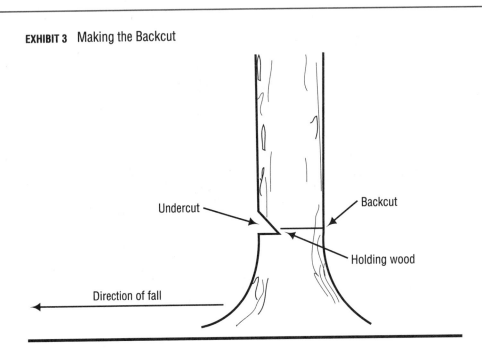

If the tree begins to bind the chainsaw, hammer a wedge into the backcut with the blunt side of the axe head. Then continue cutting.

b. As soon as the tree begins to fall, turn off the chainsaw, withdraw it, and step back immediately—the butt end of the tree could jump back toward you.

c. Move rapidly down the escape path.

**SUMMARY**

Felling is a complex and dangerous procedure. Choosing the lay, providing an escape path, making the undercut, and making the backcut are the basic steps—but trees, terrain, and other circumstances vary greatly.

For the safest operation, seek professional advice and help whenever you foresee *any* complications whatsoever.

For an uninformed audience, always begin your analysis with an introduction telling what the process is, and why, when, and where it happens. In the body, tell how it happens, analyzing each step in sequence. In the conclusion, summarize the steps, and describe one full cycle of the process.

Sections from this general outline can be adapted to any process analysis:

**I.** INTRODUCTION
    **A.** Definition, Background, and Purpose of the Process
    **B.** Intended Audience (usually omitted for workplace readers)
    **C.** Prior Knowledge Needed to Understand the Process
    **D.** Brief Description of the Process
    **E.** Principle of Operation
    **F.** Special Conditions Needed for the Process to Occur
    **G.** Definitions of Special Terms
    **H.** List of Major Steps

**II.** STEPS IN THE PROCESS
    **A.** First Major Step
        **1.** Definition and purpose
        **2.** Special conditions needed for the specific step
        **3.** Substeps (if applicable)
            **a.**
            **b.**
    **B.** Second Major Step (and so on)

**III.** CONCLUSION
    **A.** Summary of Major Steps
    **B.** One Complete Process Cycle

Adapt this outline to the structure of the process you are explaining.

Here is a document patterned after the sample outline. Our writer, Bill Kelly, belongs to an environmental group studying the problem of acid rain in their Massachusetts community. (Massachusetts is among the states most affected by acid rain.) To gain community support, the environmentalists must educate citizens about the problem. Bill's group is publishing and mailing a series of brochures. The first brochure explains how acid rain is formed.

----

**Audience-and-Use Profile**
My audience will consist of general readers. Some already will be interested in the problem; others will have no awareness (or interest). Therefore, I'll keep my explanation at the lowest level of technicality (no chemical formulas, equations). But my explanation needs to be vivid enough to appeal to less aware or less interested readers. I'll use visuals to create interest and to illustrate simply. To give an explanation thorough enough for broad understanding, I'll divide the process into three chronological steps: how acid rain develops, spreads, and destroys.

# HOW ACID RAIN DEVELOPS, SPREADS, AND DESTROYS

## Introduction

Acid rain is environmentally damaging rainfall that occurs after fossil fuels burn, releasing nitrogen and sulfur oxides into the atmosphere. Acid rain, simply stated, increases the acidity level of waterways because these nitrogen and sulfur oxides combine with the normal moisture in the air. The resulting rainfall is far more acidic than normal rainfall. Acid rain is a silent threat because its effects, although slow, are cumulative. This analysis explains the cause, the distribution cycle, and the effects of acid rain.

Most research shows that power plants burning oil or coal are the primary cause of acid rain. The fuel used to create energy is not completely expended, and some of the residue enters the atmosphere. Although this residue contains several potentially toxic elements, sulfur oxide and, to a lesser extent, nitrogen oxide are the major problem, because they are transformed when they combine with moisture. This chemical reaction forms sulfur dioxide and nitric acid, which then rain down to earth.

The major steps explained here are (1) how acid rain develops, (2) how acid rain spreads, and (3) how acid rain destroys.

## The Process

### How Acid Rain Develops

Once fossil fuels have been burned, their usefulness is over. Unfortunately, it is here that the acid rain problem begins.

Fossil fuels contain a number of elements that are released during combustion. Two of these, sulfur oxide and nitrogen oxide, combine with normal moisture to produce sulfuric acid and nitric acid. (Figure 1 illustrates how acid rain develops.) The released gases undergo a chemical change as they combine with atmospheric ozone and water vapor. The resulting rain or snowfall is more acid than normal precipitation.

Acid level is measured by pH readings. The pH scale runs from 0 through 14—a pH of 7 is considered neutral. (Distilled water has a pH of 7.) Numbers above 7 indicate increasing degrees of alkalinity. (Household ammonia has a pH of 11.) Numbers below 7 indicate increasing acidity. Movement in either direction on the pH scale, however, means multiplying by 10. Lemon juice, which has a pH value of 2, is 10 times more acidic than apples, which have a pH of 3, and is 1,000 times more acidic than carrots, which have a pH of 5.

Because of carbon dioxide (an acid substance) normally present in air, unaffected rainfall has a pH of 5.6. At this time, the pH of precipitation in the northeastern United States and Canada is between 4.5 and 4. In Massachusetts, rain and snowfall have an average pH reading of 4.1. A pH reading below 5 is considered to be abnormally acidic, and therefore a threat to aquatic populations.

### How Acid Rain Spreads

Although it might seem that areas containing power plants would be most severely affected, acid rain can in fact travel thousands of miles from its source. Stack gases escape and drift with the wind currents. The sulfur and nitrogen oxides are thus able to travel great distances before they return to earth as acid rain.

FIGURE 1 How Acid Rain Develops

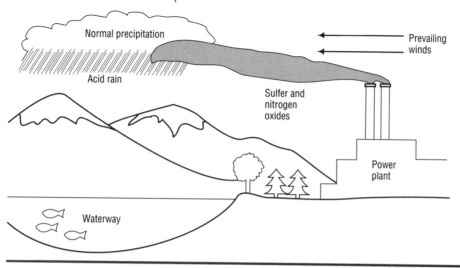

For an average of two to five days after emission, the gases follow the prevailing winds far from the point of origin. Estimates show that about 50 percent of the acid rain that affects Canada originates in the United States; at the same time, 15 to 25 percent of the U.S. acid rain problem has its origin in Canada.

The tendency of stack gases to drift makes acid rain a widespread menace. More than 200 lakes in the Adirondacks, hundreds of miles from any industrial center, are unable to support life because their water has become so acidic.

### How Acid Rain Destroys

Acid rain causes damage wherever it falls. It erodes various types of building rock such as limestone, marble, and mortar, which are gradually eaten away by the constant bathing in acid. Damage to buildings, houses, monuments, statues, and cars is widespread. Some priceless monuments and carvings already have been destroyed, and even trees of some varieties are dying in large numbers.

More important, however, is acid rain damage to waterways in the affected areas. (Figure 2 illustrates how a typical waterway is infiltrated.) Because of its high acidity, acid rain dramatically lowers the pH in lakes and streams. Although its effect is not immediate, acid rain can eventually make a waterway so acidic it dies. In areas with natural acid-buffering elements such as limestone, the dilute acid has less effect. The northeastern United States and Canada, however, lack this natural protection, and so are continually vulnerable.

The pH level in an affected waterway drops so low that some species cease to reproduce. In fact, a pH level of 5.1 to 5.4 means that fisheries are threatened; once a waterway reaches a pH level of 4.5, no fish reproduction occurs. Because each creature is part of the overall food chain, loss of one element in the chain disrupts the whole cycle.

**FIGURE 2** How Acid Rain Destroys

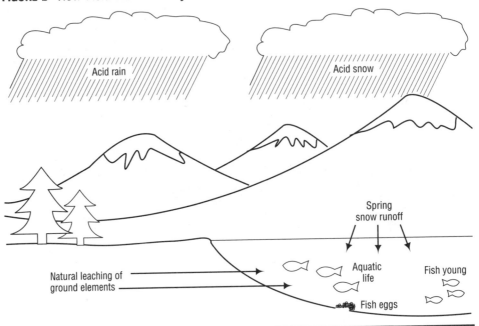

In the northeastern United States and Canada, the problem of excess acidity is compounded by the runoff from acid snow. During the cold winter months, acid snow sits with little melting, so that by spring thaw, the acid released is greatly concentrated. Aluminum and other heavy metals normally present in soil are also released by acid rain and runoff. These toxic substances leach into waterways in heavy concentrations, affecting fish in all stages of development.

**Summary**
Acid rain develops from nitrogen and sulfur oxides emitted by industrial and power plants burning fossil fuels. In the atmosphere, these oxides combine with ozone and water to form acid rain: precipitation with a lower-than-average pH. This acid precipitation returns to earth many miles from its source, severely damaging waterways that lack natural buffering agents. The northeastern United States and Canada are the most profoundly affected areas in North America.

## REVISION CHECKLIST FOR INSTRUCTIONS

(Numbers in parentheses refer to the first page of discussion.)

### Content

☐ Does the title promise exactly what the instructions deliver? (326)

☐ Is the background adequate for this audience? (328)

☐ Do explanations enable readers to understand what to do? (325)

☐ Do examples enable readers to see how to do it correctly? (329)

☐ Are the definition and purpose of each step given as needed? (328)

☐ Is all needless information omitted? (329)

☐ Are all obvious steps omitted? (329)

☐ Do notes, cautions, or warnings appear whenever needed, before the step? (331)

☐ Are visuals adequate for clarifying the steps? (326)

☐ Do visuals repeat prose information as needed? (330)

## Organization

☐ Is the introduction adequate but not excessive? (335)

☐ Do the instructions follow the exact sequence of steps? (337)

☐ Is all the information for a step close together? (337)

☐ For a complex step, does each sentence begin on a new line? (333)

☐ Is the conclusion necessary, and if necessary, adequate? (339)

## Style

☐ Do introductory sentences have enough variety to maintain interest? (333)

☐ Does each step have similar sentence structure? (333)

☐ Do steps generally have short sentences? (333)

☐ Does each step begin with the action verb? (332)

☐ Are all steps in the active voice and imperative mood? (332)

☐ Do all steps have parallel phrasing? (332)

☐ Are transitions adequate for marking time and sequence? (332)

## Format

☐ Does each heading clearly tell readers what to expect? (334)

☐ On a typed page, are steps single-spaced within, and double-spaced between? (334)

☐ Do white space and highlights set off discussion from step? (334)

☐ Are notes, cautions, or warnings set off by lined boxes or highlights? (334)

☐ Are visuals beside or near the step, and set off by white spaces? (334)

## EXERCISES

**1.** In a memo to your instructor, discuss the types of process-related explanations you would expect to write on the job. As specifically as you can, identify typical subjects, your expected audience, and their level of technical understanding.

**2.** Make these instructions more readable by rewriting them in the appropriate voice, mood, phrasing, sentence or paragraph division, and format. Insert transitions where necessary.

*What to Do Before Jacking Up Your Car*

Whenever the misfortune of a flat tire occurs, some basic procedures should be followed before the car is jacked up. If possible, your car should be positioned on as firm and level a surface as is available. The engine has to be turned off; the parking brake should be set; and the automatic transmission shift lever must be placed in "park," or the manual transmission lever in "reverse." The wheel diagonally opposite the one to be removed should have a piece of wood placed beneath it to prevent the wheel from rolling. The spare wheel, jack bar, jack stand, hook, and lug wrench should be removed from the luggage compartment.

**3.** Select part of a technical manual in your field or instructions for a general reader, and make a copy of the material. Using the revision checklist, evaluate the sample. In a memo to your instructor, discuss the strong and weak points of the instructions. Or be prepared to explain to the class why the sample is effective or ineffective.

**4.** Select a specialized process that you understand well (from your major or a subject that interests you). Choose a process that has several distinct steps. Using the process analysis on pages 346–348 as a model, write your own analysis for classmates who are unfamiliar with this process. Begin by completing an audience-and-use profile (page 33). Here are a few possible topics: how the body metabolizes alcohol, how economic inflation occurs, how the federal deficit threatens our future, how a lake or pond becomes a swamp, how exposure to the sun causes skin cancer, how cigarettes cause heart and lung disease, how a volcanic eruption occurs.

**5.** Choose a topic from this list, your major, or a subject you are interested in. Using the general outline in this chapter as a model, outline instructions for a procedure that requires at least three major steps. Address a general reader, and begin by completing an audience-and-use profile (page 33). Write one draft; modify your outline as needed; and write a second draft in accordance with the revision checklist. Exchange instructions with a classmate for revision suggestions.

| | |
|---|---|
| planting a tree | preparing logs for building a log cabin |
| hot-waxing skis | removing the rear wheel of a bicycle |
| hanging wallpaper | adding an electrical outlet |
| jump-starting a car battery | avoiding hypothermia |
| filleting a fish | loading ("booting") a software program |
| warming up before exercise | creating a file on your college's mainframe |
| serving a tennis ball | computer |
| hitting a golf ball | |

**6.** Assume you are assistant to the communications manager for a manufacturer of outdoor products. Among the company's best-selling items are its various models of gas grills. Because of fire and explosion hazards, all grills must be accompanied by detailed instructions for safe assembly, use, and maintenance.

One of the first procedures in the manual is the "leak test," to ensure that the gas supply and transport apparatus is leak free. One of the engineers has prepared the instructions in Figure 15.2. Before being published in the manual, they must be approved by communications management. Your boss directs you to evaluate the instructions for accuracy, completeness, and clarity, and to report your findings in a memo. Because of the legal implications here, your evaluation must spell out all positive and negative details of content, organization, style, and format. (Use the Revision Checklist as a guide.) The boss is a busy and impatient reader, and expects your report to be no longer than two pages. Do the evaluation and write the memo.

## TEAM PROJECTS

**1.** From your major field, select a simple but specialized process that you understand well (how gum disease develops, how a savings bank makes a profit, how the heart pumps blood, how an earthquake occurs, how steel is made, how a computer compiles and executes a program, how electricity is generated, how a corporation is formed, how a verdict is appealed to a higher court, how a bankruptcy claim is filed, how a photocopying machine produces copies, how nicotine affects the body, how paper is made). Write a brief explanation of the process. Exchange your analysis with a classmate in another major. Study your classmate's explanation for fifteen minutes and then write the same explanation *in your own words,* referring to your classmate's paper as needed. Now, evaluate your classmate's version of your original explanation for accuracy of content. Does it show that your explanation was understood? If not, why not? Discuss your conclusions in a memo to your instructor, submitted with all samples.

**2.** Draw a map of the route from your classroom to your dorm, apartment, or home — whichever is closest. Be sure to include identifying landmarks. When your map is completed, write instructions for a classmate who will try to duplicate your map from the information given in your written instructions. Be sure your classmate does not see your map! Exchange your instructions and try to duplicate your classmate's map. Compare your results with the original map. Discuss your conclusions.

**3.** Divide into small groups and visit your computer center, library, or any place on campus or at work where you can find operating manuals for computers, lab or office equipment, or the like. (Or look through the documentation for your own computer hardware or software.) Locate fairly brief instructions that could use revision for improved content, organization, style, or format. (Try to choose instructions for a procedure you have done successfully.) Make a copy of the instructions, and revise them. Submit all materials to your instructor, along with a memo explaining the improvements. Or be prepared to discuss your revision in class.

**4.** Visit your computer center and ask to borrow a software package that arrived without documentation or with very limited instructions. (Or perhaps you've received such programs as a member of a software club). Run the program, and then prepare instructions for the next users. Test your instructions by having a classmate use them to run the program. Revise as needed. Remember that you are writing for a user who has no previous experience.

**FIGURE 15.2** Instructions for Leak Testing a Gas Grill

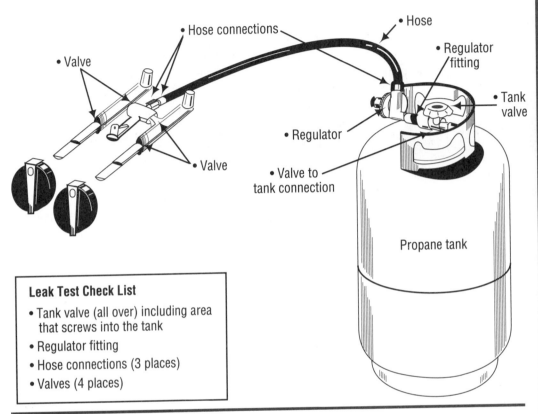

Tank must be filled prior to this step (See "Propane Tank" section for details about filling this tank)

**Leak Test Check List**

- Tank valve (all over) including area that screws into the tank
- Regulator fitting
- Hose connections (3 places)
- Valves (4 places)

**SAFETY!** All models must be Leak Tested! Take the hose, valve, and regulator assembly outdoors in a well ventilated area. Keep away from open flames or sparks. Do Not smoke during this test. Use only a soap and water solution to test for leaks.

1. Have propane tank filled with propane gas only by a reputable propane gas dealer.
2. Attach regulator fitting to tank valve.
   - Regulator fitting has left-hand threads. Turn counterclockwise to attach.
3. Tighten regulator fitting securely with wrench.
4. Place the two matching control knobs onto the valve stems.
5. Turn the control knobs to the right, clockwise. This is the **"Off"** position.
6. Make a solution of half liquid detergent and half water.

7. Turn gas supply "ON" at the tank valve (counterclockwise).
8. Brush soapy mixture on all connections listed in the **Leak Test Check List**.
9. Observe each place for bubbles caused by leaks.
10. Tighten any leaking connections, if possible.
    - If leak cannot be stopped, **Do Not Use Those Parts!** Order new parts.
11. Turn gas supply **"Off"** at the tank valve.
12. Push in and turn control knobs to the left ("HI" position) to release pressure in hose.
13. Disconnect the regulator fitting from the tank valve.
    - Regulator fitting has left-hand threads. Turn fitting clockwise to disconnect.
14. Leave tank outdoors and return to the grill assembly with valve, hose, and regulator.

Source: Instructions reprinted by permission of Thermos® Division.

# Letters and Employment Correspondence

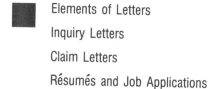

Elements of Letters

Inquiry Letters

Claim Letters

Résumés and Job Applications

Supporting Your Application

$\mathbf{A}$ report may be compiled by a team of writers for various readers, but a letter usually is written by one writer for one or more definite readers. Because a letter is more personal than a report, proper tone is essential; you want the reader to be on your side. Because your signature certifies the content of your letter (which may serve as a legal document), precision is crucial.

This chapter covers four common letter types: inquiry letters, claim letters, and letters of application, along with résumés. Other useful types—letters of transmittal and letter reports—are discussed in Chapters 17 and 18.

## ELEMENTS OF LETTERS

*Never send a letter until you genuinely feel confident about signing it.* Use the guidelines below for judging the adequacy of letters before you sign them.

### Introduction-Body-Conclusion Structure

Structure most letters to include (1) a brief *introduction* paragraph (five lines or fewer) which identifies you and your purpose; (2) one or more *body* paragraphs with the details of your message; (3) a *conclusion* paragraph which sums up and encourages action. For readability, keep your paragraphs short (usually fewer than eight lines). If your body section is long, divide it into shorter paragraphs, as in Figure 16.1. Notice that this body section is broken down into four questions for easy answering. Figure 16.2 shows the response to these questions.

## Standard Parts

Letters typically have six parts, in order from top to bottom: heading, inside address, salutation, the text (introduction, body, and conclusion), complimentary close, and signature.

**Heading.** If the stationery has a letterhead, add only the date two lines below the letterhead (Figure 16.2), at the right or left margin. On blank stationery, include your address and the date (but not your name), as in Figure 16.1 and here:

```
Street Address        154 Sea Lane
City, State Zip Code  Harwich, MA 02163
Month Day, Year       July 15, 1987
```

Avoid abbreviations (but do use the Postal Service's two-letter state abbreviations on the envelope, and in the heading itself when the state has a long name).

**Inside Address.** Two to six spaces below your heading and abutting your left margin is the inside address.

```
Reader's Name, Title, and Position  Dr. Ann Mello, Dean
Company Name                        Western University
Street Address (if applicable)      Stow, Iowa 51350
City, State Zip Code
```

Whenever possible, address your letter to a specifically named reader, using that reader's title (Attorney, Major), but don't be redundant by writing "Dr. Ann Mello, Ph.D."; use Dr. or Ph.D., Dr. or M.D. Abbreviate only titles that are routinely abbreviated (Mr., Ms., Dr.). Titles such as "Captain" are written out in full. Although some women prefer the traditional titles "Mrs." or "Miss," many prefer "Ms." When unsure of your reader's preference, use "Ms."

**Salutation.** Your salutation, two spaces below the inside address, begins with "Dear" and ends with a colon ("Dear Mr. Smith:"). Include the person's full title. If you don't know the person's name or gender, use the position title: "Dear Manager" or, preferably, an attention line (page 358). When appropriate, address your reader by first name—if that is the way you would address this individual in person.

```
Dear Ms. Jones:
Dear Professor Smith:
```

No satisfactory phrase exists for greeting several people at once. Women rightly resent the term "Gentlemen," and "Ladies and Gentlemen" sounds like the beginning of a speech. "Dear Sir or Madam" sounds formal and old-fashioned. And "To Whom It May Concern" is too vague and impersonal. Your best bet there is to eliminate the salutation completely by using an attention line.

**FIGURE 16.1** A Letter of Inquiry

154 Sea Lane
Harwich, MA 02163
July 15, 1990

A.B. Coolidge
Land Use Manager
Eastern Paper Company
Waldoboro, Maine 04572

Dear Mr. Coolidge:

Mr. Melvin Blount, your sales representative, has told me that Eastern Paper Company offers parcels of its lakefront property in northern Maine for public lease. I am interested in a lease, and would appreciate answers to these questions:

1. Does your company have any lakefront parcels in highly remote areas?

2. What is the average size of a leased parcel?

3. How long does a lease remain in effect?

4. What is the yearly leasing fee?

I would welcome any other details you might send along.

Yours truly,

*Thomas Stacy*

Thomas Stacy

P.S. I am planning a trip to Maine for August 5–11, and would be happy to stop by your office any time you are free.

FIGURE 16.2   A Letter Responding to an Inquiry

## EASTERN PAPER COMPANY
### WALDOBORO, MAINE 04572

July 25, 1990

Mr. Thomas Stacy
154 Sea Lane
Harwich, MA 02163

Dear Mr. Stacy:

In answer to your inquiry about leasing lakeside lots in Maine, we have no lands for lease in "highly remote areas." The most remote area is the north shore of Deerfoot Lake, where you can reach the camp lot by boat.

Leases on these 30,000-square-foot parcels are renewable yearly each June. The fee is $250 per year. We have a limited number of leases available, but you would have to visit our office for information about exact locations. Drop by any weekday morning before 11 o'clock.

The Land Use Regulation Commission in Augusta, Maine regulates campsite leasing, and you need to apply there for a building permit.

Thank you for your inquiry.

Yours truly,

*A.B. Coolidge*

A.B. Coolidge
Townsite Manager


ABC/de

cc: Melvin Blount

Some readers might take offense at the use of their full name. "Dear Mary Smith" or "Dear Joe Brown" can sound too distant and superior and condescending. The implied message in this kind of salutation seems to be, "I don't know you enough to use your first name only, but I don't respect you enough to use your last name only."

**Letter Text.** Begin your letter two spaces below the salutation. For letters that will fill most of the page, use single-spacing within paragraphs and double-spacing between. For short letters, double-space within paragraphs and triple-space between, to balance the page.

**Complimentary Close.** Place the complimentary close two spaces below the concluding paragraph, aligned with your heading. Any conventional closing (polite but not overly intimate) will do, but it should parallel the level of formality in your salutation, and should reflect your relationship to the reader. These possibilities are ranked in decreasing order of formality:

```
Respectfully,

Sincerely,

Cordially,

Best wishes,

Warmest regards,

Regards,

Best,
```

The complimentary close is followed by a comma.

**Signature.** Type your full name and title four spaces below your complimentary close. Sign in the space between.

```
Sincerely yours,


Martha S. Jones
Personnel Manager
```

Your signature indicates your approval of and responsibility for the letter (even if it's typed by a secretary). If you are writing as a representative of a company or group that bears legal responsibility for the correspondence, type the company's name in full caps two spaces below the complimentary close; place your typed name and title four spaces below the company name, and sign in between.

```
Yours truly,

HASBROUCK LABORATORIES


Lester Fong
Research Associate
```

## Specialized Parts

Some letters have one or more specialized parts.

**Attention Line.** Use an attention line when writing to an organization and when you want a specific person (whose name you don't know), title, or department to receive your letter.

```
ATTENTION: Research and Development Division

ATTENTION: Quality Assurance Supervisor
```

Drop two spaces below the inside address, and place the attention line either flush with the left margin or centered on the page. The attention line replaces your salutation, as on page 367.

**Subject Line.** The subject line forecasts what your letter is about, a good device for getting a busy reader's attention.

```
SUBJECT: Market Outlook for Absco Portable Computers
```

Place the subject line two spaces below the inside address (or attention line). Write the subject in caps or underline it.

**Typist's Initials.** If someone types your letter, your initials (in caps) and your typist's (in lowercase letters) should appear two spaces below the typed signature, abutting the left margin.

```
JJ/pl
```

Because they repeat the signature block, the writer's initials may be eliminated.

**Enclosure Notation.** When other documents accompany your letter, add one of these notations one space below the typist's initials, abutting the left margin:

```
Enclosure

Enclosures 2

Encl. 3
```

If the enclosures are important documents, name them:

```
Enclosures: 2 certified checks, 1 set of KBX plans.
```

**Distribution Notation.** If you will distribute copies of your letter to other readers, so indicate, one space below any enclosure notation.

```
cc: office file
    Melvin Blount

cc: S. Furlow
    B. Smith
```

**Postscript.** A postscript draws the reader's attention to a point you wish to emphasize. Do not use a postscript for a point you've forgotten in the letter. Rewrite instead. Do use a postscript to add a personal note.

```
P.S. You will love the way this model prints documents.
```

Place the postscript two spaces below any other notation, abutting your left margin. Use the postscript sparingly in professional communication. Many readers regard postscripts as sales-letter gimmicks.

## Appropriate Format

Use uniform margins, spacing, and indention: frame your letter with a 2½-inch top margin and side and bottom margins of 1 to 1¼ inches; single-space within paragraphs and double-space between; avoid hyphenating at the end of a line.

If your letter takes up more than one page, begin each additional page seven spaces from the top, with a line identifying the addressee, date, and page number:

```
Walter James, June 25, 1987, p. 2
```

Begin your text two spaces below this line. Place at least two lines of your paragraph at the bottom of page one, and at least two lines of your final text on page two.

On your 9½-by-4-inch envelope, center your reader's address, single-spaced. Use only accepted abbreviations. Place your own single-spaced address in the upper left corner.

## Accepted Letter Form

Although several letter forms are acceptable, and your company may have its own requirements, we discuss two common forms: modified block, with paragraphs not indented and the date, complimentary close, and signature aligned slightly right of page center (Figure 16.3); and block, with every line beginning at the left margin (Figure 16.4). The letters of inquiry and response in Figures 16.1 and 16.2 exemplify the modified block and block form, respectively.

**FIGURE 16.3** A Modified Block Letter

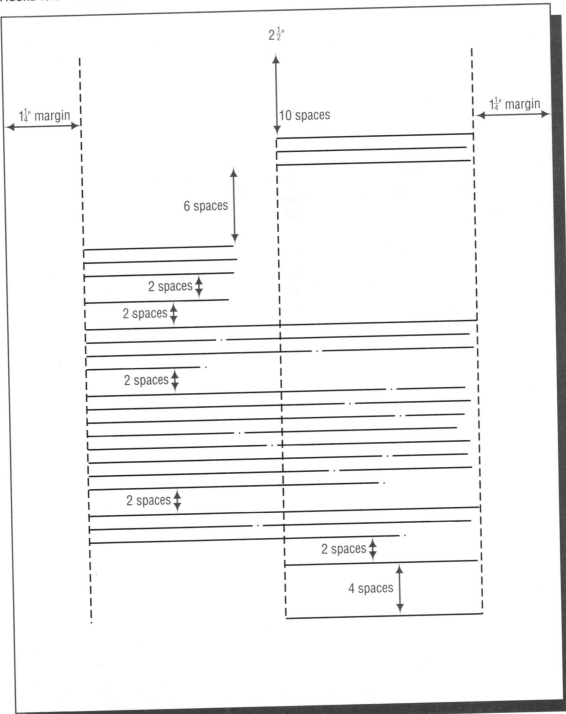

**FIGURE 16.4** A Block Letter

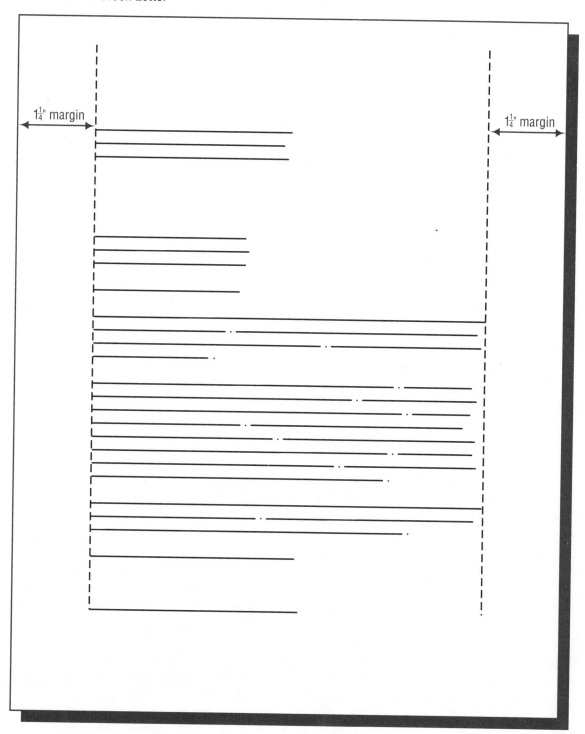

## Plain English

Workplace correspondence too often suffers from *letterese*, those tired, stuffy, and overblown phrases some writers think they need to sound important. Here is a typically overwritten closing sentence:

```
Humbly thanking you in anticipation of your kind cooperation,
I remain

                                        Faithfully yours,
```

Although no one speaks in this way, some writers lean on such heavy prose instead of writing in plain English: "We will appreciate your cooperation."

Here are a few of the many old standards that make letters seem unimaginative and boring:

| *Letterese* | *Translation into Plain English* |
| --- | --- |
| As per your request | As you requested |
| Contingent upon receipt of | As soon as we receive |
| Enclosed please find my résumé. | My résumé is enclosed. |
| I am desirous of | I want, I would like |
| At the earliest possible date | As early as possible |
| Please be advised that my new address is | My new address is |
| This writer | I |
| I humbly request that you consider | Please consider |
| In the immediate future | Soon |
| In accordance with your request | As you requested |
| Due to the fact that | Because |
| I wish to express my gratitude. | Thank you. |
| Pursuant to our agreement | As we agreed |
| In a timely manner | On time, promptly |

Be natural: write as you would speak in a classroom.

## "You" Perspective

In speaking face to face, you unconsciously modify your delivery as you read the listener's signals: a smile, a frown, a raised eyebrow, a nod. Even in a phone conversation a listener provides cues that signal approval, anger, and so on. In writing a letter, however, you face a blank page; you can too easily write only to please yourself, forgetting that a flesh-and-blood person will be reading your letter.

The "you" perspective affects your tone; by careful word choice, you show your readers respect. Put yourself in their place; consider how readers will respond to what you've just written. A letter creates a relationship between writer and reader, and, like the letter's appearance, the words themselves are another basis for that relationship. If you bury readers in letterese and clichés, they are bound to conclude that you care little about your writing task or about them.

Even a single word, carelessly chosen, can offend. In a letter complaining about the monitor on your new computer, you have the choice of saying, "Although the amber screen causes very little eyestrain, the character resolution is not sharp enough for lengthy word processing use" or "The monitor is lousy for word processing." Clearly, "lousy" is a poor choice because it insults the manufacturer or dealer, and fails to describe the problem (poor resolution).

Imagine how a parent felt after receiving a school letter calling her child a "fortunate deviate." Only after three angry phone calls to school officials did she learn that "fortunate deviate" is educational jargon for "bright child."

Put yourself in the reader's place in this example: imagine you manage the complaint department for a large mail-order company. Which of these versions requesting repair, replacement, or refund would you honor quickly?

1. I demand that you creeps send a replacement for this faulty calculator, and I only hope it won't be as big a piece of junk as the first! (the belligerent attitude)

2. This new calculator does not seem to work, and I wonder if you might kindly consider the possibility of sending me a replacement, if that is all right with you. Please accept my humble thanks in advance. (the pardon-me-for-living attitude)

3. I beg to advise you that I am appalled by the patent paucity of workmanship in this calculator and find it imperative that you refund the full price at the earliest possible date. (the phony-baloney attitude)

4. Please send me a replacement or a refund for this defective calculator as soon as possible. (the courteous and direct attitude)

Statement 4 would most likely achieve results. It is courteous and direct. The earlier versions express only the writer's need to sound off (except 2, in which the "you" perspective is carried to a grotesque extreme), and would probably end up in the reader's wastebasket.

## Clear Purpose

Like any effective writing, good letters do not just "happen." Each is the product of a deliberate process. Most effective letters are *rewritten*; words almost never tumble out for a perfect message on your first try. As you plan, write, and revise, answer these questions:

1. *What purpose do I wish to achieve?* (get a job, file a complaint, ask for advice or information, answer an inquiry, give instructions, request a favor, share good news, share bad news).

2. *What facts does my reader need?* (measurements, dates, costs, model numbers, enclosures, other details)

3. *To whom am I writing?* (Do you know the reader's name? When possible, write to a person, not a title.)

4. *What is my relationship to my reader?* (Is the reader a potential employer, an employee, a person doing a favor, a person whose products are disappointing, an acquaintance, an associate, a stranger?)

Answer these four questions *before* drafting the letter. After you have written a draft, answer the next three questions, which pertain to the *effect* of your letter. Will readers be encouraged to respond favorably?

5. *How will my reader react to what I've written?* (with anger, hostility, pleasure, confusion, fear, guilt, resistance, satisfaction)

6. *What impression of me will my reader get from this letter?* (intelligent, courteous, friendly, articulate, pretentious, illiterate, confident)

7. *Am I ready to sign my letter with confidence?* (Think about it.)

Don't mail your letter until you have answered each question to your full satisfaction. Revise as often as needed to achieve your purpose.

### Direct or Indirect Plan for Good or Bad News

The reaction you anticipate from your reader (question 5, above) should determine the organizational plan of your letter: either a *direct* or *indirect* plan. The direct plan puts the main point right in the first paragraph, followed by the explanation. Use the direct plan when you expect the reader to react with approval or when you want the reader to know immediately the point of your letter (say, in good-news, routine complaint, inquiry, or application letters—or other routine correspondence).

If you expect the reader to resist or to need persuading, consider an indirect plan: give the explanation *before* the main point (as in refusing a request, admitting a mistake, or requesting a pay raise). An indirect plan might make readers more tolerant of bad news or more receptive to your argument.

Anytime you consider using an indirect plan, think carefully about its *ethical* implications. Never try fooling the reader—or even creating an impression that you have something to hide.

### INQUIRY LETTERS

Inquiry letters may be solicited or unsolicited. You often write the first type as a consumer requesting information about an advertised product. Such letters are welcomed because the reader stands to benefit. You can be brief: "Please send me your brochure on . . ." or some such.

Other inquiries will be unsolicited, that is, not in response to an ad, but requesting information for a report or project. Here, you are asking your reader the favor of reading your letter, considering your request, collecting the information, and writing a response. Always apologize for any imposition, express appreciation, and state a reasonable request clearly and briefly (long, involved inquiries are likely to go unanswered). Begin with something more cordial and less abrupt than "I need information."

Before you can ask specific questions, do your homework. Don't expect the respondent to read your mind. A general request ("Please send me all your data on . . .") is likely to be ignored.

A typical inquiry situation: You are preparing an analytical report on the feasibility of harnessing solar energy for home heating in northern climates. You learn that a private, nonprofit research group has been experimenting in solar energy systems. After deciding to write for details, you plan and compose your inquiry.

In your introduction, tell the reader who you are and why you want information. Maintain the "you" perspective with an opening statement that sparks interest and goodwill.

In the body of your letter, write specific and clearly worded questions that are easy to understand and answer. If you have several questions, arrange them in a list. (Lists help readers organize their answers, increasing your chances of getting all the information you want.) Number each question, and separate it from the others, perhaps leaving space for responses right on the page. If you have more than five questions, consider placing them in an attached questionnaire.

Conclude by explaining how you plan to use the information and, if possible, how your reader might benefit. If you have not done so earlier, specify a date by which you need a response. Offer to send a copy of your finished report. Close with a statement of appreciation; it will encourage your reader to respond.

Your letter might resemble the one below (or Figure 16.1).

While gathering data on home solar heating, I encountered references (in *Scientific American* and elsewhere) to your group's pioneering work. Would you please allow me to benefit from your experience?

As a student at Evergreen College, I am preparing a report on the feasibility of solar energy as a major source of home heating in northern climates within a decade. Your answers to the questions below would help me complete my course project (April 15 deadline).

1. At this stage of development, have you found active or passive solar heating more practical?
2. Do you hope to surpass the 60 percent limit of heating needs supplied by the active system? If so, what efficiency do you expect to achieve, and how soon?
3. What is the estimated cost of building materials for your active system, per cubic foot of living space?
4. What metal do you use in collectors to obtain the highest thermal conductivity at the lowest maintenance costs?

Your answers, along with any recent findings you can share, will enrich a learning experience I will put into practice next summer by building my own solar-heated home. I would be glad to send you a copy of my report, along with the house plans I have designed. I will appreciate your help.

Sincerely,

Sometimes your questions will be too numerous or involved to be answered by letter or questionnaire. You might then request an informative interview (if the respondent is nearby). Karen Granger, our next writer, wanted a state representative's "opinions of the EPA's progress" in cleaning up local chemical contamination. Because she anticipated a complex answer, she requested an interview.

> As a technical writing student at Southeastern Massachusetts University, I am preparing a report analyzing the EPA's progress in cleaning up the PCBs in the New Bedford Harbor.
>
> Throughout my preliminary research, I have encountered your name again and again. Your dedicated work has had a definite influence on this situation, and I am hoping to benefit from your knowledge.
>
> I was surprised to learn that, although this contamination is considered the most serious anywhere, the EPA has still not moved beyond conducting studies. My own study questions the need for such extensive data gathering. Your opinion, as I can ascertain from *Standard Times* articles, is that the EPA is definitely moving too slowly.
>
> The EPA refutes that argument by asserting they simply do not yet have the information necessary to begin a clean-up operation.
>
> As both a writer and a New Bedford resident, I am very interested in hearing your opinions on the EPA's progress. Would you grant me an interview? With your permission, I will phone in a few days to make an appointment. This interview would be an invaluable aid to me.
>
> I would appreciate your assistance and would be glad to send you a copy of my completed report.

With your unsolicited inquiry, include a stamped, return address envelope.

## CLAIM LETTERS

Claim letters request adjustments for defective goods or poor services, or they complain about unfair treatment, or the like. Claims can be routine or arguable. Routine claims follow the direct plan, because the claim is backed by a contract, guarantee, or the company's reputation. Because they are open to interpretation, arguable claims call for an indirect plan.

### Routine Claims

In a routine claim, make your request or state the problem in your introductory paragraph; then explain in the body section. Close courteously, repeating the action you request. Do not thank the company in advance. If your claim is valid, a reputable firm will honor it.

Make the tone courteous and reasonable. Your goal is not to express dissatisfaction, but to achieve results: a refund, replacement, apology. Avoid a belligerent tone, as well as one that is apologetic or meek. Press your claim objectively yet firmly by explaining it clearly, and by stipulating the *reasonable* action that will satisfy you.

*Explain* the problem in enough detail for a reader to understand the basis for your claim. Explain that your new alarm clock never rings, instead of merely saying it's defective. Identify the faulty item clearly, giving serial and model numbers, and date and place of purchase. Then propose a fair adjustment. Conclude by expressing goodwill and confidence in the reader's integrity.

The next writer does not ask whether the ski manufacturer will honor his claim; he assumes it will. He asks directly how to return the skis for repair. He uses an attention line to direct his claim to the right department. The subject line, and its reemphasis in the first sentence, makes clear the nature of the claim.

Attention: Consumer Affairs Department

Subject: Delaminated Skis

This winter, my Tornado skis began to delaminate. I want to take advantage of your lifetime guarantee to have them relaminated.  
*States problem and action desired*

I bought the skis from the Ski House in Erving, Massachusetts, in November 1967. Although I no longer have the sales slip, I did register them with you. The registration number is P9965.  
*Provides necessary details*

I'm aware that you no longer make metal skis, but as I recall, your lifetime guarantee on the skis I bought was a major selling point. Only your company and Head were backing their skis so strongly.  
*Explains basis for claim*

Would you please let me know how to go about returning my delaminated skis for repair?  
*Ends courteously, specifically stating desired action*

Although this is a routine claim, twenty years separate the 1967 purchase from the 1987 claim. The claim was honored without question.

## Arguable Claims

When your direct request has been denied, ignored, or is in some way unusual, you must *persuade* the firm to grant your claim. Say your parked car is wrecked by a drunk driver, and the insurance company appraises the car at $1,500. But two months earlier, you had the engine rebuilt, a muffler system installed, and the front end repaired. By settling for the $1,500, you would lose $1,000. You write a claim letter, explaining your circumstances and requesting a fair adjustment.

Use the indirect plan for an arguable claim. Readers are more likely to respond favorably *after* reading your explanation. Begin your introduction with a neutral statement both parties can agree to–but which also serves as the basis for your request (e.g., "Customer goodwill is often an intangible quality, but a quality that brings tangible benefits.").

Once you've established initial agreement, explain and support your claim. Include enough information for a fair evaluation: date and place of purchase, order or policy number, dates of previous letters or calls, and background.

Conclude by requesting a *specific action* (a credit to your account, a replacement, a rebate). Ask confidently.

Notice how the writer below uses a tactful, reasonable tone and the indirect plan to achieve her goal.

**Begins with neutral statements both parties can agree to**

Your company has an established reputation as a reliable wholesaler of office supplies. And for eight years we have counted on that reliability. But a recent episode has left us annoyed and disappointed.

**Presents facts to support claim**

On November 29, 1989, we ordered (#675198) five cartons of Maxell 5¼" diskettes (#A74−866), eight cartons of Scotch 5¼" diskettes (#A74−892), and three cartons of Epson MX 70/80 ribbon cartridges (#A19−556).

**More support**

On December 5, the order arrived. But instead of double-sided, double-density Maxells ordered, we received single-sided Verbatim diskettes. Instead of Scotch 5¼" diskettes, we received 8" diskettes. And the Epson ribbons were blue, not the black we had ordered. We returned the order on the same day.

**Includes all relevant information**

Also on the 5th, we called John Fitzsimmons at your company to explain our problem. He promised delivery of a corrected order by the 12th. Finally, on the 22nd we did receive an order—the original incorrect one—with an attached note claiming that the packages had been water damaged while in our possession.

**Sticks to the facts— writer accuses no one**

Our warehouse manager insists the packages were in perfect condition when he gave them to the shipper. Because we had the packages only five hours, and had no rain on the 5th, we are certain the damage did not occur here.

**Requests a specific adjustment**

Responsibility for damages therefore rests with either the shipper or your warehouse staff. What bothers us is our outstanding bill from Hightone ($1,049.50) for the faulty shipment. We insist that the bill be canceled, and that we receive a corrected statement. Until this misunderstanding, our transactions with your company were excellent. We hope they can be again.

Hightone agreed that Office Systems had a valid claim, and adjusted the bill.

## RÉSUMÉS AND JOB APPLICATIONS

In today's job market, many applicants compete for few openings. Whether you are applying for your first professional job or changing careers in midlife, you have to wage an effective campaign to market your skills. Your résumé and letter of application must stand out among those of your competitors.

## Job Prospecting: The Preliminary Step

Begin your employment search with some prospecting: study the job market to identify realistically the careers and jobs for which you best qualify.

**Being Selective.**   Many new graduates make the mistake of applying for too broad a range of jobs, including some for which they have no real qualifications. Such a shotgun approach is ambitious, but can be discouraging. Focus on openings that suit you best. Begin by evaluating your skills, aptitudes, and goals:

- What skills have you acquired in schools, in jobs, or from hobbies and interests?
- What can you do well: write, draw, speak other languages, organize, lead, instruct, socialize, sell, solve problems, think creatively?
- Are you good at making decisions?
- Are you better at original thinking or at following directions?
- Are you looking for security, excitement, money, travel, power, prestige, or something else?

Besides helping focus your search, answers to these questions will be handy when you write your résumé.

If your goal is to join some major corporation, consult these library resources for information about prospective employers.

- *Moody's Industrial Index* or Standard and Poor's *Register of Corporations*, for data on plant locations, major subsidiaries, products, executive officers, and corporate assets.
- The *Business Index*, for articles about particular companies in publications such as the *Wall Street Journal, Fortune,* and *Forbes.*
- The yearly issue of *Fortune* Magazine that lists "The Fortune 500" companies.
- Annual reports, for data on a company's assets, innovations, recent performance, and prospects. (Many libraries collect annual reports in a separate file cabinet. Ask your librarian.)

Once you have a general direction, you can begin the actual job search.

**Launching Your Search.**   Launch your job campaign early. Don't wait for the job to come to you. Take the initiative by observing these suggestions.

1. Scan the Help Wanted section in Sunday newspapers. Here you will find descriptions, salaries, and qualifications for countless jobs.
2. Ask your reference librarian for occupational handbooks, government publications, newsletters, and job listings in your field.
3. Visit your college placement service; here, openings are posted, interviews are scheduled, and counselors give job-hunting advice.

4. Ask people in your field for an inside view and practical advice.

5. Sign up at your placement office for interviews with company representatives who visit the campus.

6. Seek the advice of faculty in your major who do outside consulting or who have worked in business, industry, or government.

7. Look for a summer job in your field; this experience may count as much as your education.

8. Establish contacts; don't be afraid to ask for advice. Make a list of names, addresses, and phone numbers of people willing to help.

9. Many professional organizations invite student memberships (at reduced fees). Such affiliations can generate excellent contacts, and they look good on the résumé. If you do join a professional organization, try to attend meetings of the local chapter.

By taking these steps well before your senior year, you may learn that specific courses make you more marketable.

If you're changing jobs or careers, employers will be more interested in what you've accomplished *since* college. Be prepared to show how your work experience is relevant to the new job. Capitalize on the contacts you've made over the years. Collect on favors owed.

Whether you're a beginner or a veteran, you might register with an employment agency. Of course, a fee is payable after you're hired, but many employers pay the fee. Ask about fee arrangements *before* you sign up.

Once you have a picture of where you fit into the job market, you will set out to answer the big question asked by all employers: *What do you have to offer?* Your answer must be a highly polished presentation of yourself, your education, work history, interests, and special skills—your résumé.

## The Résumé

The résumé summarizes your experience and qualifications. Written before your application letter, it provides background information to support your letter. In turn, the letter will emphasize specific parts of your résumé, and will discuss how your background is suited to that job. The résumé gets you the interview, not the job.

Employers generally spend less than sixty seconds scanning a résumé. They look for an obvious and persuasive answer to this question: *What can you do for us?* Employers are impressed by a résumé that

1. looks good (conservative, tasteful, and uncluttered, on quality paper)

2. reads easily (headings, typeface, spacing, and punctuation that provide clear signals)

3. provides information the employer needs for making an interviewing decision

Most employers discard immediately résumés that are mechanically flawed, cluttered, sketchy, or hard to follow. Don't leave readers guessing or annoyed; make you résumé perfect.

Organize your information within these categories:

- name and address
- job and career objectives
- education
- work experience
- personal data
- interests, activities, awards, and skills
- references

Select and organize your material to emphasize what you can offer. Don't just list *everything;* be selective. (We're talking about *communicating* instead of merely delivering information.) Don't abbreviate, because some readers may not know the referent. Use punctuation to clarify and emphasize, not to be "artsy." Try to limit your résumé to a single page, as most employers prefer.[1]

Begin your résumé well before you begin your job search. You will need that much time to do a first-class job. Your final version can be duplicated for various similar targets—but each new type of job requires a new résumé that is tailored to fit the advertised demands of that job.

*CAUTION:* Never *invent* credentials just to qualify for a job. Your résumé should make you look as good as the facts allow. But distorting the facts is both unethical and counterproductive. Companies routinely investigate claims made in a résumé. And people who have lied are fired.

**Name and Address.** Under the first heading, include your full name, mailing address, and phone number (many interview invitations and job offers are made by phone). If your school and summer address differ, include both, indicating dates on which you can be reached at each.

**Job and Career Objectives.** From your collected information (steps 1–9, listed earlier), you should have a clear idea of the *specific* jobs for which you *realistically* qualify.[2] Resist the impulse to be all things to all people. The key to a successful résumé is the image of *you* it projects—disciplined and purposeful, yet flexible. State your specific job and career goals:[3]

Intensive-care nursing in a teaching hospital, ultimately supervising and instructing.

---

[1] Of course, if you are changing jobs or careers, or if your résumé looks cramped, you might need a second page.

[2] Be prepared to have different statements of objectives for different résumés.

[3] To save space, you can omit your statement of career objectives from the résumé and include it in your letter instead.

Your statement of objectives should show a clear sense of purpose. *Do not* borrow a trite or long-winded statement from a placement brochure. If applying to a company with various branches, state your willingness to relocate.

**Educational Background.** If your education is more impressive than your work experience, place it first. Begin with your most recent school and work backward, listing degrees, diplomas, and schools *beyond* high school (unless prestige, program, or your achievement warrant its inclusion). List the courses that have directly prepared you for the job you seek. If your class rank and grade point average are in the upper 30 percent, list them. Include any schools attended or courses completed while you were in military service. If you financed part of all of your education by working, say so, indicating the percentage.

**Work Experience.** If you have solid experience, place it before your education. Beginning with your most recent job and working backward, list and clearly identify each job, giving dates and names of employers. Tell whether the job was full-time, part-time (hours weekly), or seasonal. Tell exactly what you did in each job, indicating promotions. If the job was major (and related to this one), describe it in detail; otherwise, describe it briefly. Include any military experience. If you have no real experience, show that you have potential by emphasizing your preparation, and by writing an enthusiastic letter.

Do not use complete sentences in your job descriptions; they take up room best left for other items. But do use action verbs throughout (*supervised, developed, built, taught, opened, managed, trained, solved, planned, directed*, and so on). Such verbs emphasize your vitality, and help you stand out.

**Personal Data.** An employer cannot legally discriminate on the basis of sex, religion, color, age, national origin, physical features or marital status. Therefore, you aren't required to provide this information or a photograph. But if you believe that any of this information could advance your prospects, by all means include it.

**Personal Interests, Activities, Awards, and Skills.** List hobbies, sports, and other pastimes; memberships in teams and organizations; offices held; and any recognition you have received. Include dates and types of volunteer work. Employers know that persons who seek well-rounded lives are likely to take an active interest in their jobs. Be selective in this section. List only items that show the qualities employers seek.

**References.** Your list of references names four or five people *who have agreed* to write strong, positive assessments. Often a reference letter is the key to getting an employer to want to meet you; choose your references carefully.

Select references who can speak with authority about your ability and character. Avoid members of your family and close friends not in your field. Choose instead among professors, previous employers, and community figures who know you well enough to write *concretely* on your behalf. In asking for a reference, keep these points in mind:

**1.** *A mediocre letter of reference is more damaging than no letter at all.* Don't merely ask, "Could you please act as one of my references?" This question leaves the person little chance to say no. He or she might not know you well or may be unimpressed by your work but, instead of refusing, might write a watery letter that will do more harm than good. Instead, make an explicit request: "Do you feel you know me and my work well enough to write me a strong letter? If so, would you act as one of my references?" This second version gives people the option to decline gracefully or to promise a positive letter.

**2.** *Letters are time-consuming.* References have no time to write individual letters to every prospective employer. Ask for one letter only, with no salutation. Your reference keeps a copy; you keep the original for your personal dossier (so that you can reproduce it as necessary); and a copy goes to the placement office for your placement dossier. Because the law permits you to read all material in your dossier, this arrangement provides you with your own copy of your credentials.[4] (The dossier is discussed later in this chapter.)

If the people you select as references live far away, you may wish to make your request by letter, like this one:

> From September 1985 to August 1987, I worked at Teo's Restaurant as waiter, cashier, and then assistant manager. Because I enjoyed my work, I decided to study for a career in the hospitality field.
>
> In three months I will graduate from San Jose City College with an A.A. degree in Hotel and Restaurant Management. Next month I begin my job search. Do you remember me and my work enough to write me a strong letter of recommendation? If so, would you be one of my references?
>
> To save your time, please omit the salutation from the letter. If you could send me the original, I will forward a copy to my college placement office.
>
> To update you on my recent activities, I've enclosed my résumé. I appreciate your help and support.

Opinion is divided about whether names and addresses of references should be in a résumé. If saving space is important, simply state, "References available on request," keeping your résumé only one page long, but if your other résumé items take up more than one page, you probably should include names and addresses of references. (An employer might recognize a name, and thus notice *your* name along the crowd of applicants.) If you are changing careers, a full listing of references is especially important.

---

[4]Under some circumstances you may—and may wish to—waive the right to examine your recommendations. Some applicants, especially those applying to professional schools as in medicine and law, do waive this right. They do so in concession, one supposes, to a general feeling that a letter writer who is assured of confidentiality is more likely to provide a balanced, objective, and reliable assessment of a candidate. In your own case, you might seek the advice of your major adviser or a career counselor.

## Composing the Résumé

With data collected and references lined up, you are ready to compose your actual résumé. Imagine you're a twenty-four-year-old student about to graduate from a community college with an A.A. degree in Hotel and Restaurant Management. Before college, you worked at related jobs for more than three years. You now seek a junior management position while you continue your education part-time. You have spent two weeks compiling information for your résumé, and obtaining commitments from four references. Figure 16.5 shows your résumé. Notice that this résumé mentions nothing about salary. Wait until this matter comes up in your interview, or later. The information, specific but concise, describes what you have to offer, and can be read in a moment. (For a résumé composed in search of summer employment, see Figure 16.6.)

As a final check, look at your résumé as a personnel director might, and analyze your presentation for weaknesses. Here is Personnel Director Mary Smith's assessment in a memo to colleagues.

May 10, 1986

TO:       Hiring Committee
FROM:     Mary Smith, Personnel Director
SUBJECT:  Follow-up on James Purdy's application (copy enclosed)

I recommend pursuing James Purdy's candidacy. This applicant shows strong purpose and responsible planning. His background provides strong support for his stated plans.

The fact that he financed his own education yet achieved a high cumulative average indicates both diligence and intelligence. His courses, along with experience in sales, food service, and hospitality, suggest his career choice is based on sound knowledge of the hospitality field and an obvious talent for direct customer contact.

His history of job promotions indicates that his work has impressed employers repeatedly. His skiing ability and interest in cooking and sailing make him a strong candidate for one of our northern resorts. His activities and awards suggest he works well with others (including youngsters), is respected by his peers, and has leadership qualities. Finally, his ability to speak French could be an asset to our customer relations division.

Overall, James Purdy seems a well-rounded person who knows what he wants and who can offer youth, enthusiasm, and experience. He promises to be a responsible employee who will continue to improve personally and professionally.

When fully satisfied with your résumé, consider having your prototype printed. The prototype should be attractive, neat, and free of mechanical or grammar errors. A perfect copy is your responsibility. *Never* send out carbon, thermofax, or mimeographed copies. If you are applying for various types of jobs, have your résumé prepared on a word processor with a laser printer. This way, you can adapt each résumé to the particular job.

**FIGURE 16.5** A Résumé for an Entry-level Candidate

**James David Purdy**
203 Elmwood Avenue
San Jose, California 95139
Tel.: 214–316–2419

**Career
Objective**

Customer relations for a hospitality chain, eventually leading to market management.

**Education**

1987–1989

*San Jose City College, San Jose, California*
Associate of Arts Degree in Hotel/Restaurant Management, June 1989. Grade point average: 3.25 of a possible 4.00. All college expenses financed by scholarship and part-time job (20 hours weekly).

**Employment**

1987–1989

*Peek-a-Boo Lodge, San Jose, California*
Began as desk clerk and am now desk manager (part-time) of this 200-unit resort. Responsible for scheduling custodial and room service staff, convention planning, and customer relations.

1985–1987

*Teo's Restaurant, Pensacola, Florida*
Beginning as waiter, advanced to cashier and finally to assistant manager. Responsible for weekly payroll, banquet arrangements, and supervising dining room and lounge staff.

1984–1985

*Encyclopaedia Britannica, Inc., San Jose, California*
Sales representative (part-time). Received top bonus twice.

1983–1984

*White's Family Inn, San Luis Obispo, California*
Worked as bus boy, then waiter (part-time).

**Personal**

*Awards*
Captain of basketball team, 1988; Lion's Club Scholarship, 1987.

*Special Skills*
Speak French fluently; expert skier.

*Activities*
High school basketball and track teams (3 years); college student senate (2 years); Innkeepers' Club—prepared and served monthly dinners at the college (2 years).

*Interests*
Skiing, cooking, sailing, oil painting, and backpacking.

**References**

Available from Placement Office, San Jose City College, San Jose, California 90462

A final suggestion: avoid using a résumé-preparing service. Although professionals can use your raw materials to produce an impressive résumé, employers often recognize the source by its style, and could conclude that you are incapable of communicating on your own.

### The Job Application Letter

**Your Image.** Your job application letter is one of the most important documents you ever will write. Although it elaborates on your résumé, you must emphasize personal qualities and qualifications in a convincing way. Your résumé merely presents raw facts; your application letter relates these facts to the company to which you are applying. The tone and insight you bring to your discussion suggest a good deal about who you are. The letter is your chance to explain how you see yourself fitting into the organization, to interpret your résumé, and to show how valuable you will be.

**Targets.** Never send a photocopied letter. You can base different letters on one model—with appropriate changes—but prepare each letter fresh.

The immediate purpose of your letter is to secure an interview. The letter itself should make the reader want to meet you. Make your statements engaging, precise, and *original*. Borrowed phrases from textbook examples and "letterese" will not do the job.

Sometimes you will apply for jobs advertised in print or by word of mouth (solicited applications). At other times you will write prospecting letters to organizations that have not advertised but might need someone like you (unsolicited applications). Either letter should be tailored to the situation.

**The Solicited Letter.** Imagine you are James Purdy. In *Innkeeper's Monthly*, you read this advertisement and decide to apply:

**Resort Management Openings**

Liberty International, Inc., is accepting applications for several junior management positions at our new Lake Geneva resort. Applicants must have three years of practical experience, along with formal training in all areas of hotel/restaurant management. Please apply by June 1, 1989, to

> Elmer Borden
> Personnel Director
> Liberty International, Inc.
> Lansdowne, Pennsylvania 24135

Now plan and compose your letter.

*Introduction.* Create a confident tone by directly stating your reason for writing. Name the job, and remember you are talking *to* someone; use the pronoun "you" instead of awkward or impersonal constructions such as "One can see from the enclosed résumé. . . ." If you can, establish a connection by mention-

FIGURE 16.6  A Résumé for an Summer-internship Candidate

**Karen P. Granger**
82 Mountain Street
New Bedford, MA 02740
Telephone (617) 864–9318

| | |
|---|---|
| **Objective** | A summer internship documenting microcomputer software. |
| **Education** | Attending Southeastern Massachusetts University, North Dartmouth, MA 02747; will receive B.A. degree in January 1989. Major: English/ Writing. Concentration in computer science. Dean's List, five semesters. GPA: 3.54. Class rank 110 of 1792. |
| **Experience** | |
| Intern Technical Writer | **Conway Communications, Inc.,** 39 Wall Street, Marlboro, MA Learned local area network (LAN) technology and Conway's product line. Wrote and tested five hardware upgrade manuals. Produced a hardware installation/maintenance manual from another writer's outline. Updated documentation. Edited other writers' work. Specified and approved all illustrations for my projects. Summers 1986, 1987. |
| Writing Tutor | **Writing/Reading Center,** SMU Tutored writing and word processing for individuals and groups. Edited WRC student newsletter. Trained new tutors. Co-wrote and acted in a video about the WRC. Fall 1985–present. |
| Reporter and Editor | **Crimson Courier,** New Bedford High School Wrote features, news, and a humorous column titled *The Lone Granger.* As Editor, organized staff meetings, generated story ideas, edited articles, and was responsible for front page layout and pasteup. Fall 1982–Spring 1984. |
| **Computer Skills** | Familiar with IBM PC, Apple IIe, and Macintosh microcomputers; DEC 20 mainframe; VAX 11/780 minicomputer; HBJ Writer, Bankstreet Writer, MacWrite, and MASS11 software; and LOGO, Pascal, and BASIC languages. |
| **Achievements** | Two writing samples published in Dr. John M. Lannon's *Technical Writing,* 4th ed. (Boston: Little, Brown, 1987); Massachusetts State Honors Scholarship, 1984–1987; Advanced Placement Literature/ Composition (awarded six credits). |
| **Activities** | Student member, Society for Technical Communication, 1986; student representative, College Curriculum Committee, 1986; SMU Literary Society, 1986–1988. |
| **References** | Available from SMU Placement Office. |

ing a mutual acquaintance; say, you learn that your nutrition professor, H. V. Garlid, is a former colleague of Elmer Borden; mentioning his name—with permission—will call attention to your letter. Finally, after referring your reader to your enclosed résumé, discuss your qualifications.

*Body.* Concentrate on the experience, skills, and aptitudes you can bring to *this* job. Follow these suggestions:

- Don't come across as a jack-of-all-trades. Relate your qualifications specifically to this job.
- Avoid flattery ("I am greatly impressed by your remarkable company").
- Be specific. Replace "much experience," "many courses," or "increased sales" with "three years of experience," "five courses," or "a 35 percent increase in sales between June and October 1983."
- Support your claims with *evidence*, and show how your qualifications will benefit this employer. Instead of saying, "I have leadership skills," say, "I was student senate president during my senior year and captain of the lacrosse team."
- Create a dynamic tone by using *active* voice and action verbs:

  **Weak**   Management responsibilities were steadily given to me.

  **Strong**   I steadily assumed management responsibilities.

- Trim the fat from your sentences:

  **Flabby**   I have always been a person who enjoys a challenge.

  **Lean**   I enjoy a challenge.

- Express self-confidence:

  **Unsure**   It is my opinion that I have the potential to become a successful manager because . . .

  **Confident**   I will be a successful manager because . . .

- Never be vague:

  **Vague**   I am familiar with the 1022 interactive database management system, and RUNOFF, the text processing system.

  **Definite**   As a lab grader for one semester, I kept grading records on the 1022 database management system, and composed lab procedures on the RUNOFF text processing system.

- Avoid letterese. Write in plain English.

- Show enthusiasm. An enthusiastic attitude can be as important as your background.

*Conclusion.* Restate your interest and emphasize your willingness to retrain or relocate (if necessary). If your reader is nearby, request an interview; otherwise, request a phone call, stating times you can be reached. Leave your reader with the impression that you are worth knowing.

**Revision.** *Never* settle for a first draft—or a second or third! This letter is your model for letters serving in varied circumstances. Make it perfect.

After several revisions, James Purdy finally signed the letter shown below.

203 Elmwood Avenue
San Jose, California 10462
April 22, 1989

Mr. Elmer Borden
Personnel Director
Liberty International, Inc.
Lansdowne, Pennsylvania 24135

Dear Mr. Borden:

Please consider my application for a junior management position at your Lake Geneva resort. I will graduate from San Jose City College on May 30 with an Associate of Arts degree in Hotel/Restaurant Management. Dr. H. V. Garlid, my nutrition professor, described his experience as a consultant for Liberty International, and encouraged me to apply.

For two years I worked as a part-time desk clerk, and I am now the desk manager at a 200-unit resort. This experience, along with customer relations work described in my résumé, has given me a clear and practical understanding of customers' needs and expectations.

As an amateur chef, I know of the effort, attention, and patience required to prepare fine food. Moreover, my skiing and sailing background might well be assets to your resort's recreation program.

I have confidence in my hospitality management skills. My experience and education have prepared me to work well with others and to respond creatively to changes, crises, and added responsibilities.

If my background meets your needs, please phone me any weekday after 4 p.m. at 214-555-2419.

Sincerely,


James D. Purdy

Enclosure


Purdy wisely emphasizes practical experience because his background is varied and impressive. An applicant with less experience would emphasize education instead, discussing courses and activities.

As an additional example, here is the letter composed by Karen Granger in her quest for a summer internship.

Dear Mr. White:

I read in Internships 1989 that your company offers a summer documentation internship. Because of my education and previous technical writing employment, I am highly interested in such a position.

In January 1989, I will graduate from Southeastern Massachusetts University with a B.A. in English/Writing. I have prepared specifically for a computer documentation career by taking computer science, mathematics, and technical writing courses.

In one writing course, the Computer Documentation Seminar, I wrote three software manuals. One manual uses a tutorial to introduce beginners to the Apple Macintosh and MacWrite. The other manuals describe two IBM PC applications that arrived at SMU with no documentation.

The enclosed résumé describes my work as the intern technical writer with Conway Communications, Inc., for two summers. I learned local area networking (LAN) by documenting Conway's LAN hardware and software. I was responsible for several projects simultaneously and spent much of my time talking with engineers and testing procedures. If you would like samples of my writing, please let me know.

Although Conway has invited me to return next summer and to work full-time after graduation, I would like more varied experience before committing myself to permanent employment. I know I could make a positive contribution to Birchwood Group, Inc. May I telephone you next week to arrange a meeting?

**The Unsolicited Letter.**   Ambitious job seekers do not limit their search to advertised openings. The unsolicited, or "prospecting," letter is a good way to uncover other possibilities. Such letters have advantages and disadvantages.

*Disadvantages.*   The unsolicited approach has two drawbacks: (1) you might waste time writing to organizations that have no openings; and (2) because you don't know what the opening is—even if there is one—you cannot tailor your letter to the specific requirements.

*Advantages.*   This cold-canvassing approach does have one big advantage: for an advertised opening you will compete with legions of qualified applicants, whereas your unsolicited letter might arrive just when an opening has materialized. If it does, your application will receive immediate attention and you just might get the job. Even when there is no immediate opening, most companies welcome unsolicited letters, and may keep an impressive application on file until an opening does occur. Or the application may be passed along to a company that has an opening.

There are often good reasons for going beyond the Help Wanted columns. Unsolicited letters generally are a sound investment if your targets are well chosen and your expectations are realistic.

*Reader Interest.* Because your unsolicited letter is unexpected, attract the reader's attention immediately. Don't begin: "I am writing to inquire about the possibility of obtaining a position with your company." By now, your reader is asleep. If you can't establish a connection through a mutual acquaintance, use a forceful opening:

> Does your hotel chain have a place for a junior manager with a college degree in hospitality management, a proven commitment to quality service, and customer relations experience that extends far beyond mere textbook learning? If so, please consider my application for a position.

Address your letter to the person most likely in charge of hiring. (Consult the business directories listed on page 103 for names of company officers.)

**The Prototype.** Many of your letters, solicited or unsolicited, can be versions of one model, or prototype. Your prototype must therefore present you in the best possible light.

Take plenty of time to compose the model letter and résumé. Employers will regard the quality of your application as an indication of the quality of work you will do. Businesses spend much money and time projecting favorable images. The image you project, in turn, must measure up to their standards.

## SUPPORTING YOUR APPLICATION

### Your Dossier

Your dossier is a folder containing your credentials: college transcript, recommendation letters, and any other items (such as a notice of scholarship award or commendation letter) that document your achievements. In your letter and résumé, you talk about yourself; in your dossier, others talk about you. An employer impressed by what you say about yourself will want to read what others think, and will request your dossier. By collecting recommendations in one folder, you spare your references from writing the same letter over and over.

Your college placement office will keep your dossier on file and send copies to employers who request them. Always keep your own copy as well. Then, if an employer requests your dossier, you can make a photocopy and mail it, advising your reader that the placement office copy is on the way. This step is not needless repetition! Most employers establish a specific timetable for (1) advertising an opening, (2) reading letters and résumés, (3) requesting and reviewing dossiers, (4) interviewing, and (5) making an offer. Obviously, if your letter and résumé do not arrive until screening is at step 3, you are out of luck. The same is true if your dossier arrives when screening is at the end of step 4. Timing is crucial. Too often, dossier requests from employers sit and gather dust in some "incoming" box in the placement office. Weeks can pass before your dossier is mailed. The only loser is you.

## Interviews

An employer impressed by your credentials will arrange to have an interview. You are a finalist. But now you must show the interviewer(s) you are as impressive in person as you seem on paper—and the best person for the job.

You might meet with one interviewer, a group, or several groups in succession. You might be interviewed alone or with several candidates at once. Interviews can last one hour or less, a full day, or even several consecutive days. The character of the interview can range from a pleasant, informal chat to grueling quiz sessions. Some interviewers may antagonize you deliberately to observe your reaction ("Whatever made you imagine that you could fill this position?"). Take as many interviews as you can. Your confidence and competence will increase with each.

Prepare for the interview by learning about the company (its products or services, history, prospects, branch locations) in trade journals and industrial indexes. If time permits, request company literature and annual reports. Prepare specific answers to the obvious questions:

- *Why do you wish to work here?*
- *What do you know about our company?*
- *What do you see as your biggest weakness? Your biggest strength?*
- *What would you like to be doing in ten years?*

Plan informative and direct answers to questions about your background, training, experience, and salary requirements. Prepare your own list of questions about the job and the organization; you will be invited to ask questions, and the questions you ask can be as revealing as the answers you give.

If you've prepared, your interview should be a chance for you and the employer to learn much about each other. The purpose of the interview is to confirm the impressions an employer has from your application. Be yourself. You have specific skills and a unique personality to offer. Busy people are taking the time to speak with you because they recognize your worth.

Know the exact time and location of the interview. Come dressed as if you already work for the company. Maintain eye contact most of the time; if you stare at your shoes, the interviewer will not be impressed. Relax in your chair, but don't slouch. Don't smoke, even if invited. Don't pretend to know more than you do; if you can't answer a question, say so. Avoid abrupt yes or no answers, as well as life stories. Make your answers concrete but to the point.

*Don't be afraid to allow silence.* An interviewer simply may *stop* talking, just to observe your reaction to silence. If you really have nothing more to say or ask, don't feel compelled to speak; caught off-guard, you're likely to ramble. Let the interviewer make the next move.

When your interviewer hints that the meeting is ending (perhaps by checking a wristwatch), take the cue. Restate your interest; ask when you can expect further word; thank the interviewer; and leave.

## The Follow-up Letter

Within a few days after your interview, jog the employer's memory with a letter restating your interest. Here is Purdy's follow-up to his interview with Elmer Borden.

> Thank you again for your hospitality during my visit to your Lansdowne offices. The trip and scenery were delightful.
>
> After meeting you and your colleagues and touring the resort, I know I could be a productive staff member.

## The Letter of Acceptance

If all goes well, you will receive a job offer by phone or letter. If by phone, request a written offer, and respond with a formal letter of acceptance. This letter may serve as part of your contact; be sure to spell out the terms you are accepting. Here is Purdy's letter of acceptance.

> I am happy to accept your offer of a position as assistant recreation supervisor at Liberty International's Lake Geneva Resort, with a starting salary of $28,500.
>
> As you requested, I will phone Ms. Druid in your personnel office for instructions on reporting date, physical exam, and employee orientation.
>
> I look forward to a long and satisfying career with Liberty International.

## The Letter of Refusal

You may find yourself in the enviable position of having to refuse offers. Even if you refuse by phone, write a prompt and cordial letter of refusal, explaining your reasons, and leaving the door open for future possibilities. You may find later that you dislike the job you accepted, and wish to explore old contacts. Here is how Purdy handled a refusal.

> Although I was impressed by the challenge and efficiency of your company, I must decline your offer of a position as assistant desk manager of your London hotel. I have taken a position with Liberty International because it will allow me to work full-time to complete the requirements for my B.S. degree in hospitality management.
>
> If any later openings should materialize, however, I would again appreciate your considering me as a candidate.
>
> Thank you for your interest and courtesy.

## REVISION CHECKLIST FOR LETTERS

Use this checklist to refine the content, arrangement, and style of your letters. (Numbers in parentheses refer to the first page of discussion.)

### Content

☐ Is your letter addressed to a specifically named person? (354)

☐ Does the letter have all the standard parts? (354)

☐ Does it have all needed specialized parts? (358)

☐ Have you given the reader all necessary information? (363)

☐ Have you identified the name and position of your reader? (354)

### Arrangement

☐ Does the introduction immediately engage the reader and lead naturally to the body? (353)

☐ Are transitions between letter parts clear and logical? (518)

☐ Does the conclusion encourage the reader to act? (353)

☐ Is the format correct? (359)

☐ Does the letter follow an accepted form? (359)

### Style

☐ Is the letter in conversational language (free of letterese)? (362)

☐ Does the letter reflect a "you" perspective? (362)

☐ Does the tone reflect your relationship with your reader? (362)

☐ Is the reader likely to react favorably to this letter? (364)

☐ Are all sentences clear, concise, and fluent? (199)

☐ Is the letter in correct English? (Appendix A)

☐ Does the letter's appearance enhance the writer's image?

## EXERCISES

**1.** Bring to class a copy of a business letter addressed to you or a friend. Compare letters. Choose the most and least effective.

**2.** Write and mail an unsolicited letter of inquiry about the topic you are investigating for an analytical report or research assignment. In your letter you might request brochures, pamphlets, or other informative literature, or you might ask specific questions. Submit to your instructor a copy of your letter and the response.

**3.** *a.* As a student in a state college, you learn that your governor and legislature have cut next year's operating budget for all state colleges by 20 percent. This cut will cause the firing of many young and popular faculty members, drastic reduction in student admissions, reduction in financial aid, cancellation of new programs, and deterioration in college morale and quality of instruction. Write a complaint letter to your governor or representative, expressing your strong disapproval and justifying a major adjustment in the proposed budget.

b. Write a complaint letter to a politician about some issue affecting your school or community.

c. Write a complaint letter to an appropriate school official to recommend action on a campus problem.

**4.** Write a 500- to 700-word essay explaining your reasons for applying to a specific college for transfer or for graduate or professional school admission. Be sure your essay covers two general areas: (1) what you can bring to this school by way of attitude, background, and talent; and (2) what you expect to gain from this school in personal and professional growth.

**5.** Write a letter applying for a part-time or summer job, in response to a specific ad. Choose an organization that can offer you experience related to your career goal. Identify the exact hours and calendar period during which you are free to work.

**6.** a. Identify the job you hope to have in a few years. Using newspaper ads, library (see your reference librarian for assistance), placement office, and personal sources, write your own full description of the job: duties, responsibilities, work hours, salary range, requirements for promotion, highest promotion possible, unemployment rate in the field, employment outlook for the next decade, need for further education (advanced degrees, special training), employment rate in terms of geography, optimum age bracket (as in football, does one fade around thirty-five?).

b. Using these sources, construct a profile of the ideal employee for this job. If you were the personnel director screening applicants, what specific qualifications would you require (education, experience, age, physical ability, appearance, special skills, personality traits, attitude, outside interests, and so on)? Try to locate an actual newspaper ad describing job responsibilities and qualifications. Or assume you're a personnel officer, and compose an ad for the job.

c. Assess your own credentials against the ideal employee profile. Review the plans you have made to prepare for this job: specific courses, special training, work experience. Assume that you have completed your preparation. How do you measure up to the requirements in (b)? Are your goals realistic? If not, why not? What alternative plan should you formulate?

d. Using your list from part (c) as raw material, construct your résumé. Revise until it is perfect.

e. Write a letter of application for the job described in part (a). Revise until you feel good about signing the letter.

f. Write a follow-up letter thanking your fictional employer for your interview and restating your interest.

g. Write a letter accepting a job offer from this same employer.

h. Write a letter graciously declining this job offer.

i. Submit each of these items, in order, to your instructor.

*Note:* Use the sample letters in this chapter for guidance, but don't borrow specific expressions.

**7.** Most of these sentences need to be overhauled before being included in a letter. Identify the weakness in each statement, and revise as needed.

  *a*. Pursuant to your ad, I am writing to apply for the internship.

  *b*. I need all the information you have about methane-powered engines.

  *c*. You idiots have sent me a faulty disk drive!

  *d*. It is imperative that you let me know of your decision by January 15.

  *e*. You are bound to be impressed by my credentials.

  *f*. I could do wonders for your company.

  *g*. I humbly request your kind consideration of my application for the position of junior engineer.

  *h*. If you are looking for a winner, your search is over!

  *i*. I have become cognizant of your experiments and wish to ask your advice about the following procedure.

  *j*. You will find the following instructions easy enough for an ape to follow.

  *k*. I would love to work for your wonderful company.

  *l*. As per your request I am sending the county map.

  *m*. I am in hopes that you will call soon.

  *n*. We beg to differ with your interpretation of this leasing clause.

  *o*. I am impressed by the high salaries paid by your company.

**8.** Write a complaint letter about a problem you've had with goods or service. State your case clearly and objectively, and request a specific adjustment.

**9.** Last October 25, you bought an internal frame backpack (the Mountaineer, Item #51–6131) from Maple Mountain Outfitters, a direct-mail outfitter specializing in expedition equipment. The pack cost $159.95, plus $5.65 for shipping. You've been pleased with other gear bought from Maple Mountain (their Staysnug down sleeping bag and two-person lightweight tent), but the backpack is worthless. Because you take good care of your equipment, you were looking forward to years of use from the pack. But on your third trip, a two-week hike along the Sawtooths in Idaho, you had lots of problems. The zippers on two of the outside pockets and on the main compartment came unstitched; the ice-ax loop ripped off the first time your ax snagged, and worst of all, on your fifth day out, one of the carrying straps ripped out—an unpleasant experience while lugging 70 pounds of gear up a 60-degree slope. You endured the trip by promising yourself revenge on whoever constructed the pack. That shoddy pack turned a beautiful trip into a monumental headache. With the trip over, you have decided you would be satisfied with a new pack—definitely not the same model—and an apology. You will accept a Maple Mountain Expedition Pack as a replacement. It costs $40 more, but you figure the company should pay the difference to make up for the problems you've had. Write a diplomatic letter, explaining the situation and requesting the replacement. The company's address is 10 Mountain Way, Jim's Creek, WA 83190.

**10.** As purchasing agent for Greyfox Electronics, you decided to save money by ordering invoice forms, letterhead, and envelopes from a new vendor, Express Forms. The shipment arrived, but the company's name is misspelled: Greyfox is spelled Greyfor. Your purchase orders have Greyfox Electronics prominently displayed, and so the vendor is responsible. The problem is compounded because the shipment arrived two weeks late, and you're almost out of invoices. You need corrected invoices within seven working days—or your great money-saving idea may turn into a disaster, which certainly won't help your chances for promotion. Write Express Forms, explaining the problem and the urgency. Mention that repeat business depends on their meeting the deadline. Express Forms, 82 Ranger Road, Smithville, AR 51001.

**11.** Two weeks ago, Jane Swan, your third-shift supervisor, called to say the plant had no heat. You immediately called Acme Heating and requested emergency service. Acme sent a repairperson, and by 3:00 A.M., the heating system was running. A day later, the system again shut down; you called Acme, and again they got the system running. When the system shut down a third time, you called Harry Foreman at Acme to find out why they hadn't repaired the system the first or second time. He explained that the problem was excessive condensation in the fuel tank. As a result, water was getting into the fuel lines, causing the system to fail. The only solution was to drain the tank to get rid of the water. You approved this plan. Today you received Acme's bill for $648. The breakdown shows $225 for three emergency calls. Because the same problem occurred three times, and Acme didn't explain until the final occurrence, you don't believe you should have to pay for two of those calls. Besides, you're one of Acme's biggest customers, buying at least 12,000 gallons of fuel oil each winter. Request an adjustment on your bill. The address: Box 229, Hillcrest Ave., Warwick, RI 05131.

## TEAM PROJECT

Assume that this advertisement has appeared in your school newspaper:

**STUDENT CONSULTANT WANTED**

The office of Dean of Students invites applications for the position of student consultant to the Dean for the coming academic year. Duties will include (1) meeting with fellow students as individuals and groups to discuss issues, opinions, questions, complaints, and recommendations about all areas of college policy, (2) presenting oral and written reports to the Dean of Students regularly, and (3) attending college planning sessions as student spokesperson. Time commitment: 15 hours weekly both semesters. Salary: $4,000.

Candidates should be full-time students with at least one year of student experience at this college. The ideal applicant will be skilled in report writing and oral communication, will work well in groups, and will demonstrate a firm commitment to our college. Application deadline: May 15.

*a*. Compose a résumé and a letter of application for this position.

*b*. Divide your class into screening, interview, and hiring committees.

*c*. Exchange your group's letters and résumés with those of another group.

*d*. As an individual committee member, read and evaluate each of the applications you have received. Rank each application privately, on paper, according to the criteria in this chapter, before discussing them with your group. *Note:* While screening applicants, you will be competing for selection by another committee, which is reviewing your own application.

*e*. As a committee, select the three strongest applications, and invite the applicants for interviews. Interview each selected applicant for ten minutes, after you have prepared a list of standardized questions.

*f*. On the basis of these interviews, rank your preferences privately, on paper, giving specific reasons for your final choice.

*g*. Compare your conclusions with those of your colleagues and choose the winning candidate.

*h*. As a committee, compose a memo to your instructor, giving specific reasons for your final recommendations.

# Memos and
# Short Reports

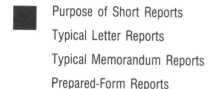

Purpose of Short Reports

Typical Letter Reports

Typical Memorandum Reports

Prepared-Form Reports

**R**eports present ideas and facts to decision makers. In the professional world, superiors and colleagues rely on short reports as a basis for making *informed* decisions on matters as diverse as the most comfortable office chairs to buy or the best recruit to hire for management training.

## PURPOSE OF SHORT REPORTS

The short report communicates concisely in one of these formats: the letter report, the memorandum, or the prepared-form report.

On the job, personnel must communicate speedily and precisely. Your success may well depend on how rapidly and clearly you share information. Here are some of the kinds of reports you might write on any workday:

- a request for assistance on a project
- a requisition for parts and equipment
- a proposal outlining reasons and suggesting plans for a new project
- a set of instructions
- a cost estimate for planning, materials, and labor on a new project
- a report of your progress on a specific assignment
- an hourly or daily account of your work activities
- a report of your inspection of a site, item, or process
- a statement of reasons for equipment or project failure
- a record of a meeting
- a report of your survey to select the best prices or products

The length of these reports may vary from a few lines to several pages.

Whether you report your data in a letter, memo, or on a prepared form, will depend on the situation and your company's reporting policies. Quite possibly, the same information you cast in a memo to a superior will be incorporated in a letter to a client.

## TYPICAL LETTER REPORTS

Letter reports typically go to people outside the organization. For the introductory and closing elements of your letter, follow the standard letter format discussed in Chapter 16. The two format additions in a letter report are (1) a subject heading, placed two spaces below the salutation, and (2) other headings, as needed, to segment your letter into specific areas (see Frymore's letter report, below). Be sure your letter report keeps the personal "you" perspective.

Letter reports serve many purposes. Here we show letter reports designed for two broad purposes: to *inform* and to *recommend*.

### The Informational Report

The informational letter report typically provides readers with information about a firm's products, services, or operations. The following report is addressed to Wilmington's fire marshal, Mary Pine. While inspecting Frymore's plant, the marshal found its fire evacuation procedures haphazard, and gave the firm one month to establish uniform and sensible evacuation procedures.

In responding, the writer, Chris Alverez, numbers items and underlines job titles to delineate everyone's responsibilities. The marshal can see at a glance who is responsible for each procedure. Likewise, employees receiving the same report (as a directive in memo form) can quickly determine their responsibilities and see where they should exit.

SUBJECT: Frymore's Revised Evacuation Procedures

Dear Ms. Pine:

I'm pleased to send you a description of our revised evacuation procedures. As you instructed, we have held two fire drills using the new procedures and were able to evacuate within 5 minutes. With practice, we *will* achieve the required 3½-minute evacuation.

Specific procedures:

1. The Packaging Manager will stand in front of his department near the flow tracks. He will direct employees out through the employee entrance or through the main office, depending on location of the fire.
2. The Processing Manager will shut down all equipment before leaving the building. Press operators will shut down fryers and extruders.
3. The Casing Manager will go to the junction at the packaging line and kitchen lane to keep traffic moving toward exits.
4. During the first shift, the Quality Control Manager will direct traffic at the entrance of the cafeteria hallway to prevent running and keep employees from

going to the locker rooms. On the second and third shifts, the <u>Shift Manager</u> will cover the area.

5. The <u>Plant Services Manager</u> will evacuate office personnel through the front door and away from the building.

6. <u>Maintenance employees</u> who are part of the plant's fire-fighting department will report to the <u>Maintenance Manager</u> to help locate and contain the fire until the fire department arrives.

Please let us know if these procedures are satisfactory. Should you have any questions or further suggestions, please call me at 555-8315, ext. 332. We look forward to your next inspection in September.

## The Recommendation Report

Beyond merely providing information, the writer of the *recommendation* report interprets data, draws conclusions, and offers recommendations, usually in response to a specific request from a reader.

The following recommendation report from an investment advisor is cast as a letter. It differs from Frymore's report in that headings are included. (The itemized list format—most appropriate for the diverse contents of Frymore's report—precluded the need for headings.) This writer uses technical language (dividend, stock split) because she knows her reader is familiar with such terms. Following a direct plan, the report begins with a brief introduction to the topic and the report's purpose, then gives its recommendation, an overview, and the topics of analysis. The writer's tone is formal and courteous, befitting the advisor/client relationship. Notice how the recommendation is persuasively supported by the facts.

SUBJECT: <u>The Feasibility of Investing in Orbital Computer, Inc.</u>

We are pleased to submit the report you requested on the feasibility of a common stock investment in Orbital Computer.

<u>Recommendation</u>
Orbital Computer's profit margin makes Orbital stock an excellent investment. We recommend you invest at least $10,000 in Orbital Class B stock.

<u>Overview of Orbital Computer</u>
Orbital has been in business for more than twenty years, successfully competing with IBM, Digital Equipment Corporation, and Hewlett-Packard Corporation, to name a few. Orbital's share of the market has remained steady for the past four years. Total services and sales ranked 457th in the industrial United States, with orders increasing in the last year from $700 million to more than $1 billion. Net income placed Orbital 280th in the country, and 11th in return to its investors.

<u>Types of Common Stock</u>
Investors are offered two types of common stock listed on the American Stock Exchange. The assigned par value of both classes is $.50 per share. Class B stock entitles the investor to an additional $.025 per share dividend,

but restricts voting privileges to one vote for every ten shares held. Class C stock is not entitled to the additional dividend, but carries full voting rights.

Profit Margin
The margin for profit on sales is 9 percent, and it has been steady for the last three years. Earnings in 1979 were $.09 per share; in 1989 they were $1.36 per share. Included in these ten-year earnings is a two-for-one stock split issued on October 31, 1988.

Orbital stock would add strength and diversity to your portfolio.

## TYPICAL MEMORANDUM REPORTS

The major form for internal written communication in organizations, memos leave a "paper trail" of directives, inquiries, instructions, requests, recommendations, and so on. Organizations rely heavily on memos to trace decisions, track progress, recheck data, or identify the person(s) responsible. Therefore, any memo you write can have far-reaching ethical and legal implications. Always make sure your memo includes the date, your initials for verification, and that your information is specific, accurate, and unambiguous.

The standard memo (Figure 17.1) has a heading that names the organization and identifies the sender, recipient, subject (often in caps or underlined for emphasis), and date. (As with company letterheads, placement of these items differs slightly among firms.) Memos simply end after the final point; no summary or closing remarks are necessary. Figure 17.1 shows format guidelines for memos.

Memo reports cover every conceivable topic necessary to run an operation. The most common types include informational, recommendation, justification, progress, periodic, and survey reports.

### Internal Recommendation Reports

Page 391 showed a recommendation report cast as a letter to a client. The following recommendation report is addressed to the writer's boss. This sample typifies the problem solving carried on daily by professionals and documented in memo form.

**A Problem-Solving Recommendation**
Assume you are the writer, Bruce Doakes, in the situation that generated the Trans Globe memo below:

You are assistant manager of occupational health and safety for a major airline that employs over two hundred reservation and booking agents. Each agent spends eight hours daily seated at a workstation that has a computer, telephone, and other electronic equipment. Many agents have complained of chronic discomfort from their work: headache, eyestrain and irritation, blurred or double vision, backache, and stiff neck and joints.

**FIGURE 17.1** Guidelines for Formatting Memorandums

NAME OF ORGANIZATION

MEMORANDUM

Date:      (also serves as a chronological record for future reference)
To:        Name and title (the title also serves as a record for reference)
From:      Your name and title (your initials for verification)
Subject:   GUIDELINES FOR FORMATTING MEMOS

*Subject Line*

Summarize the memo's purpose and contents on the subject line, to help readers focus on the subject and gauge its importance. A precise title also makes filing by subject easier.

*Introductory Paragraph*

Unless you have reason for being indirect, state your main point immediately.

*Topic Headings*

When discussing a number of subtopics, include headings (as we do here). Headings help you organize, and help readers locate information quickly.

*Paragraph Spacing*

Do not indent the first line of paragraphs. Single space within, and double space between paragraphs.

*Second-page Notation*

When the memo exceeds one page, begin the second and subsequent pages with (1) the recipient's name, (2) the date, (3) the page number. For example: Ms. Baxter, June 12, 19xx, page 2. Place this information three lines from the page top and begin your text three lines below.

*Memo Verification*

Do not sign your memos. Initial the "From" line, after your name.

*Copy Notation*

When sending copies to people not listed on the "To" line, include a copy notation two spaces below the last line, and list, by rank, the names and titles of those receiving copies. For example,

Copies:    J. Spring, V.P., Production
           H. Baxter, General Manager, Production

Your boss asks you to study the problem and recommend improvements in the work environment. You survey employees and consult ophthalmologists, chiropractors, orthopedic physicians, and the latest publications on ergonomics (tailoring work environments for employees' physical and psychological well-being). After completing the study, you compose the report.

Like most short reports, this one begins by stating directly the purpose and main point, giving the reader an immediate orientation.

**TRANS GLOBE AIRLINES**
Memorandum
August 15, 1990

TO: R. Ames, Vice President, Personnel

FROM: B. Doakes, Health and Safety

SUBJECT: <u>Recommendations for Reducing Agents' Discomfort</u>

During our July 20 staff meeting, we discussed the problem of physical discomfort among reservation and booking agents, who spend the better part of eight hours daily at automated workstations. Our agents routinely complain of headaches, eyestrain and irritation, blurred or double vision, backaches, and stiff joints. This report outlines the apparent causes, and recommends immediate and nondisruptive ways of reducing discomfort.

Causes of Agents' Discomfort
I have ruled out the computer display screens as the cause of headaches and eye problems for two reasons:
1. Our new display screens have excellent contrast and no flicker.
2. Based on the best current data, low-level radiation from the screens seems to pose no health hazard.
The headaches and eye problems seem to result from the excessive glare on display screens from background lighting.
    Other discomforts, such as backaches and stiffness, apparently result from the agents' sitting in one position for up to two hours between breaks.

Recommended Changes
We can eliminate much discomfort by improving background lighting, workstation conditions, and work routines and habits.

<u>Background Lighting.</u> To reduce the glare on display screens, I recommend these changes in background lighting:
1. Decrease all overhead lighting by installing lower-wattage bulbs.
2. Keep all curtains and adjustable blinds on the south and west windows at least half-drawn, to block direct sunlight.
3. Install shades to direct the overhead light straight downward, so that it is not reflected by the screens.

<u>Workstation Conditions.</u> I recommend these changes in the workstations:
1. Reposition all screens so light sources are neither at front nor back.
2. Wash the surface of each screen weekly.

3. Adjust each screen so that the top is slightly below the operator's eye level.
4. Adjust all keyboards so that they are 27 inches from the floor.
5. Replace all fixed chairs with adjustable, armless secretarial chairs.

Work Routines and Habits. I recommend these changes in all agents' work routines and habits:

1. Allow frequent rest periods (10 minutes each hour instead of 30 minutes twice daily).
2. Provide yearly eye exams for all terminal operators, as part of our routine health-care program.
3. Train employees to adjust screen contrast and brightness whenever the background lighting changes.
4. Offer workshops on improving posture.

These changes will give us time to consider more complex options such as installing hoods and antiglare filters on terminal screens, replacing fluorescent lighting with incandescent, covering surfaces with nonglare paint, or other disruptive procedures.

## Justification Reports

Justification reports (a type of recommendation report) are in a unique class in that they often are initiated by the writer rather than requested by the readers. Writers use such reports to suggest and justify changes in policy or procedures. Justification reports overlap with proposals (Chapter 19), but usually require less persuasion, for the "justification" should be obvious. Such reports therefore typically begin rather than end with recommendations, unlike proposals. A justification report provides an excellent chance for junior staff to show their initiative and be recognized by superiors. Such reports answer this key question for readers: *Why should we?*

Justification reports should include specific benefits such as savings in time, increased productivity, better performance, or increased profits. The benefits need not always be tangible, however; they could include such intangibles as customer goodwill or improved employee morale.

Typically, justification reports follow a version of this arrangement:

1. State the problem and your recommendations for solving it.
2. Point out the cost, savings, and benefits of your plan.
3. If needed, explain how your suggestion can be implemented. If people need release time for training workshops, state how many people are involved, how much release time is required, and how long the sessions will last.
4. Conclude on a note that encourages the reader to act.

Our next writer uses a version of the preceding arrangement: he begins with the problem and recommended solution, spells out the costs and benefits, and concludes by reemphasizing the major benefit. The tone is confident yet diplomatic—appropriate for an unsolicited recommendation to a superior.

GREENTREE BIONOMICS, INC.
Memorandum

April 18, 1990

TO:      D. Spring, Personnel Director
         Greentree Bionomics, Inc. (GBI)
FROM:   M. Noll, Biology Division M.N.
ABOUT:  The Need to Hire Additional Personnel

Introduction and Recommendation
With twenty-six active employees, GBI has been unable to keep up with scheduled contracts. As a result, we have a contract backlog of roughly $500,000. This backlog is caused by understaffing in the biology and chemistry divisions.

To increase production and ease the work load, I recommend that GBI hire three general laboratory assistants.

The lab assistants would be responsible for cleaning glassware and general equipment; feeding and monitoring fish stocks; preparing yeast, algae, and shrimp cultures; preparing stock solutions; and assisting scientists in various tests and procedures.

Cost and Benefits
While costing $28,080 yearly (at $4.50/hour), three full-time lab assistants would create several distinct advantages:

1. Uncleaned glassware no longer would pile up, and the fish-holding tanks could be cleaned daily (as they should be) instead of weekly.
2. Because other employees would no longer need to work more than forty hours weekly, morale would improve.
3. Research scientists would be freed from general maintenance work (cleaning glassware, feeding and monitoring the fish stock, etc.). With more time to perform client tests, the researchers could eliminate our backlog.
4. With our backlog eliminated, clients would no longer have cause for impatience.

Conclusion
Increased production at GBI is essential to maintaining good client relations. These additional personnel would allow us to continue a reputation of prompt and efficient service, thus ensuring our steady growth and development.

## Progress Reports

Large organizations depend on progress (or status) reports to keep track of activities, problems, and progress on various projects. Daily progress reports are vital in a business that assigns crews to many projects. Management uses progress reports to evaluate the project and its supervisor, and to decide how to allocate funds. Progress reports typically are informational, but they may recommend appropriate action as well.

Often, a progress report is one of a series. Together, the project proposal (Chapter 19), progress reports (the number varies with the scope and length of the project), and the final report (Chapter 20) provide a record and history of the project.

To give management the answers it needs, progress reports must, at a minimum, answer these questions:

1. *How much has been accomplished since the last report?*
2. *Is the project on schedule?*
3. *If not, what went wrong?*
   a. *How was the problem corrected?*
   b. *How long will it take to get back on schedule?*
4. *What else needs to be done?*
5. *What is the next step?*
6. *Are there any unexpected developments?*
7. *When do you anticipate completion? Or, on a long project, when do you anticipate completion of the next phase?*

If the report is part of a series, you might refer to prior problems or developments, but the bulk of the report must cover the project's current status.

Many organizations have forms for organizing progress reports, and so no one format is best. But all reports in a series should follow the same organization. The following memorandum illustrates how one writer organized her report. Like all examples, use it as a guide; don't follow it slavishly. The format and organization you choose must fit *your* purpose, audience, and situation.

SUBJECT: Progress Report: Equipment for New Operations Building

Work Completed

Our training group has met twice since our May 12 report in order to answer the questions you posed in your May 16 memo. In our first meeting, we identified the types of training we anticipate.

Types of Training Anticipated

- Divisional Surveys
- Loan Officer Work Experience
- Divisional Systems Training
- Divisional Clerical Training (Continuing)
- Divisional Clerical Training (New Employees)
- Divisional Management Training (Seminars)
- Special/New Equipment Training

In our second meeting, we considered various areas for the training room.

Training Room

The frequency of training necessitates our having a training room available daily.

The large training room in the Corporate Education area (10th floor) would be ideal. Before submitting our next report, we need your confirmation that this room can be assigned to us.

To support the training programs, we purchased this equipment:

- Audioviewer
- 16mm projector
- Videocassette recorder and monitor
- CRT
- Mini/micro computer, for computer-assisted instruction
- Slide projector
- Tape recorder

This equipment will allow us to administer training in a variety of modes, ranging from programmed and learner-controlled instruction to group seminars and workshops.

Work Remaining

To support the training, we should furnish the room appropriately. Because the types of training will vary, the furniture should provide a flexible environment. Outlined here are our anticipated furnishing needs.

- Tables and chairs that can be set up in many configurations. These would allow for individual or group training and large seminars.
- Portable room dividers. These would provide study space for training with programmed instruction, as well as allow for simultaneous training.
- Built-in storage space for audiovisual equipment and training supplies. Ideally, this storage space should be multipurpose, providing work or display surfaces.
- A flexible lighting system, important for audiovisual presentations and individualized study.
- Independent temperature control, to ensure that the training room remains comfortable regardless of group size and equipment used.

The project is on schedule. As soon as we receive your approval of these specifications, we will proceed to the next step: sending out bids for room dividers, and having plans drawn for the built-in storage space.

cc: R.S. Pike, SVP
    G.T. Bailey, SVP

As you work on a longer report or term project, your instructor might require a progress report. In this next memo, Karen Granger reports her progress on her term project: an evaluation of the Environmental Protection Agency's effectiveness in cleaning a heavily contaminated harbor.

**PROGRESS REPORT**

| | |
|---|---|
| TO: | Dr. John Lannon |
| FROM: | Karen P. Granger |
| DATE: | April 17, 1990 |
| SUBJECT: | Evaluation of the EPA's *Remedial Action Master Plan* |

#### Work Completed

February 23: Began general research on the PCB contamination of the New Bedford Harbor.

March 8: Decided to analyze the *Remedial Action Master Plan* (RAMP) in order to determine whether residents are being "studied to death" by the EPA.

March 9–19: Drew a map of the harbor to show areas of contamination. Obtained the RAMP from Pat Shay of the EPA.

Interviewed Representative Grimes briefly by phone; made an appointment to interview Grimes and Sharon Dean on April 13.

Interviewed Patricia Chase, President of the New England Sierra Club, briefly, by phone.

March 24: Obtained *Public Comments on the New Bedford RAMP*, a collection of reactions to the plan.

April 13: Interviewed Grimes and Dean; searched Grimes' files for information. Also searched the files of Raymond Soares, New Bedford Coordinator, EPA.

#### Work in Progress

Contacting by telephone the people who commented on the RAMP.

#### Work to Be Completed

April 25: Finish contacting commentators on the RAMP.

April 26: Interview an EPA representative about the complaints that the commentators raised on the RAMP.

Date for Completion: May 3, 1990

#### Complications

The issue of PCB contamination is complicated and emotional. The more I uncovered, the more difficult I found it to remain impartial in my research and analysis. As a New Bedford resident, I expected to find that we are indeed being studied to death; because my research seems to support my initial impression, I am not sure I have remained impartial.

Lastly, the people I want to talk to do not always have the time to find the answers for my questions. Everyone I have spoken with, however, has been interested and encouraging, if not always informative.

Page 545 shows another sample progress report on a term project.

## Periodic Reports

The periodic activity report is similar to the progress report in that it summarizes activities over a specified period. But unlike the progress report, which summarizes *project* accomplishments, periodic reports summarize daily, weekly, or monthly general activities. Many manufacturers requiring periodic reports have prepared forms, because most of their tasks are quantifiable (e.g., Units produced: 672). But most white-collar jobs do not lend themselves to prepared-form reports. You will probably have to develop your own activity report, as Fran DeWitt did in this monthly report.

DeWitt's report answers her boss's primary question: *What did you accomplish last month?* Her response has to be detailed and informative.

TO: N. Morgan, Assistant Vice President
FROM: F. C. DeWitt
SUBJECT: Coordinating Meetings for Training

For the past month, I've been working on a cooperative project with the Banking Administration Institute, Computron Corporation, and several banks. My purpose has been to develop training programs, specific to banking, appropriate for computer-assisted instruction (CAI).

We have focused on three major areas: Proof/Encoding Training, Productivity Skills for Management, and Banking Principles.

I will be hosting two meetings for this task force. On June 16, we will discuss Proof/Encoding Training, and on June 17, Productivity Training. The objective for the Proof/Encoding meeting is to compare ideas, information, and current training packages available on this topic. We will then design a training course.

The objective of the meeting on Productivity is to discuss skills that increase productivity in banking (specifically Banking Operations). Discussion will include instances in which computer-assisted instruction is appropriate for teaching productivity skills. Computron will also discuss computer applications used to teach productivity.

On June 10 I will attend a meeting in Washington, D.C., to design a course in basic banking principles for high-level clerical/supervisory-level employees. We will also discuss the feasibility of adapting this course to CAI. This type of training, not currently available through Corporate Education, would meet a definite supervisor/management need in the division.

My involvement in these meetings has two benefits. First, structured discussions with trainers in the banking industry provide an exchange of ideas, methods, and experiences. This involvement expedites development of our training programs because it saves me time on research. Second, microcomputers will continue to affect future training. With a working knowledge of these systems and their applications, I will be able to assist my group in designing programs specific to our needs.

### Survey Reports

Brief survey reports often are used to examine conditions that affect an organization (consumer preferences, available markets). The following memo, from the research director for a midwestern grain distributor, gives clear and specific information directly. To simplify interpretation of data, the writer arranged them in a table.

SUBJECT: Food-Grain Consumption, U.S., 1985–88

Here are the data you requested on April 7, as part of your division's annual marketing survey.

### U.S. PER CAPITA CONSUMPTION OF FOOD GRAINS
### IN POUNDS, 1985–88

|  | 1985 | 1986 | 1987 | 1988 |
|---|---|---|---|---|
| Corn Products |  |  |  |  |
|   Cornmeal and other | 16.4 | 16.4 | 16.6 | 16.7 |
|   Corn syrup and sugar | 22.3 | 22.7 | 23.3 | 23.8 |
| Oat Products | 3.8 | 4.1 | 4.9 | 4.8 |
| Barley Products | 1.2 | 1.2 | 1.2 | 1.2 |
| Wheat |  |  |  |  |
|   Flour | 116.2 | 116.4 | 117.1 | 117.8 |
|   Breakfast cereals | 2.9 | 2.9 | 2.9 | 2.9 |
| Rye, Flour | 1.2 | 1.2 | 1.2 | 1.2 |
| Rice, Milled | 8.3 | 6.7 | 7.7 | 7.0 |

Source: U.S. Dept. of Agriculture

If you require additional information, please call Ms. Smith at 316.

CC: C. B. Schultz, Vice President, Marketing

## Minutes

Minutes are official records of meetings. Copies of minutes are distributed to all members and interested superiors as a way of keeping track of proceedings. The person appointed secretary records the minutes.

Minutes are filed as part of an official record, and so must be precise, clear, highly informative, and free of the writer's personal commentary ("As usual, Ms. Jones disagreed with the committee") or judgmental words ("good," "poor," "irrelevant"). When you record minutes, answer these questions:

- *Which group held the meeting?*
- *When and where was it held?*
- *What was its purpose?*
- *Who chaired the meeting?*
- *Who else was present?*
- *Were the minutes of the last meeting approved (or disapproved)?*
- *Was anything resolved?*

Besides answering these questions, summarize the discussion on each agenda item. Name the person who makes the motion, and the person who seconds it. Record the votes on each motion, and describe the motion itself.

SUBJECT: Minutes of Managers' Meeting, October 5, 1990

**Members Present**
Harold Tweeksbury, Jeannine Boisvert, Sheila DaCruz, Ted Washington, Denise Walsh, Cora Parks, Cliff Walsh, Joyce Capizolo.

**Agenda**
1. The meeting was called to order on Wednesday, October 5, at 10 A.M. by Cora Parks.

2. The minutes of the September meeting were approved unanimously.

3. The first order of new business was to approve the following policies for the Christmas season:
   a. Temporary employees should list their ID numbers in the upper left corner of their receipt envelopes in order to help verification. Discount Clerical assistant managers will be responsible for seeing that this procedure is followed.
   b. When temporary employees turn in their envelopes, personnel from Discount Clerical should spot-check them for completeness and legibility. Incomplete or illegible envelopes should be corrected, completed, or rewritten. Envelopes should not be sealed.

4. Jeannine Boisvert moved that we also hold one-day training workshops for temporary employees in order to teach them our policies and procedures. Denise Walsh seconded the motion. Joyce Capizolo disagreed, saying that on-the-job training (OJT) was enough. The motion for the training session carried 6–3. The first workshop, which Jeannine agreed to arrange, will be held October 25.

5. Joyce Capizolo requested that temporary employees be sent a memo explaining the temporary employee discount procedure. The request was converted to a motion and seconded by Cliff Walsh. The motion passed by a 7–2 vote.

6. Cora Parks adjourned the meeting at 11:55 A.M.

## PREPARED-FORM REPORTS

To streamline communications and keep track of data, many companies use prepared forms for short reports. Such forms are useful in two ways:

1. A prepared form provides clear guidance for recording data. If you complete the form correctly, you are sure to satisfy the readers' needs.

2. A prepared form standardizes data reported from various sources. Identical categories of data are recorded in identical order. This fixed format allows for rapid processing and tabulating of data.

Countless forms are prepared for countless purposes, and so we reproduce no examples here. The sample questionnaire (pages 122–123) is an example of a prepared form: its data can be reviewed easily and tabulated rapidly.

## REVISION CHECKLIST FOR SHORT REPORTS

Use this checklist as a guide to refining and revising your short reports. (Numbers in parentheses refer to the first page of discussion.)

☐ Have you chosen the best report form for your purpose? (390)

☐ Does the memo have a complete heading? (392)

☐ Does the subject line clearly forecast the memo's or letter report's contents? (393)

☐ Have you given readers enough information for an *informed* decision? (389)

☐ Are your conclusions and recommendations clear? (392)

☐ Have you single-spaced within paragraphs and double-spaced between? (393)

☐ Do headings, charts, or tables appear whenever needed? (240)

☐ If more than one reader is receiving copies, does the memo include a distribution notation (cc) to identify other readers? (393)

☐ Are all sentences clear, concise, and fluent? (199)

☐ Does the document's appearance create a favorable impression? (279)

## EXERCISES

**1.** We all would like to see changes in our schools' policies or procedures, whether they are changes in our majors, school regulations, social activities, grading policies, registration procedures, or the like. Find some area of your school that needs obvious changes, and write a justification report to the person who might initiate that change. Explain why the change is necessary and describe the benefits. Follow the format on page 396.

**2.** You would like to see some changes in this course to better reflect your career plans. Perhaps you feel too much emphasis is placed on letters, too little on reports. Or maybe there's too much lecturing and not enough discussion. Write a justification report to your instructor, justifying the reasons for the changes you propose. Remember, you must illustrate specific benefits resulting from your plan. Do *not* try to justify spending less time doing course work.

**3.** Think of an idea you would like to see implemented in your job (e.g., a way to increase productivity, improve service, increase business, or improve working conditions). Write a justification memo, persuading your audience that your idea is worthwhile.

**4.** Write a memo to your boss, justifying reimbursement for this course. *Note:* You might have written another version of this assignment for exercise 7 in Chapter 1. If so, compare early and recent versions for content, organization, style, and format.

**5.** Assume you work in a large electronics company for a division manager named Bentley. You and your coworkers would like to institute an in-house day-care center because, collectively, you have eighteen preschool children. All of you presently have babysitters or take your children to centers or nursery schools scattered throughout the city. With an in-house center, employees would have less scurrying in the morning, cutting back on lateness resulting from traffic, bad weather, and so on. Employees also have to leave work early sometimes to pick up their children. Employees would pay for the two supervisors and do all the necessary paperwork to ensure that the center meets state regulations. (The center would actually be cheaper than hiring individual sitters as you presently do.) The unused lounge on the second floor would be an ideal location.

An in-house center would have many benefits for employees, supervisors, and your employer. Some supervisors might want to include their children. Brainstorm for ideas and benefits.

Because Bentley would have to give initial approval for the center, decide what kinds of questions she would ask. Draw up a list of ten questions you might have to answer.

Write your memo to Bentley. Submit all materials to your instructor.

**6.** Identify a dangerous or inconvenient area or situation on campus or in your community (endless cafeteria lines, a poorly lit intersection, slippery stairs, a poorly adjusted traffic light). Observe the problem for several hours during a peak-use period. Write a justification report to a *specifically identified* decision maker, describing the problem, listing your observations, making recommendations, and encouraging reader support or action.

**7.** Assume that you have received a $10,000 scholarship, $2,500 yearly. The only stipulation for receiving installments is that you send the scholarship committee a yearly progress report on your education, including courses, grades, school activities, and cumulative average. Write the report.

**8.** In a memo to your instructor, outline your progress on your term project. Describe your accomplishments, plans for further work, and any problems or setbacks. Conclude your memo with a specific completion date.

**9.** In a memo to your instructor, outline your progress for the course as a whole, including work completed, current assignment, plans for future projects, and setbacks. Also assess improvement in your writing skill. Be specific.

**10.** Assume that you've begun your career. Your boss wants weekly memo reports on projects and accomplishments. In short, what did you do last week to earn your salary? Write your weekly report. Unless you have worked in the field, this assignment requires research. Check your job title in *The Dictionary of Occupational Titles*. You might also want to check the *Occupational Outlook Handbook*. Both can be found in the library's reference section. If possible, interview someone in your field. Write a realistic report on a typical week's activities. Or, do the same for an actual part-time job.

**11.** Keep accurate minutes for one class session (preferably one with debate or discussion). Submit the minutes in memo form to your instructor.

**12.** Conduct a brief survey (e.g., of comparative interest rates from various banks on a car loan, comparative tax and property evaluation rates in three local towns, or comparative prices among local retailers for an item). Arrange your data and report your findings to your instructor in a memo that closes with specific recommendations for the most economical choice.

13. Although the campus store (or cafeteria) is a convenient place to buy small items (or food), you believe students are overcharged for the convenience. To prove your point, you decide to do a comparative analysis of the cost of five items at the campus store (or cafeteria) and two local stores (or restaurants). Be sure to compare like sizes, weights, brands. Conduct your survey, analyze your findings, and submit the results in memo form to the student senate. Your instructor might want you to include recommendations.

14. *Informational Report Topics* (choose *one* for a memo or letter).
   *a.* You are legal consultant to the leadership of a large auto-workers' union. Before negotiating its next contract, the union needs to anticipate effects of robotics technology on assembly-line auto workers within ten years. Do the research and write the report.
   *b.* You are a consulting engineer to an island community of 200 families suffering a severe water shortage of fresh water. Some islanders have raised the possibility of producing drinking water from salt water (desalination). Write a report for the town council, summarizing the process and describing instances in which desalination has been used successfully or unsuccessfully. Would desalination be economically feasible for a community this size?
   *c.* You are health officer in a town less than one mile from a massive radar installation. Citizens are disturbed about the effects of microwave radiation. Do they need to worry? Find the facts and write your report.
   *d.* You are health officer in a town where a power company easement allows high-voltage lines to run immediately adjacent to the elementary school playground. Are the children endangered by ionic disturbances? Parents and school committee want to know. Find out, and write your report.
   *e.* Dream up a scenario of your own, in which information would make a real difference.

15. *Recommendation Report Topics* (choose *one* for a memo or letter).
   *a.* You are an investment broker for a major firm. A longtime client calls to ask your opinion. He's thinking of investing in a company that is fast becoming a leader in fiber optics communication links. "Should I invest in this technology?" your client wants to know. Find out, and give him your recommendations in a short report.
   *b.* The buildings in the condominium complex you manage have been invaded by carpenter ants. Can the ants be eliminated by any insecticide *proven* nontoxic to humans or pets? (Many dwellers have small children and pets.) Find out, and write a report making recommendations to the maintenance supervisor.
   *c.* The "coffee generation" wants to know about the properties of caffeine and the chemicals used on coffee beans. What are the effects of these substances on the body? Write your report, making specific recommendations about precautions that coffee drinkers can take.

*d.* As a consulting dietitian to the school cafeteria in Blandville, you've been asked by the school board to report on the most dangerous chemical additives in foods. Parents want to be sure that foods containing these additives are eliminated from school menus, insofar as possible. Write your report, making general recommendations about modifying school menus.

*e.* Dream up a scenario of your own in which information and recommendations would make a real difference. (Perhaps the question could be one you've always wanted answered.)

## TEAM PROJECT

Organize into groups of four or five and choose a topic upon which all group members can take the same position. Here are some possibilities:

- Should your college abolish core requirements?
- Should every college student pass a writing proficiency exam before graduating?
- Should courses outside one's major be graded pass/fail at the student's request?
- Should your school drop or institute student evaluation of teachers?
- Should all students be required to become computer literate before graduating?
- Should campus police carry guns?
- Should dorm security be improved?
- Should students with meal tickets be charged according to what and how much they eat, instead of paying a flat fee?

As a group, decide your position on the issue, and brainstorm collectively to justify your recommendation to a stipulated primary audience in addition to your classmates and instructor. Complete an audience-and-use profile (page 33), and compose a justification report. Appoint one member to present the report in class.

# Supplements
# to Documents

■ Purpose of Supplements
Composing Report Supplements

**S**upplements are reference items generally added to a long report or proposal to make the document more accessible. The title page, letter of transmittal, table of contents, and abstract give summary information about the content of the document. The glossary, appendixes, and list of works cited can either provide supporting data or help readers follow technical sections. According to their needs, readers can refer to one or more or these supplements or skip them altogether. All supplements, of course, are written only *after* the document itself has been completed.

## PURPOSE OF SUPPLEMENTS

No matter how useful its information, a document usually has to be *accessible* to varied readers for many purposes. Supplements address these workplace realities:

- *Confronted by a long document, many readers will try to avoid reading the whole thing.* Instead they will look for the least information they need to complete the task, make the decision, and take action.

- *Different readers often use one document for different purposes.* Some look for an overview; others want the details; others are interested only in the conclusions and recommendations, or the "bottom line". Technical personnel might focus on the body of a highly specialized report and on the appendixes for supporting data (maps, formulas, calculations). Executives and managers might read only the transmittal letter and the abstract. If the latter audience reads any of the report proper, they are likely to focus on the conclusions and recommendations.

By carefully planning and designing supplements for your documents, you can accommodate the needs of diverse audiences.

## COMPOSING REPORT SUPPLEMENTS

Report supplements can be classified in two groups:

1. *supplements that precede your report* (front matter): cover, title page, letter of transmittal, table of contents (and figures), and abstract

2. *supplements that follow your report* (end matter): glossary, footnotes, endnote pages, appendix(es)

### Cover

Use a sturdy, plain cover with page fasteners. With the cover on, the open pages should lie flat. Use covers only for long documents.

Center the report title and your name four to five inches from the upper edge:

FEASIBILITY ANALYSIS OF A CAREER IN TECHNICAL MARKETING
by
Richard B. Larkin, Jr.

### Title Page

The title page lists the report title, author's name, name of person(s) or organization to whom the report is addressed, and date of submission.

**Title.** Your title promises what the report will deliver by stating the report's purpose and subject. The previous title (given as an example for the cover) is clear, accurate, comprehensive, and specific. But even slight changes can distort this title's signal.

**An unclear title**     A CAREER IN TECHNICAL MARKETING

This above version does not state clearly the report's purpose. Its signal is confusing. Is the report *describing* the career, *proposing* such a career, *giving instructions* for preparing for this career, or *telling one person's story* of his/her career? Insert descriptive words ("analysis," "instructions," "proposal," "feasibility," "description," "progress,") that accurately state your purpose.

To be sure that your title promises what the report delivers, write its final version *after* completing the report.

**Placement of Items on Title Page.** Do not number your title page, but count it as page i of your prefatory pages. Center the title horizontally on the page, three to four inches below the upper edge, using all capital letters. If the title is longer than six to eight words, center it on two or more single-spaced lines. Place other items in the spacing, order, and typeface shown in Figure 18.1, the title page to a report shown in Chapter 20 (pages 470–478). Or work out your own system, as long as your page is balanced.

## Letter of Transmittal

For college reports, your letter of transmittal usually comes immediately after the title page and is bound as part of your report. For workplace reports, the letter usually precedes the title page. Include a letter of transmittal with any formal report or proposal addressed to a specific reader. Your letter adds a note of courtesy and gives you a place for personal remarks or opinions. Depending on the situation, your letter might

- acknowledge those who helped with the report
- refer to sections of special interest: unexpected findings, key visuals, major conclusions, special recommendations, and the like
- discuss the limitations of your study, or any problems gathering data
- discuss the need and approaches for follow-up investigations
- describe any personal (or "off-the-record") observations
- suggest some special uses for the information
- urge the reader to immediate action

The letter of transmittal can be tailored to a particular reader, as is Richard Larkins' letter in Figure 18.2. If a report is being sent to a number of people who are variously qualified and bear various relationships to the writer, the letters of transmittal may vary. The letter itself has an introduction-body-conclusion structure.

**Introduction.** Open with a reference to the reader's original request. Briefly review the reasons for your report. Maintain a confident and positive tone throughout the letter. Indicate pride and satisfaction in your work. Avoid implied apologies, such as "I hope this information is adequate," or "I hope this report meets your expectations."

**Body.** In the letter body, include items from our prior list of possibilities (acknowledgements, special problems). Although your informative abstract will summarize major findings, conclusions, and recommendations, your letter gives a brief and personal overview of the *entire project*.

**Conclusion.** State your willingness to answer questions or discuss findings. End positively with something like "I believe that the data in this report are accurate, that they have been analyzed rigorously and impartially, and that the recommendations are sound."

## Table of Contents

Your table of contents serves as a road map for readers and a checklist for you. Derived from your outline, this supplement is easy to compose: simply assign page numbers to headings in your outline. Keep in mind, however, that not

**FIGURE 18.1** Title Page for a Formal Report

# Feasibility Analysis
# of a Career
# in Technical Marketing

for

Professor J. M. Lannon
Technical Writing Instructor
Southeastern Massachusetts University
North Dartmouth, Massachusetts

by

Richard B. Larkin, Jr.
English 266 Student

May 1, 1990

**FIGURE 18.2** Letter of Transmittal for a Formal Report

165 Hammond Way
Hyannis, MA 02457
April 29, 1990

John Fitton
Placement Director
Southeastern Massachusetts University
North Dartmouth, MA 02747

Dear Mr. Fitton:

Here is my analysis to determine the feasibility of a career in technical marketing and sales. In preparing my report, I've learned a great deal about this career, and I hope my information will help other students as well.

Although committed to their specialities, some technical and science graduates seem interested in careers in which they can apply their technical knowledge to customer and business problems. Technical marketing may be an attractive choice of career for those who know their field, who can relate to different personalities, and who are good communicators.

Technical marketing is competitive and demanding, but highly rewarding. If fact, it is an excellent route to upper-management and executive positions. Specifically, marketing work enables one to develop a sound technical knowledge of a company's products, to understand how these products fit into the marketplace, and to perfect sales techniques and interpersonal skills. This is precisely the kind of background that paves the way to top-level jobs.

I've enjoyed my work on this project, and would be happy to answer any questions.

Sincerely,

Richard B. Larkin, Jr.

all levels of headings from your outline will appear in your table of contents or your report. Too many headings can fragment the discussion.

Follow these guidelines for a table of contents:

- List front matter (transmittal letter, abstract), numbering the pages with small roman numerals. (The title page, though not listed, is counted as page i.) List glossary, appendix, and endnotes; number these pages with arabic numerals, continuing the page sequence of your report proper, where page 1 is the first page of the report text.
- Include no headings in the table of contents not listed as headings or subheadings in the report; your report text may, however, contain subheadings not listed in the table of contents.
- Phrase headings in the table of contents exactly as in the report.
- To reflect their rank, list headings at various levels in varying typeface and indention.
- Use horizontal dots ( . . . . . . . . ) to connect heading to page number. Make sure the rows of dots line up vertically, one over the other.

Figure 18.3 shows the table of contents for Richard Larkin's feasibility analysis.

### Table of Tables and Figures

Following your table of contents is a table of tables and figures, if needed. When your report has more than four or five visuals, place this table on a separate page. Figure 18.4 shows the table of tables and figures for Richard Larkin's report.

### Informative Abstract

You can expect that some readers will have neither the time nor the inclination to read your entire report. For these readers, the informative abstract is the most important part of your report. The abstract is always written *after* your report proper.

Writing the abstract gives you the chance to measure your own control over the material. Because your are compressing your message, you must establish explicit connections. If you cannot effectively summarize your report, it probably needs revision.

Follow these guidelines for your abstract:

- Make your abstract able to stand alone in meaning—a mini-report.
- Write for the general reader. Readers of the abstract are more likely to vary in expertise, perhaps more than those who read the report itself; translate all technical data.
- Add no new information. Simply summarize the report.
- Present your information in this sequence:
  a. Begin by identifying the issue or need that led to the report.

**FIGURE 18.3** Table of Contents for a Formal Report

iii

## CONTENTS

page

**FIGURE 18.4**  A Table of Tables and Figures for a Formal Report

    b. Offer the major findings from the report body.

    c. Include a condensed conclusion and recommendations, if any.[1]

The informative abstract in Figure 18.5 accompanies Richard Larkin's report in Chapter 20.

### Glossary

A glossary is an alphabetical listing of specialized words and their definitions, immediately following your report. Many specialized reports contain glossaries, especially when written for both technical and nontechnical readers. A glossary makes key definitions available to non-technical readers without interrupting technical readers. If fewer than five terms need defining, place them instead in the report introduction as working definitions (as discussed in Chapter 8 – or use footnote definitions. If you use a separate glossary, inform readers of its location: "(see the glossary at the end of this report)."

    Follow these guidelines for a glossary:

- Define all terms unfamiliar to a general reader (an intelligent layperson).

- Define all terms that have a special meaning in your report (e.g., "In this report, a small business is defined as . . .").

- Define all terms by giving their class and distinguishing features (Chapter 8), unless some terms need expanded definitions.

- List your glossary and its first page number in your table of contents.

[1] My thanks to Professor Edith K. Weinstein for these suggestions.

**FIGURE 18.5** An Informative Abstract

## ABSTRACT

My own interest in technical marketing and sales led me to analyze the feasibility of such a career, and to compare the various options for college graduates who wish to enter the field.

Technical marketing is a highly feasible career for anyone who is motivated, who can communicate well, and who knows how to get along. Although this career offers diverse jobs and excellent potential for income, it entails almost constant travel, competition, and stress.

College graduates enter technical marketing through one of four options: entry-level positions that offer hands-on experience, formal training programs in large companies, prior experience in one's specialty, or graduate programs. The relative advantages and disadvantages of each option can be measured in resulting immediacy of income, rapidity of advancement, and long-term potential.

Anyone considering a technical marketing career should follow these recommendations:

- Speak with people who work in the field.
- Weigh carefully the implications of each entry option.
- Consider combining two or more options.
- Choose options for personal as well as professional benefits.

- List all terms in alphabetical order. Underline each term and use a colon to separate it from its single-spaced definition.
- Define only terms that need explanation. In doubtful cases, overdefining is safer than underdefining.
- On first use, place an asterisk in the text by each term defined in the glossary.

Figure 18.6 shows part of a glossary for a comparative analysis of two techniques of natural childbirth, written by a nurse practitioner for expectant mothers and student nurses.

## Endnote or Works-Cited Pages

The endnote pages list each of your outside references in the same numerical order as they are cited in your report proper. See Chapter 6 for a discussion of documentation.

## Appendix

An appendix comes at the very end of your report. The appendix expands items

FIGURE 18.6  A Partial Glossary

**GLOSSARY**

**Analgesic:** a medication given to relieve pain during the first stage of labor.

**Cervix:** the neck-shaped anatomical structure that forms the mouth of the uterus.

**Dilation:** cervical expansion occurring during the first stage of labor.

**Episiotomy:** an incision of the outer vaginal tissue, made by the obstetrician just before the delivery, to enlarge the vaginal opening.

**First stage of labor:** the stage in which the cervix dilates and the baby remains in the uterus.

**Induction:** the stimulating of labor by puncturing the membranes around the baby or by giving an oxytoxic drug (uterine contractant), or both.

discussed in the report without cluttering the report text. Typical items in an appendix include

- details of an experiment
- statistical or other measurements
- maps
- complex formulas
- long quotes (one or more pages)
- photographs
- texts of laws and regulations
- related correspondence (letters of inquiry, and so on)
- interview questions and responses
- sample questionnaires and tabulated responses
- sample tests and tabulated results
- some visuals occupying more than one full page
- anything essential to secondary readers but less relevant to primary readers.

The appendix is a catch-all for items that are important but difficult to integrate in your text. Figure 18.7 shows an appendix to the budget proposal in Chapter 19 (pages 440–449).

Do not stuff appendixes with needless information. And it is unethical to use them for burying bad or embarrassing news that belongs in the report proper. Readers should not have to turn to appendixes to understand the report itself.

**FIGURE 18.7** An Appendix

**APPENDIX A**

Table 1 Allocations and Performance of Five Massachusetts College Newspapers

| | Stonehorse College | Alden College | Simms University | Fallow State | SMU |
|---|---|---|---|---|---|
| Enrollment | 1,600 | 1,400 | 3,000 | 3,000 | 5,000 |
| Fee paid (per year) | $65.00 | $85.00 | $35.00 | $50.00 | $65.00 |
| Total fee budget | $88,000 | $119,000 | $105,000 | $150,000 | $334,429.28 |
| Newspaper budget | $10,000 | $6,000 | $25,300 | $37,000 | $21,500 |
| | | | | | $25,337.14[a] |
| Yearly cost per student | $6.25 | $4.29 | $8.43 | $12.33 | $4.06 |
| | | | | | $5.20[a] |
| Format of paper | Weekly | Every third week | Weekly | Weekly | Weekly |
| Average no. of pages | 8 | 12 | 18 | 12 | 20 |
| Average total pages | 224 | 120 | 504 | 336 | 560 |
| | | | | | 672[a] |
| Yearly cost per page | $44.50 | $50.00 | $50.50 | $110.11 | $38.25 |
| | | | | | $38.69[a] |

[a]These figures are next year's costs for the SMU *Torch*.

*Source:* Figures were quoted by newpaper business managers in April 1989.

Follow these guidelines:

- Include only material that is relevant.
- Use a separate appendix for each major item.
- Title each appendix clearly: "Appendix A: Projected Costs."
- Do not use too many appendixes. Four or five appendixes in a ten-page report would indicate a poorly organized report.
- Limit an appendix to a few pages, unless length is imperative.
- Mention your appendix early in your introduction, and refer readers to it at appropriate points in the report: "(see Appendix A)."

Use an appendix for any item which is essential, but which would harm the unity and coherence of your report. Remember that readers should be able to understand your report without having to turn to the appendix. Distill the essential facts from your appendix and place them in your report text.

| Improper reference | The whale population declined drastically between 1976 and 1977 (see Appendix B for details). |
|---|---|
| Proper reference | The whale population declined *by 16 percent* from 1976 to 1977 (see Appendix B for statistical breakdown). |

## EXERCISES

**1.** These titles are intended for investigative, research, or analytical reports. Revise each inadequate title to make it clear and accurate.

   *a.* The Effectiveness of the Prison Furlough Program in Our State
   *b.* Drug Testing on the Job
   *c.* The Effects of Nuclear Power Plants
   *d.* Woodburning Stoves
   *e.* Interviewing
   *f.* An Analysis of Vegetables  (for a report assessing the physiological effects of a vegetarian diet)
   *g.* Wood as a Fuel Source
   *h.* Oral Contraceptives
   *i.* Lie Detectors and Employees

**2.** Prepare a title page, a letter of transmittal (for a definite reader who can *use* your information in a definite way), a table of contents, and an informative abstract for a report you have written earlier.

**3.** Find a short but effective appendix in one of your textbooks, in a journal article in your field, or in a report from your workplace. In a memo to your instructor and classmates, explain how the appendix is used, how it relates to the main text, and why it is effective. Attach a copy of the appendix to your memo. Be prepared to discuss your evaluation in class.

# Proposals

**19**

**A** proposal is an offer to do something or a suggestion that something be done. The general purpose of a proposal is to *persuade* readers to improve conditions, authorize work on a project, accept a service or product (for payment), or otherwise support a plan for solving a problem or doing a job.

Your own proposal may be a letter to your school board to suggest changes in the English curriculum; it may be a memo to your firm's vice president to request funding for a training program for new employees; or it may be a 1,000-page document to the Defense Department to bid for a missile contract (competing with proposals from other firms). You might write the proposal alone or as part of a team. It might take hours or months.

## PURPOSE OF PROPOSALS

Whether in science, business, industry, government, or education, proposals are written for decision makers: managers, executives, directors, clients, trustees, board members, community leaders, and the like. Inside or outside your organization, these are the people who decide whether your suggestions are worthwhile, whether your project will ever get off the ground, whether your service or product is useful. If your job depends on funding from outside sources, proposals might well be your most important writing activity. To succeed, a proposal must be convincing.

## THE PROPOSAL PROCESS

The basic proposal process can be summarized simply: someone offers a plan for something that needs to be done. In business and government, this process has three stages:

1. Client *X* needs a service or product.
2. Firms *A*, *B*, *C*, propose ways of meeting the need.
3. Client *X* awards the job to the firm offering the best proposal.

The complexity of events within each phase will of course depend on the situation. Here is a typical situation:

Assume that you manage a mining engineering firm in Tulsa, Oklahoma. On Wednesday, February 19, you spot this announcement in the *Commerce Business Daily:*[1]

**R—Development of Alternative Solutions to Acid Mine Water Contamination from Abandoned Lead and Zinc Mines** near Tar Creek, Neosho River, Ground Lake, and the Boone and Roubidoux aquifers in northeastern Oklahoma. This will include assessment of environmental impacts of mine drainage followed by development and evaluation of alternate solutions to alleviate acid mine drainage in receiving streams. An optional portion of the contract, to be bid upon as an add-on and awarded at the discretion of the OWRB, will be to prepare an Environmental Impact Assessment for each of three alternative solutions as selected by the OWRB. The project is expected to take 6 months to accomplish, with anticipated completion date of September 30, 19xx. The projected effort for the required task is 30 person-months. Requests for proposals will be issued and copies provided to interested sources upon receipt of a written request. Proposals are due March 1, 19xx. (044)

> Oklahoma Water Resources Board
> P.O. Box 53585
> 1000 Northeast 10th Street
> Oklahoma City, OK 73152
> (405) 555-2541

Your firm has the personnel, experience, and time to do the job, and so you decide to compete for the contract. Because the March 1 deadline is fast approaching, you write immediately for a copy of the request for proposal (RFP). The RFP will give you the guidelines for developing the proposal—guidelines for spelling out your plan to solve the problem (methods, timetables, costs).

You receive a copy of the RFP on February 21—only one week before deadline. You get right to work, with the two staff engineers you have appointed to your proposal team. Because the credentials of your staff could affect the

---

[1]This daily publication lists the government's latest needs for *services* (salvage, engineering, maintenance) and for *supplies, equipment, and materials* (guided missiles, engine parts, building materials, and so on). The *Commerce Business Daily* is an essential reference tool for anyone whose firm seeks government contracts.

*The Grantsmanship Center News*, published six times a year, contains a wealth of information about federal grants for nonprofit organizations in areas such as rural development, media, science, energy and environment, and education. This publication also lists notices of training programs in proposal writing, and offers books on proposal writing and evaluation, along with bibliographies of essential publications about the proposal process.

client's acceptance of the proposal, you ask team members to update their résumés (for inclusion in an appendix to the proposal).

Several other firms will be competing for this contract. The client will award it to the firm submitting the best proposal; clients use these criteria (and perhaps others) to review various proposals:

- understanding of the client's needs, as described in the RFP
- soundness of the firm's technical approach
- quality of the project's organization and management
- ability to complete the job by the deadline
- ability to control costs
- specialized experience of the firm in this type of work
- qualifications of staff to be assigned to the project
- the firm's record for similar projects

A client's specific evaluation criteria often are listed (in order of importance or on a point scale) in the RFP. Although these criteria may vary, every client expects a proposal that is *clear, informative,* and *realistic.*

Some clients will hold a preproposal conference for the competing firms. During this briefing, the firms are told of the client's needs, expectations, specific start-up and completion dates, criteria for evaluation, and any other details that might help guide the competing firms.

The sample situation above is only one among countless possibilities. You might encounter proposal situations that are quite different, but the process will always be similar: you will propose a plan to fill a need. And in most cases, the merit of your proposed plan will be judged *solely* by what you have *on paper.* This reality is summed up in the advice below from a government publication that solicits proposals.

> [Proposal] reviewers will base their conclusions only on information contained in the proposal. It cannot be assumed that readers are acquainted with the firm or key individuals or any referred-to experiments.[2]

Make any proposal able to stand alone in meaning and persuasiveness.

## TYPES OF PROPOSALS

Despite their variety, proposals are classified in three ways: according to *origin, audience,* or *intention.* Based on its origin, a proposal is either *solicited* or *unsolicited*—that is, requested by someone or initiated on your own because you have recognized a need. Business and government proposals usually are solicited, and they originate from a customer's request (as shown in the sample situation on pages 420–421.

[2]*Small Business Innovation Research Program* (Washington: U.S. Department of Defense, 1983): 9.

Based on its audience, a proposal may be *internal* or *external*—written for members of your organization or for clients and funding agencies. (The situation on pages 420–421 calls for an external proposal.)

Based on its intention, a proposal may be a *planning, research,* or *sales* proposal. These last categories by no means account for all variations among proposals. Some proposals may in fact fall under all three categories, but these are the types you are most likely to have to write. A discussion of each type follows.

## The Planning Proposal

A planning proposal suggests ways of solving a problem or bringing about improvement. It might be a request for funding to expand the campus newspaper, an architectural plan for new facilities at a ski area, or a plan to develop energy alternatives to fossil fuels. The successful planning proposal always answers this main question for readers:

- *What are the benefits of following your suggestions?*

The planning proposal that follows is external and solicited. The XYZ Corporation has contracted a team of communication consultants to design in-house writing workshops. The consultants here need to persuade the reader that their methods are likely to succeed. In their proposal, addressed to the company's education officer, the consultants offer concrete and specific solutions to clearly identified problems.

After a brief introduction summarizing the problem, our writers develop their proposal under two headings, "Assessment of Needs" and "Proposed Plan."

Under "Proposed Plan," subheadings offer an even more specific forecast. The "Limitations" section shows that our writers are careful to promise no more than they can deliver.

Because this proposal is external, it is cast as a letter. Notice, however, that the word choice ("thanks," "what we're doing," "Jack and Terry") creates an informal, familiar tone. Such a tone is appropriate in this external document because the writers and reader have spent many hours in conferences, luncheons, and phone conversations.

Like any document that gains reader acceptance, this one is the product of careful decisions about content, organization, and style.

Dear Mary:

Thanks for sending the writing samples from your technical support staff. Here's what we're doing to design a realistic approach.

Assessment of Needs
After conferring with technicians in both Jack's and Terry's groups, and analyzing their writing samples, we identified this hierarchy of needs:

- improving readability
- achieving precise diction

- summarizing information
- organizing a set of procedures
- formulating various memo reports
- analyzing audiences for upward communication
- writing persuasive bids for transfer or promotion
- writing persuasive suggestions

### Proposed Plan
Based on the needs listed above, we have limited our instruction package to eight carefully selected and readily achievable goals.

### Course Outline
Our eight 2-hour sessions are structured as follows:

1. achieving sentence clarity
2. achieving sentence conciseness
3. achieving fluency and precise diction
4. writing summaries and abstracts
5. outlining procedures and manuals
6. editing manuals and procedures
7. formulating various reports for various purposes
8. analyzing the audience and writing persuasively

### Classroom Format
The first three meetings will be lecture-intensive with weekly exercises to be done at home and edited collectively in class. The remaining five weeks will combine lecture and exercises with group editing of work-related documents. We plan to remain flexible so we respond to needs that arise.

### Limitations
Given our limited contact time, we cannot realistically expect to turn out a batch of polished communicators. By the end of the course, however, our students will have begun to appreciate writing as a deliberate process.

If you have any suggestions for refining this plan, please let us know.

## The Research Proposal

Research (or grant) proposals request approval (and often funding) for a research project. A chemistry professor might address a research proposal to the Environmental Protection Agency for funds to identify toxic contaminants in local groundwater. Research proposals are solicited by many government and private agencies: National Science Foundation, National Institutes of Health, Department of Agriculture, Carnegie Foundation, and others. Each granting agency has its own requirements for the format and content of proposals, but any successful research proposal answers these questions:

- *Why is this project worthwhile?*
- *What qualifies you to undertake the project?*
- *What are its chances of success?*

In college, you might submit proposals for independent study, field study, or a thesis project. Here is the title of a research proposal submitted to the thesis committee in a geology department:

PROPOSAL FOR A MASTER'S THESIS PROJECT
TO INVESTIGATE THE TERTIARY GEOLOGY
OF THE ST. MARIES RIVER DRAINAGE
FROM ST. MARIES TO CLARKIA, IDAHO

A technical writing student might submit an informal proposal requesting the instructor's approval for a term paper (which, in turn, may be a formal proposal). The introduction of the next proposal describes the problem and justifies the need for the study. The body outlines the scope, method, and sources for the proposed investigation. The conclusion describes the goal of the investigation and encourages reader support. This proposal is convincing because it answers questions about *what, why, how, when,* and *where.* Because this proposal is internal, it is cast informally as a memo.

March 16, 19xx

TO:     Dr. John Lannon

FROM:     T. Sorrells Dewoody

SUBJECT: Proposal for Determining the Feasibility of Marketing Dead Western White Pine

Introduction

Over the past four decades, huge losses of western white pine have occurred in the northern Rockies, primarily attributable to white pine blister rust and the attack of the mountain pine beetle. Estimated annual mortality is 318 million board feet. Because of the low natural resistance of white pine to blister rust, this high mortality rate is expected to continue indefinitely.

If white pine is not harvested while the tree is dying or soon after death, the wood begins to dry and check (warp and crack). The sapwood is discolored by blue stain, a fungus carried by the mountain pine beetle. If the white pine continues to stand after death, heart cracks develop. These factors work together to cause degradation of the lumber and consequent loss in value.

Statement of Problem

White pine mortality reduces the value of white pine stumpage, because the commercial lumber market will not accept it. The major implications of this problem are two: first, in the face of rising demand for wood, vast amounts of timber lie unused; second, dead trees are left to accumulate in the woods, where they are rapidly becoming a major fire hazard here in northern Idaho and elsewhere.

Proposed Solution

One possible solution to the problem of white pine mortality and waste is to search for markets other than the conventional lumber market. The last few years have seen a burst of popularity and growing demand for weathered barn boards and wormy pine for interior paneling. Some firms around the country are marketing defective wood as specialty products. (These firms call the wood from which their products come "distressed," a term I will use hereafter to refer to dead and defective white pine.) Distressed white pine quite possibly will find a place in such a market.

## Scope

To assess the feasibility of developing a market for distressed white pine, I plan to pursue six areas of inquiry:

1. What products are presently being produced from dead wood, and what are the approximate costs of production?
2. How large is the demand for distressed-wood products?
3. Can distressed white pine meet this demand as well as other species meet it?
4. Is there room in the market for distressed white pine?
5. What are the costs of retrieving and milling distressed white pine?
6. What prices for the products can the market bear?

## Methods

My primary data sources will include consultations with Dr. James Hill, Professor of Wood Utilization, and Dr. Sven Bergman, Forest Economist—both members of the College of Forestry, Wildlife, and Range. I will also inspect decks of dead white pine at several locations, and visit a processing mill to evaluate it as a possible base of operations. I will round out my primary research with a letter and telephone survey of processors and wholesalers of distressed material.

Secondary sources will include publications on the uses of dead timber, and a review of a study by Dr. Hill on the uses of dead white pine.

## My Qualifications

I have been following Dr. Hill's study on dead white pine for two years. In June of this year I will receive my B.S. in forest management. I am familiar with wood milling processes and have firsthand experience at logging. My association with Drs. Hill and Bergman gives me the opportunity for an in-depth feasibility study.

## Conclusion

Clearly, action is needed to reduce the vast accumulations of dead white pine in our forests. The land on which they stand is among the most productive forest land in northern Idaho. By addressing the six areas of inquiry mentioned earlier, I can determine the feasibility of directing capital and labor to the production of distressed white pine products. With your approval I will begin my research at once.

## The Sales Proposal

A sales proposal offers a service or product. Sales proposals may be solicited or unsolicited. If they are solicited, several firms may compete with proposals of their own. Because sales proposals are addressed to readers outside your organization, they are cast as letters (if they are brief). But long sales proposals, like long reports, are formal documents with supplements (cover letter, title page, table of contents).

The sales proposal, a major marketing tool in business and industry, will be successful if it answers this question:

- *How will you serve our needs better than your competitors?*

or

- *Why should we hire you instead of someone else?*

The following solicited proposal offers a service. Because the writer is competing with other firms, he explains specifically *why* his machinery is best for the job, *how* the job can best be completed, *what* his qualifications are for getting the job done, and *how much* the job will cost. He will be legally bound by his estimate. To protect himself, he points out possible causes of increased costs. *Never underestimate costs by failing to account for all variables* — a sure way to lose money or clients.

The introduction describes the subject and purpose of the proposal. The conclusion reinforces the confident tone throughout and encourages readers' acceptance by ending with — and thus emphasizing — two vital words: "economically" and "efficiently."

SUBJECT: <u>Proposal to Dig a Trench and Move Boulders at Site Ten Miles West of Bliss</u>

Dear Mr. Haver:

I've inspected your property and would be happy to undertake the landscaping project necessary for the development of your farm.

The backhoe I use cuts a span 3 feet wide and can dig as deep as 18 feet — more than an adequate depth for the mainline pipe you wish to lay. Because this backhoe is on tracks rather than tires, and is hydraulically operated, it is particularly efficient in moving rocks. I have more than twelve years of experience with backhoe work and have completed many jobs similar to this one.

After examining the huge boulders that block access to your property, I am convinced they can be moved only if I dig out underneath and exert upward pressure with the hydraulic ram while you push forward on the boulders with your D-9 Caterpillar. With this method, we can move enough rock to enable you to farm that now inaccessible tract. Because of its power, my larger backhoe will save you both time and money in the long run.

This job should take 12 to 15 hours, unless we encounter subsurface ledge formations. My fee is $100 an hour. The fact that I provide my own dynamiting crew at no extra charge should be an advantage to you because you have so much rock to be moved.

Please phone me at any time for more information. I'm sure we can do the job economically and efficiently.

The proposal categories (planning, research, and sales) discussed in this section are neither exhaustive nor mutually exclusive. A research proposal, for example, may request funds for a study that will lead to a planning proposal.

The Vista proposal partially shown on pages 427–431 is a combined planning and sales proposal: if clients accept the writer's preliminary plan, they will hire the firm to install the automated system.

## PROPOSAL GUIDELINES

Readers will evaluate your proposal according to how clearly, informatively, and realistically you answer these questions:

- *What are you proposing?*
- *What problem will you solve?*
- *Why is your plan worthwhile?*
- *What is unique about your plan?*
- *What are your (or your firm's) credentials?*
- *How will the plan be implemented?*
- *How long will it take to accomplish?*
- *How much will it cost?*
- *How will we benefit if we accept your plan?*

In addition to answering the questions above, successful proposal writers adhere to the following guidelines.

### Create an Accessible and Appealing Format

Format is the *look* of a document. It includes such features as

- the layout of words and graphics
- typeface, type size, and white space
- highlights and lists
- headings

See chapter 13 "Designing Effective Formats," for a discussion of these features.)
   Design your proposal to reflect your attention to detail, and your reader will pay attention. A poorly formatted and assembled proposal suggests to readers the writer's careless attitude toward the project.

### Signal the Proposal's Intent with a Clear Title

Decision makers are busy people who have no time for guessing-games. Begin with a title that is absolutely clear as to the proposal's purpose and content.

> Unclear    PROPOSED OFFICE PROCEDURES FOR VISTA FREIGHT, INC.

What kinds of office procedures are being proposed? The title is too broad; it gives readers no specific idea as to the proposal's focus.

Don't write "Recommended Improvements" when you mean "Recommended Wastewater Treatment." A specific and comprehensive title signals the proposal's intent.

## Include Supporting Material and Appropriate Supplements

Both short and long proposals often include supporting materials (maps, blueprints, specifications, calculations, and so forth). Place supporting material in an appendix (pages 415–417) so that it won't interrupt the discussion.

Depending on your readers, appropriate supplements (pages 407–418) for a long proposal might include a title page, cover letter, table of contents, summary, abstract, and appendixes. Readers with various responsibilities will be interested in different parts of your proposal: Some know about the problem and will read only your plan; some look only at the summary; others will study recommendations or costs; and still others need all the details. If you're unsure as to which supplements to include in an internal proposal, ask the intended reader(s) or study other proposals. If you're preparing a solicited proposal (one written for an outside agency) follow the agency's instructions *exactly*.

## Focus the Proposal's Subject and Purpose

Focus on one major question you can answer exhaustively and make your approach original enough to get the reader's attention and support. A fatal mistake is to begin writing too soon, before you have zeroed in on your subject and purpose.

Readers want specific suggestions for filling specific needs. By spelling out your subject and purpose, you show them immediately that you understand their problem or have an innovative plan for improving their products, sales, or services.

Notice how this proposal writer focuses on Vista's inefficient office procedures, and then outlines specific solutions.

Subject
Vista provides two services. (1) It finds freight carriers for its clients. The carriers, in turn, pay Vista a 6 percent commission for each referral. (2) Vista handles all shipping paperwork for its clients. For this auditing service, clients pay Vista a monthly retainer.

Although Vista's business has increased steadily for the past three years, record-keeping, accounting, and other paperwork still are done *manually*. These inefficient procedures have caused a number of problems, including late billings, lost commissions, and poor account maintenance. Vista will lose clients unless its office procedures improve.

Purpose

This proposal offers a realistic and efficient way for Vista to streamline office procedures. We first identify the problems imposed on your staff by the current system, and then we show how you can reduce inefficiency, eliminate client complaints, and improve your cash flow by automating most office procedures.

## Treat Potential Problems Realistically

Do not underestimate the project's complexity. Identify problems your readers might not anticipate, and propose realistic methods for solving them. Here is how the proposal to automate Vista's office procedures treats potential problems:

> As outlined below, Vista can realize many benefits by automating office procedures. But, as many firms have learned, *imposing* automated procedures on their staffs can create severe morale problems—particularly among senior staff. To diminish employee resistance, encourage your staff to study and comment on this proposal. To help avoid hardware and software problems once the system is operational, we have included recommendations and a budget for staff training. (Again, as firms have learned, cutting training costs from their automation budgets leads to inefficiency and negligible benefits.)

If the best available solutions have limitations, let the readers know. Otherwise, you and your firm could be liable in the case of project failure. Avoid overstatement. Notice how the above solutions are qualified ("diminish" and "help avoid" instead of *eliminate*) so as not to promise more than the writer can deliver. Ethical communication is essential.

## Explain the Benefits of Implementing the Proposal

A proposal's goal is to *persuade* readers to improve conditions, authorize work on a project, accept a service or product (for payment), or otherwise support a plan for solving a problem or doing a job. An effective proposal shows readers how they (or their organization) will benefit by adopting your plan.

The proposal written for Vista outlines these benefits. (Each benefit will be described at length in the body of the proposal.)

Once your automated system is operational, you will be able to

- identify cost-effective carriers,
- coordinate shipments (which will ensure substantial discounts for clients),
- print commission bills,
- track shipments by weight, miles, fuel costs, and destination,
- send clients weekly audit reports on their shipments
- bill clients on a 25-day cycle
- produce weekly or monthly reports

Additional benefits include eliminating repetitive tasks, improving cash flow, and increasing staff productivity.

## Include Concrete and Specific Information

Vagueness is a fatal flaw in a proposal. Before you can persuade readers, you must inform them; therefore, you need to *show* as well as *tell*. Instead of writing, "We will install state-of-the-art equipment," write,

> To meet your automation requirements, we will install 12 IBM PS/2 computers with 20 megabyte hard drives. The IBM's will be connected (networked) for rapid file transfer between offices. The plan also includes interconnection with 4 Epson high-speed dot matrix printers, and 1 Hewlett-Packard laser printer.

To avoid any misunderstanding and to reflect your ethical commitment, a proposal must elicit *one* interpretation only.

## Include Effective Visuals

If they enhance your proposal, use visuals, properly introduced and discussed. For example,

> As the flowchart illustrates (Figure 19.1), your routing and billing system creates a good deal of redundant work for your staff. The routing sheet alone is handled at least six times. Such extensive handling leads to errors, misplaced sheets, and tardy billing.

## Use a Tone That Connects with Your Readers

Your tone is your personal mark—the voice and personality that appears between the lines. While writing a proposal, keep your tone confident, encouraging, and diplomatic. Show readers you believe in your plan; urge them to act, and anticipate how they will react to your suggestions. But do not come across with a bossy or insulting tone, as in this example:

> *Had Vista's managers been better trained,* they could have streamlined office procedures by eliminating some of the more obvious paper-shuffling problems. Moreover, the clerical staff could have lessened the number of errors *had they been more vigilant.*

Here is a more diplomatic version:

> Vista's manual office procedures have resulted in an efficient flow of information. The problems include repetitive tasks, late billings, lost documents, and poor account maintenance.

## Analyze the Needs of Your Audience

Proposals address diverse audiences. A research proposal might be read by experts, who would then advise the granting agency whether to accept or reject it. Planning and sales proposals might be read by colleagues, superiors, and clients (often laypersons). Informed and expert readers will be most interested

**FIGURE 19.1** Flowchart of Vista's Manual Routing and Billing System

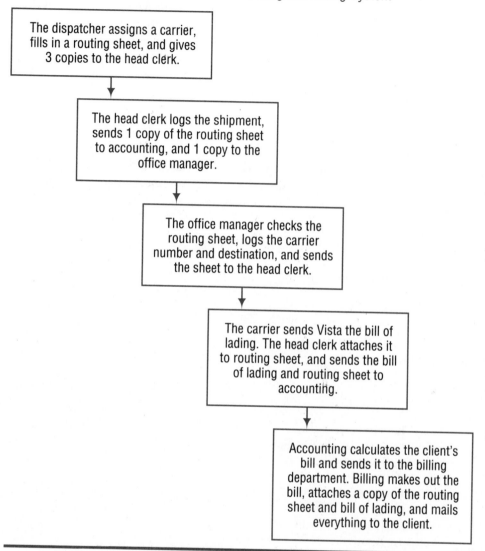

The dispatcher assigns a carrier, fills in a routing sheet, and gives 3 copies to the head clerk.

The head clerk logs the shipment, sends 1 copy of the routing sheet to accounting, and 1 copy to the office manager.

The office manager checks the routing sheet, logs the carrier number and destination, and sends the sheet to the head clerk.

The carrier sends Vista the bill of lading. The head clerk attaches it to routing sheet, and sends the bill of lading and routing sheet to accounting.

Accounting calculates the client's bill and sends it to the billing department. Billing makes out the bill, attaches a copy of the routing sheet and bill of lading, and mails everything to the client.

in the technical details of the project. Nontechnical readers will be more interested in the expected results, but they will need an explanation of technical details as well. Learn all you can about the needs, interests, and biases of your audience.

Unless your proposal gives all readers what they need, it is not likely to move anyone to action. This is where supplements are useful (especially abstracts, glossaries, and appendixes). Let your knowledge of the audience guide your decisions about supplements. Who is your secondary audience? Who else will be evaluating your proposal?

If the primary audience is expert or informed, keep the proposal itself technical. For uninformed secondary readers (if any), provide an informative abstract, a glossary, and appendixes explaining specialized information. If the primary audience has no expertise and the secondary audience does, follow this pattern: write the proposal itself for laypersons, and provide appendixes with the technical details (formulas, specifications, calculations) that the informed readers will use to evaluate your plan.

## COMPOSING A PROPOSAL

Like all informative writing, proposals are organized into these main sections: introduction, body, and conclusion. Depending on project complexity, each section has some or all of the subsections listed in the following general outline:

I. INTRODUCTION
   A. Subject and Purpose (or Objective)
   B. Statement of Problem
   C. Background
   D. Need
   E. Benefits
   F. Qualifications of Personnel
   G. Data Sources
   H. Limitations
   I. Scope

II. BODY
   A. Methods
   B. Timetable
   C. Materials and Equipment
   D. Personnel
   E. Available Facilities
   F. Needed Facilities
   G. Cost
   H. Expected Results
   I. Feasibility

III. CONCLUSION
   A. Summary of Key Points
   B. Request for Action

Subsection headings can be rearranged, combined, divided, or deleted as needed. Not every proposal has all subsections, but each major section must answer specific reader's questions, as illustrated next.

### Introduction

The introduction answers all these questions—or all those which apply to the situation:

- *What problem do you propose to solve?*
- *In general, what solution are you proposing?*
- *Why are you proposing it?*
- *What are the benefits?*
- *What are your qualifications for this project?*

From the beginning, your goal is to sell your idea, to convince readers that the job needs doing and you are the one to do it. If your introduction is long-winded, evasive, or vague, readers might not read on. Make it concise, specific, and clear.

Spell out the problem, to make it absolutely clear to the audience—and to show that you understand it fully. Explain the benefits of solving the problem or launching the project. Identify any sources of data. In a research or sales proposal, state your qualifications for doing the job. If your plan has limitations, explain them. Finally, give the scope of your plan by listing the subsections to be discussed in the body section.

Here is the introduction for a planning proposal titled "A Proposal for Solving the Noise Problem in the University Library." Jill Sanders, a library work-study student, addresses her unsolicited proposal to the chief librarian and the administrative staff. Because this proposal is unsolicited, it must first make the problem vivid through details that arouse concern and interest. This introduction is longer than it would be in a solicited proposal, whose readers would already agree on how severe the problem was.

## INTRODUCTION

### Subject
During the October 19xx Convocation at Margate University, many students and faculty members complained about noise in the library. Shortly afterward, areas were designated for "quiet study," but the library continues to receive complaints about noise. To create a scholarly atmosphere, the library staff should take immediate action to decrease noise.

### Purpose
In this proposal I examine the noise problem from the viewpoint of students, faculty, and library staff. I then offer a plan to make areas of the library quiet enough for serious study and research.

### Sources
My data come from a university-wide questionnaire, interviews with students, faculty, and library staff, inquiry letters to other college libraries, and my own observations for three years on the library staff.

### Statement of Problem
In this subsection I examine the severity and causes of the noise.

*Severity.* Since the 19xx Convocation, the library's fourth and fifth floors have been reserved for quiet study, but students hold group-study sessions at the large tables and disturb others working alone. The constant use of computer terminals on both floors adds to the noise, especially when students working in pairs discuss the program. Moreover, people often chat as they enter or leave study areas.

On the second and third floors, designed for reference, staff help patrons locate materials, causing constant shuffling of people and books, as well as loud conversation. At the computer service desk on the third floor, conferences between students and instructors create more noise.

The most frequently voiced complaint from the faculty members interviewed was about the second floor, where people using the Reference and Government Documents services converse loudly. Students complain about the lack of a quiet spot to study, especially in the evening, when even the "quiet" floors are as noisy as the dorms.

More than 80 percent of respondents (530 undergraduates, 30 faculty, 22 graduate students) to a university-wide questionnaire (Appendix A) insisted that excessive noise discourages them from using the library as often as they would otherwise. Of the student respondents, 430 cited quiet study as their primary reason for wishing to use the library.

The library staff recognizes the problem but has insufficient personnel to cope with it. Because all staff members have assigned tasks, they have no time to monitor noise in their sections.

*Causes.* Respondents complained specifically about these causes of noise (in descending order of frequency):

1. Loud study groups that often lapse into social discussions.
2. General disrespect for the library, with some students' attitudes characterized as "rude," "inconsiderate," or "immature."
3. The constant clicking of computer terminals on all five floors, and of typewriters on the first three.
4. Vacuuming by the evening custodians.

All complaints converged on lack of enforcement by the library staff.

Because the day staff works on the first three floors, quiet-study rules are not enforced on the fourth and fifth floors. Work-study students on these floors have no authority to enforce rules not enforced by the regular staff. Small, black-and-white "Quiet Please" signs posted on all floors go unnoticed, and the evening security guard provides no deterrent.

### Needs

Excessive noise in the library is keeping patrons away. By addressing this problem immediately, we can help restore the library's credibility and utility as a campus resource. We must reduce noise on the lower floors and eliminate it from the quiet-study floors.

### Scope

The proposed plan includes a detailed assessment of methods, costs and materials, personnel requirements, feasibility, and expected results.

## Body

The body (or plan) section of your proposal will receive most attention from readers. It answers all these questions that are applicable:

- *How will it be done?*
- *When will it be done?*
- *What materials, methods, and personnel will it take?*
- *What facilities are available?*
- *How long will it take?*
- *How much will it cost, and why?*
- *What results can we expect?*
- *How do we know it will work?*
- *Who will do it?*

Here you spell out your plan in enough detail for readers to evaluate its soundness. If this section is vague, your proposal stands no chance of being accepted. Be sure your plan is realistic and promises no more than you can deliver. The main goal of this section is to prove that your plan will work.

### PROPOSED PLAN

This plan takes into account the needs and wishes of our campus community, as well as the available facilities in our library.

#### Methods

Noise in the library can be reduced in three complementary steps: (1) improving publicity, (2) shutting down and modifying our facilities, and (3) enforcing the quiet rules.

*Improving Publicity.* First, the library must publicize the noise problem. This assertive move will demonstrate the staff's interest. Publicity could include articles by staff members in the campus newspaper, leaflets distributed on campus, and a freshman library orientation acknowledging the noise problem and asking cooperation by new students. All forms of publicity should detail the steps the library is taking to solve the problem.

*Shutting Down and Modifying Facilities.* After notifying campus and local newspapers, you should close the library for one week. To minimize disruption, the shutdown should occur between the end of summer school and the beginning of the fall term.

During this period, you can convert the fixed tables on the fourth and fifth floors to cubicles with temporary partitions (six cubicles per table). You could later convert the cubicles to shelves as the need increases.

Then you can take all unfixed tables from the upper floors to the first floor, and set up a space for group study. Plans already are under way for removing the computer terminals from the fourth and fifth floors.

***Enforcing the Quiet Rules.*** Enforcement is the essential, long-term element in this plan. No one of any age is likely to follow all the rules all the time—unless the rules are enforced.

First, you can make new "Quiet" posters to replace the present, innocuous notices. A visual-design student can be hired to draw up large, colorful posters that attract attention. Either the design student or the university print shop can take charge of poster production.

Next, through publicity, library patrons can be encouraged to demand quiet from noisy people. To support such patron demands, the library staff can begin monitoring the fourth and fifth floors, asking study groups to move to the first floor, and revoking library privileges of those who refuse. Patrons on the second and third floors can be asked to speak in whispers. Staff members should set an example by regulating their own voices.

## Costs and Materials

- The major cost would be for salaries of new staff members who would help monitor. Next year's library budget, however, will include an allocation for four new staff members.

- A design student has offered to make up four different posters for $200. The university printing office can reproduce as many posters as needed at no additional cost.

- Prefabricated cubicles for 26 tables sell for $150 apiece, for a total cost of $3,900.

- Rearrangement on various floors can be handled by the library's custodians.

The Student Fee Allocations Committee and the Student Senate routinely reserve funds for improving student facilities. A request to these organizations would yield at least partial funding for the plan.

## Personnel

The success of this plan ultimately depends on the willingness of the library administration to implement it. You can run the program itself by committees made up of students, staff, and faculty. This is yet another area where publicity is essential to persuade people that the problem is severe and that you need their help. To recruit committee members from among students, you can offer Contract Learning credits.

The proposed committees include an Antinoise Committee overseeing the program, a Public Relations Committee, a Poster Committee, and an Enforcement Committee.

## Feasibility

On March 15, 19xx, I mailed survey letters to 25 New England colleges, inquiring about their methods for coping with noise in the library. Among the respondents, 16 stated that publicity and the administration's attitude toward enforcement were main elements in their success.

Improved publicity and enforcement could work for us as well. And slight modifications in our facilities to concentrate group study on the busiest floors would automatically lighten the burden of enforcement.

*Expected Results*

Publicity will improve communication between library and the campus. An assertive approach will show that the library is aware of its patrons' needs and willing to meet those needs. Offering the program for public inspection will draw the entire community into improvement efforts. Publicity, begun now, will pave the way for the formation of committees.

The library shutdown will have a dual effect: it will dramatize the problem to the community, and also provide time for the physical changes. (An antinoise program begun with carpentry noise in the quiet areas would hardly be effective.) The shutdown will be both a symbolic and a concrete measure, leading to reopening of the library with a new philosophy and a new face.

Continued strict enforcement will be the backbone of the program. It will prove that staff members care enough about the atmosphere to jeopardize their own friendly image in the eyes of some users, and that the library is not afraid to enforce its rules consistently.

## Conclusion

The conclusion relates the need for the project and persuades readers to act. It answers the questions readers will ask:

- *How badly do we need this change?*
- *Why should we accept your proposal?*
- *How do we know this is the best plan?*

End on a strong note, with a conclusion that is assertive, confident, and encouraging—and keep it short.

### CONCLUSION AND RECOMMENDATION

The noise in Margate University library has become embarrassing and annoying to the whole campus. Forceful steps are needed to improve the academic atmosphere.

Aside from the intangible question of image, close inspection of the proposed plan will show that it will work if you take the recommended steps and—most important—if daily enforcement of quiet rules becomes a part of the library's services.

In long proposals, especially those beginning with a comprehensive abstract, the conclusion can be omitted.

When a few lines or short paragraphs can answer the readers' questions in each section, a short proposal will suffice, but a complex plan calls for a long, formal proposal. In the following section we will illustrate a formal proposal accompanied by all necessary supplements.

## APPLYING THE PRINCIPLES

The formal planning proposal in Figure 19.2 typifies the kind of specialized proposal that justifies a funding request.

### Audience-and-Use Profile

A university newspaper is struggling to meet rising costs. The paper's yearly budget is funded by a college allocations committee that disburses money to all student organizations. Because of revenue cuts throughout the university, the newspaper has received no funding increase in three years.

Bill Trippe, author of this proposal, is the newspaper's business manager. His task is to justify a request for a 17 percent budget increase for the coming year. Before drafting his proposal, Bill constructs a detailed profile of his audience (based on the worksheet, page 33).

---

**Audience Identity and Needs**

My primary audience includes all members of the allocations committee. My secondary audience is the newspaper staff, who will implement the proposed plan—if it is approved by the allocations committee.

The primary audience will use my document as perhaps the sole basis for deciding whether to grant the additional funds. Most of these readers have overseen the newspaper budget for years, and so they already know quite a bit about our overall operation. But they still need an item-by-item explanation of the conditions created by our problems with funding and ever-increasing costs. Probable questions I can anticipate are these:

- Why should the paper receive priority over campus organizations?
- Just how crucial is the problem?
- Are present funds being used efficiently?
- Can any expenses be reduced?
- How would additional funds be spent?
- How much will this increase cost?
- Will the benefits justify the cost?

**Attitude and Personality**

My primary audience often has expressed interest in this topic. But they are likely to object to any request for more money by arguing that everyone has to economize in these difficult times. I guess I could characterize their attitude as both receptive and hesitant. (Almost every campus organization is trying to make a case for additional funds.)

I do know most committee members pretty well, and they seem to respect my management skills. But I still need to spell out the problem and propose a realistic plan, showing that the newspaper staff is sincere in its intention to eliminate nonessential operating costs. At a time when everyone is expected to make do with less, I need to make an especially strong case for salary increases (to attract talented personnel).

### Expectations about the Document

My audience has requested (solicited) this proposal, and so I know it will be carefully read—but also scrutinized and evaluated for its soundness! Especially in a budget request, my audience expects no shortcuts; I'll have to itemize every expense. The Costs sections then could be the longest part of the proposal.

And to further justify the requested budget, I can demonstrate just how well the newspaper manages its present funds. Under Feasibility, I'll give a detailed comparison of funding, expenditures, and size of newspaper in relation to the four other local colleges. This section should be the "clincher" because these facts are most likely to persuade the committee that my plan is cost-effective. To avoid clutter, I'll add an appendix with a table of figures for the comparisons above (page 417).

To organize my document, I'll (1) identify the problem, (2) establish need, (3) propose a solution, (4) show that the plan is cost-effective, and (5) conclude with a request for action. My audience here expects a confident and businesslike—but not stuffy—tone. I want to be sure that everything in this proposal encourages readers to support our budget request.

---

(Figure 19.2 follows.)

FIGURE 19.2  A Sample Budget Proposal

# A Budget Proposal
# for
# the SMU *Torch*
# (1990–91)

Prepared for
The Student Fee Allocation Committee
Southeastern Massachusetts University
North Dartmouth, Massachusetts

by
William Trippe
*Torch* Business Manager

May 1, 1990

**FIGURE 19.2** A Sample Budget Proposal *Continued*

# The SMU *Torch*

Old Westport Road
North Dartmouth Massachusetts

May 1, 1990

Charles Marcus, Chair
Student Fee Allocation Committee
Southeastern Massachusetts University
North Dartmouth, MA 02747

Dear Dean Marcus:

No one needs to be reminded about the effects of increased costs on our campus community. We are all faced with having to make do with less.

Accordingly, we at the *Torch* have spent long hours devising a plan to cope with increased production costs—without compromising the newspaper's tradition of quality service. I think you and our colleagues will agree that our plan is realistic and feasible. Even the "bare-bones" operation that will result from our proposed spending cuts, however, will call for a $3837.14 increase in our 1990–91 budget.

We have received no funding increase in three years. Our present need is absolute. Without additional funds, the *Torch* simply cannot continue to function as a professional newspaper. I therefore submit the following budget proposal for your consideration.

Respectfully,

*William Trippe*

William Trippe
Business Manager, SMU *Torch*

**FIGURE 19.2** A Sample Budget Proposal *Continued*

## TABLE OF CONTENTS

Page

**FIGURE 19.2** A Sample Budget Proposal *Continued*

iv

## INFORMATIVE ABSTRACT

The student newspaper at Southeastern Massachusetts University is crippled by inadequate funding, having received no budget increase in three years. Increased costs and inadequate funding are the major problems facing the *Torch*. Increases in costs of printing, layout, and photographic supplies have called for a decrease in production. Moreover, our low salaries are inadequate to attract and retain qualified personnel. A nominal increase would make salaries more competitive.

Our staff plans to cut costs by reducing page count, saving on photographic paper, reducing circulation, and hiring a new printer. The only proposed cost increase (for staff salaries) is essential.

A detailed breakdown of projected costs establishes the need for a $3837.14 budget increase to keep the paper a weekly with adequate page count and distribution to serve our campus.

Compared with other college newspapers, the *Torch* makes much better use of its money. This comparison illustrates the cost effectiveness of our proposal.

**FIGURE 19.2** A Sample Budget Proposal *Continued*

1

### INTRODUCTION

Our campus newspaper faces the contradictory challenge of surviving ever-increasing production costs while maintaining its quality. The following proposal offers a realistic plan for meeting the crisis. The plan's ultimate success, however, depends on the Fee Allocation Committee's willingness to approve a long-overdue increase in our 1990–91 budget.

In its ten years, the *Torch* has grown in size, scope, and quality. Roughly 6000 copies (24 pages/issue) are printed weekly each 14-week semester.

Each week the *Torch* prints news, features, editorials, sports articles, announcements, notices, classified ads, columns, and letters to the editor. A vital part of university life, the paper disseminates information, ideas, and opinions—all with highest professionalism.

With much of its staff about to graduate, the *Torch* faces next year with rising costs in every phase of production, and the need to replace equipment that is outdated and worn out.

Our newspaper also suffers from a lack of student involvement. Few students can be expected to work without some kind of salary. Most staff members do receive minimal weekly salaries: from $10 for the distributor to $45 for the Editor-in-Chief, but salaries averaging less than $2 per hour cannot possibly compete with the minimum wage. Since most SMU students must work part-time, the *Torch* will have to make salaries more competitive.

The newspaper's operating expenses can be divided into four categories: composing costs, salaries, printing costs, and miscellaneous (office supplies, mail, etc.). The first three categories account for nearly 90 percent of the budget. In the past year, costs in all categories have increased from as little as 3 percent for darkroom chemicals to as much as 60 percent for photographic paper and film. Printing costs (roughly one-third of our total budget) rose by 9 percent in the past year, and another price hike of 10 percent has just taken effect.

Despite the steady increase in production costs, the *Torch* has received no increase in its yearly allocation ($21,500) in 3 years.

The plan following includes:
1. methods for reducing production costs while maintaining the quality of our staff
2. projected costs for equipment, material, salaries, and services in 1990–91
3. a demonstration of feasibility, showing our cost effectiveness
4. a summary of the attitudes shared by our staff

**FIGURE 19.2** A Sample Budget Proposal *Continued*

2

## PROPOSED PLAN

The following plan is designed to trim operating costs without compromising quality.

**Methods**

We can respond successfully to our budget and staffing crisis by taking the following steps:

**Reducing Page Count.**   By condensing free notices for campus organizations, abolishing "personals," and limiting press releases to one page, we can reduce page count per issue from 24 to 20. This reduction will save nearly 17 percent in production costs.

**Saving on Photographic Paper.**   A newspaper with fewer pages will call for fewer photographs. Also, we can further save money by purchasing a year's supply of photographic paper within the next month, before the upcoming 15 percent increase becomes effective.

**Reducing Circulation.**   By reducing circulation from 6000 to 5000 copies weekly, we will barely cover the number of full-time day students, but we will save nearly 17 percent in printing costs.

**Hiring a New Press.**   We can save money by hiring Sadus Press for printing. Bids from other presses (including our present printer) were at least 25 percent higher than Sadus's price. Also, no other company offers the delivery service we will get from Sadus.

**Increasing Staff Salaries.**   Although our staff seeks talented students who expect little money and much good experience, our salaries for all positions *must* increase by an average of $5 weekly. Otherwise, any of our staff could make as much money elsewhere by working only one-third the time. In fact, many students could make more than the minimum wage by working for local newspapers. To illustrate: *The Standard Times* pays $20 to $30 for a news article and $10 for a photo, while the *Torch* pays nothing for articles and $2 for a photo.

A most striking example of our low salaries is the $3.35 we pay our typesetters. On the outside, trained typesetters like ours make from $8 to $12 an hour. Thus, our present typesetting cost of $3038 easily could be as much as $7000—or even higher if we had the typesetting done by an outside firm, as many colleges do.

Without this nominal salary increase, we cannot possibly attract qualified personnel.

**Budget Request for 1990-91**

Our proposed budget is itemized below, but the main point is clear: if the *Torch* is to remain a viable newspaper, increased funding is essential for meeting projected costs.

**FIGURE 19.2** A Sample Budget Proposal *Continued*

3

| **Projected Costs** | | | **Yearly Total** |
|---|---|---|---|
| Equipment Leasing | | | |
| (We must continue to meet the obligations of a leasing arrangement with Chase Manhattan Leasing Corporation for Compugraphic equipment.) | | | |
| Execuwriter | $ 106.25/month | | |
| 7200 L and hardware | 127.38/month | | |
| Execuwriter II | 138.63/month | | |
| | 372.26/month | = | $3,350.34 |
| | | | |
| Composing Chemicals | | | |
| Activator | $ 105.00 | | |
| Stabilizer | 180.00 | | |
| Processor Cleaner | 55.00 | | |
| | $ 340.00 | = | $ 340.00 |
| | | | |
| Photo Paper for Execuwriters | | | |
| 3" x 150' | $ 711.00 | | |
| 7" x 150' | 556.37 | | |
| Headline paper | 732.63 | | |
| 8" x 150' | 227.00 | | |
| | $2,227.00 | = | $2,227.00 |
| | | | |
| Layout Supplies | | | |
| Font strips | $160.00 | | |
| Layout boards | 160.00 | | |
| Exacto knives and blades | 80.00 | | |
| Non-repro pens | 30.00 | | |
| Rulers | 12.00 | | |
| Scotch tape | 50.00 | | |
| Folders | 30.00 | | |
| Reduction wheels | 20.00 | | |
| Wax | 45.00 | | |

**FIGURE 19.2** A Sample Budget Proposal *Continued*

4

| Projected Costs | | | Yearly Total |
|---|---|---|---|
| Construction paper | 10.00 | | |
| Border tape | 50.00 | | |
| | $647.00 | = | $647.00 |
| | | | |
| **Photography Supplies** | | | |
| Paper | $ 730.00 | | |
| Film | 250.00 | | |
| Darkroom supplies | 100.00 | | |
| | $1,080.00 | = | $1,080.00 |
| | | | |
| **Typing Staff** | | | |
| 35 hrs. at $3.35/hr | $ 117.25 /wk | = | $3,283.00 |
| | | | |
| **Salaries** | | | |
| Editor-in-Chief | $1,400.00/year | | |
| News Editor | 840.00 | | |
| Asst. News Editor | 420.00 | | |
| Features Editor | 840.00 | | |
| Head Writer (Features) | 420.00 | | |
| Sports Editor | 840.00 | | |
| Head Writer (Sports) | 420.00 | | |
| Advertising Manager | 1,050.00 | | |
| Advertising Designer | 700.00 | | |
| Free Ad Designer | 280.00 | | |
| Layout Editor | 840.00 | | |
| Art Director | 560.00 | | |
| Photo Editor | 840.00 | | |
| Business Manager | 840.00 | | |
| | $10,290.00 | = | $10,290.00 |
| | | | |
| **Fixed Printing Costs** | | | |
| 5000 copies/week x 28 | $12,399.80 | = | $12,399.80 |

FIGURE 19.2  A Sample Budget Proposal *Continued*

5

| Projected Costs | | Yearly Total |
|---|---|---|
| Miscellaneous Costs | | |
| Graphics by SMU art students: | | |
| 3/wk. at $5 each | $ 420.00 | |
| Mail costs | 550.00 | |
| Telephone costs | 500.00 | |
| Print shop costs | 200.00 | |
| Copier fees | 50.00 | |
| | $1,720.00 = | $1,720.00 |
| | | |
| TOTAL YEARLY COSTS | | $35,337.14 |
| Expected Ad Revenue | | |
| (1000/mo. x 10) | | $10,000.00 |
| Total Costs minus Ad Revenue | | $25,337.14 |
| | | |
| TOTAL BUDGET REQUEST | | $25,337.14 |

## Feasibility

Beyond exhibiting our need, we feel that the feasibility of this proposal can be measured through an objective assessment of our cost effectiveness: Compared with other school newspapers, how well does the *Torch* use its funds?

In a survey of the four area colleges, we found that the *Torch*—by a sometimes huge margin—makes the best use of its money. Table 1 in Appendix A [page 417] shows that, of the five newspapers, the *Torch* costs the students least, runs the most pages a week, and spends the least money per page, *despite a circulation two to three times the size of the other papers.*

The most striking comparison is between the *Torch* and the newspaper at Fallow State College (Appendix A). Each student at FSC pays $12.33 yearly for a paper averaging 12 pages per issue. Here at SMU, each student pays $4.06 yearly for a paper averaging 20 pages an issue. Thus, for 33 percent of FSC's cost, SMU students are getting 66 percent more newspaper.

The *Torch* has the lowest yearly cost of all five papers, despite the largest circulation. With the budget increase requested above, the cost would rise only by $0.44, for a yearly cost of $4.50 to each student. Although Alden College's paper costs each student $4.29, it is

**FIGURE 19.2** A Sample Budget Proposal *Continued*

6

published only every third week, averages 12 pages an issue, and costs nearly $50.00 yearly per page to print—as opposed to our yearly printing cost of $38.25 per page.

As the figures in Appendix A demonstrate, our cost management is responsible and effective.

**Personnel**

Students on the *Torch* staff are unanimous in their determination to maintain the highest professionalism. Many are planning careers in journalism, writing, editing, advertising, photography, or public relations. In any issue, the balanced, enlightened coverage is evidence of our judicious selection and treatment of articles and our shared concern for quality.

**CONCLUSION**

As a forum for ideas and opinions, the *Torch* continues to reflect a seriousness of purpose and a commitment to free expression. Its place in the campus community is more vital than ever in these troubled times.

There are increases and decreases in student allocations every year. Last year, for example, eight allocations were increased by an average of $2,166. The *Torch* has received no increase since 1986–87. Presumably, various increases materialize as priorities change and as special circumstances arise. The *Torch* staff urges the Allocation Committee to respond to the paper's legitimate and proven needs by increasing our 1990-91 allocation to $25,337.14.

## REVISION CHECKLIST FOR PROPOSALS

Use this list to revise your proposal. (Numbers in parentheses refer to the first page of discussion.)

### Format

☐ Is the short internal proposal in memo form and the short external proposal in letter form? (422)

☐ Does the long proposal have adequate supplements to serve the different needs of different readers? (428)

☐ Is the format professional in appearance?

☐ Are headings logical and adequate? (427)

☐ Does the title forecast the proposal's subject and purpose? (427)

### Content

☐ Is the subject original, concrete, and specific? (428)

☐ Is the purpose clear and worthwhile? (428)

☐ Does everything in the proposal support its stated purpose? (428)

☐ Does the proposal *show* as well as *tell*? (430)

☐ Is the proposed plan, service, or product beneficial? (429)

☐ Are the proposed methods practical and realistic? (435)

☐ Are all related problems identified? (429)

☐ Is the proposal free of overstatement? (429)

☐ Are visuals used effectively and whenever appropriate? (430)

☐ Is the proposal's length appropriate to the subject? (437)

### Arrangement

☐ Is there an introduction, body, and conclusion? (432)

☐ Are all *relevant* headings from the general outline included? (432)

☐ Does the introduction explain *what* and *why?* (432)

☐ Does the body explain *how, where,* and *how much?* (435)

☐ Does the conclusion encourage acceptance of the proposal? (437)

☐ Are there clear transitions between related ideas? (518)

### Style

☐ Is the level of technicality appropriate for primary readers? (430)

☐ Do supplements follow the appropriate style guidelines? (428)

☐ Will the informative abstract be understood by all readers? (412)

☐ Does the tone connect with readers? (430)

☐ Is each sentence clear, concise, and fluent? (199)

☐ Is the language convincing and precise? (221)

☐ Is the proposal in correct English? (Appendix A)

Now revise all material that has not been checked off.

## EXERCISES

**1.** After consulting instructors or people on the job, write a memo to your instructor identifying the kinds of proposals most often written in your field. Are such proposals usually short or long, internal or external, solicited or unsolicited? Who are the decision makers in your field? Provide a scenario of a work situation that calls for a proposal. Include your sources of information in your memo.

**2.** Assume the head of your high school English Department has asked you, as a recent graduate, for suggestions about revising the English curriculum to prepare students for writing. Write a proposal, based on your experience since high school. (Primary audience: the English Department head and faculty; secondary audience: school committee.) In your external, solicited proposal, identify problems, needs, and benefits and spell out a realistic plan. Review the outline on page 432 before selecting specific headings.

**3.** From a faculty member in your major, obtain a copy of a recent proposal for changes in the department's course offerings or staffing, or for solving a departmental problem (e.g., instituting a minor, adding new courses, dropping courses, or hiring new faculty). Using the revision checklist, evaluate the proposal. Submit your findings in a memo to your instructor, or discuss your evaluation in class, using an opaque projector.

**4.** After identifying your primary and secondary audience, compose a short planning proposal for improving an unsatisfactory situation in the classroom, on the job, or in your dorm or apartment (e.g., poor lighting, drab atmosphere, health hazards, poor seating arrangements). Choose a problem or situation whose resolution is more a matter of common sense and lucid observation than of intensive research. Be sure to (a) identify the problem clearly, give brief background, and stimulate the readers' interest; (b) state clearly the methods proposed to solve the problem; and (c) conclude with a statement designed to gain readers' support for your proposal.

**5.** Write a research proposal to your instructor (or an interested third party) requesting approval for the final term project (an analytical report or formal proposal). Identify the subject, background, purpose, and benefits of your planned inquiry, as well as the intended audience, scope of inquiry, data sources, methods of inquiry, and a task timetable. Be certain that adequate primary and secondary sources are available. Convince your reader of the soundness and usefulness of the project.

**6.** As an alternate term project to the formal analytical report (see Chapter 20), develop a long proposal for solving a problem, improving a situation, or satisfying a need in your school, community, or job. Choose a subject sufficiently complex to justify a formal proposal, a topic requiring research (mostly primary). Identify an audience (other than your instructor) who will use your proposal for a specific purpose. Compose an audience-and-use profile, using the sample on pages 438–439 as a model. Here are possible subjects for your proposal:

- improving living conditions in your dorm or fraternity/sorority
- creating a student-operated advertising agency on campus
- creating a day-care center on campus
- creating a new business or expanding a business
- saving labor, materials, or money on the job
- improving working conditions
- supplying a product or service to clients or customers
- improving campus facilities for the handicapped
- developing a community waste-recycling program
- developing a bicycle safety program in your community
- increasing tourist trade in your town
- eliminating traffic hazards in your neighborhood
- developing plans to handle parking problems (e.g., plans to handle campus parking and traffic patterns for sports events)
- reducing energy expenditures on the job
- improving security in dorms or in the college library
- improving in-house training or job orientation programs
- deciding how a chapel might best use a bequest of money
- creating a jogging path through the campus
- creating a bike path or special bike lane on main streets
- creating a one-credit course in job hunting or stress management for students
- improving tutoring conditions in the learning center
- making the course content in your major more relevant to student needs
- creating a new student organization within the government (say, one to handle ethnic affairs)
- finding ways for an organization to raise money
- improving faculty advising for students
- purchasing new equipment
- improving food service on campus
- funding purchase of new uniforms for an athletic or musical organization
- easing freshman students through the transition to college
- improving bomb-alert procedures in the student union building
- making word processing available to all advanced writing students
- establishing a college-wide computer fluency program
- persuading local manufacturers to disclose their toxic-waste dump sites
- changing the grading system at your school
- leasing microcomputers to students
- establishing more equitable computer terminal use
- offering a service to potential customers

# Analytical Reports

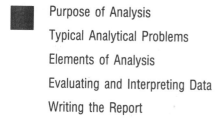

Purpose of Analysis

Typical Analytical Problems

Elements of Analysis

Evaluating and Interpreting Data

Writing the Report

Applying the Principles

The formal analytical report, like the short recommendation reports in Chapter 17, leads to recommendations. The formal report is used instead of the memo when the topic requires lengthy discussion.

Analysis is basic to our thinking, and in this sense all the assignments you have written have involved analysis. A summary requires analyzing the original for essential points; expanded definition often includes analyzing parts; description or process explanation requires analyzing the item or process. In each analysis, we divide the subject into its parts (partition) and group these parts within specific categories according to their similarities (classification).

When you analyze data, however, your subject is not an item that needs describing or a process that needs explaining. Instead it is a question that needs answering, or a problem that needs solving. Your analysis might influence a major decision. Say, for example, you receive this assignment from your supervisor:

Recommend the best method for removing the heavy-metal contamination from our company dump site.

Clearly, to get this job done, you will have to do more than observe your subject. Because high-level decisions will depend on your findings, you must *seek out* and *interpret* all data that will help you make the best recommendations. This is where you apply critical thinking skills along with the research activity discussed in Chapter 6. As part of your research, you might do a literature search to discover whether anyone has been able to solve a problem like yours, or to learn about the newest technologies for toxic cleanup. You then have to decide how much, if any, of what others have done can apply to your situation.

## PURPOSE OF ANALYSIS

On the job, you apply analytical skills to problems, proposals, and planning. A structural engineer analyzes material and design in evaluating plans for a suspension bridge or skyscraper. Before buying stock, the wise investor analyzes economic trends, treasurers' reports, performance, dividend rates, and market conditions. A legal defense is built on the attorney's analysis and logical reconstruction of events in a case. Medical diagnosis relies on the precise identification and classification of symptoms, along with the communication of findings, interpretations, and recommendations for treatment.

As an employee you may be asked to evaluate a new assembly technique on the production line, or to locate and purchase the best equipment at the best price. You might have to identify the cause behind a monthly drop in sales, the reasons for low employee morale, the causes of an accident, or the reasons for equipment failure. You might need to assess the feasibility of a proposal for a company's expansion or investment. The list is endless but the procedure remains the same: (1) making a plan, (2) finding the facts, (3) interpreting the findings, and (4) drawing conclusions and making recommendations.

## TYPICAL ANALYTICAL PROBLEMS

The analytical report is far more than an encyclopedia presentation of information. Your aim is to show how you arrived at your conclusions and recommendations. Your approach will depend on your subject, purpose, and reader's needs. Here are some typical analytical problems.

### "Will X Work for a Specific Purpose?"

Analysis can answer practical questions. Say your employer is concerned about the effects of stress on personnel. He or she asks you to investigate the claim that low-impact aerobics has therapeutic benefits—with an eye toward such a program for employees. You design your analysis to answer this question: "Do low-impact aerobics programs significantly reduce stress?" The analysis follows a *questions-answers-conclusions* structure. Because the report could lead to action, you include recommendations based on your conclusions.

### "Is X or Y Better for a Specific Purpose?"

Analysis is essential in comparing machines, processes, business locations, computer systems, or the like. Assume that you manage a ski lodge and need to answer this question: Which of the two most popular types of ski binding is best for our rental skis? In a comparative analysis of the Salomon 555 and the Americana bindings, you might assess the strengths and weaknesses of each in a point-by-point comparison: toe release, heel release, ease of adjustment, friction, weight, and cost. Or you might use an item-by-item comparison, discussing the first whole binding, then the next.

The comparative analysis follows a *questions-answers-conclusions* structure and is designed to help the reader make a choice. Examples are found in magazines such as *Consumer Reports* and *Consumer's Digest*.

## "Why Does X Happen?"

The problem-solving analysis is designed to answer questions like this: Why do independent television service businesses have a high failure rate? This kind of analysis follows a variation of the questions-answers-conclusions structure: namely, *problem-causes-solution*. Such an analysis follows this sequence:

1. identifying the problem
2. examining possible and probable causes, and isolating definite ones
3. recommending solutions

An analysis of low employee morale would follow this structure.

## "How Can X Be Improved?"

Another form of problem solving identifies needs, and suggests ways of filling these needs. This type of analysis answers questions like these:

How can we improve safety procedures in our plant? How can we operate our division more efficiently? How can we improve campus security? The sample report on pages 463–467 tackles this question: How can our university's communications program better serve students and area businesses?

## "What Are the Effects of X?"

An analysis of the consequences of an event or action would answer questions like this: How has air quality been affected by the local power plant's change from burning oil to coal?

Another kind of problem-solving analysis is done to predict an effect: "What are the consequences of my changing majors?" Here, the structure is *proposed action-probable effects-conclusions and recommendations*.

## "Is X Practical in This Situation?"

The feasibility analysis assesses the practicality of an idea or plan: Will the consumer interests of Hicksville support a microcomputer store? In a variation of the questions-answers-conclusions structure, a feasibility analysis uses *reasons for-reasons against*, with both sides supported by evidence. Business owners often use this type of analysis.

## Combining Types of Analyses

Types of analytical problems overlap considerably. Any one study may in fact require answers to two or more of the questions above. The sample report on pages 470–478 is both a feasibility analysis and a comparative analysis. It is designed to answer these questions: Is technical marketing the right career for me? If so, how do I enter the field?

## ELEMENTS OF ANALYSIS

The analytical report incorporates many writing strategies used in earlier assignments, along with the guidelines that follow.

### Clearly Identified Problem or Question

Know what you're looking for. If your car's engine fails to turn over when you switch on the ignition, you would wisely check battery and electrical connections before dismantling parts of the engine. Apply a similar focus to your report.

Earlier, a hypothetical employer posed this question: *Will a low-impact aerobics program significantly reduce stress among my employees?* The aerobics question obviously requires answers to three other questions: *What are the therapeutic claims for aerobics? Are they valid? Will aerobics work in this situation?* How aerobic exercise got established, how widespread it is, who practices, and other such questions are not relevant to this problem (although some questions about background might be useful in the report's introduction). Always begin by defining the main questions and thinking through any subordinate questions they may imply. Only then can you determine the data or evidence you need.

With the main questions identified, the writer of the aerobics report can formulate her statement of purpose:

> This report examines for the general reader some of the claims about therapeutic benefits made by practitioners of low-impact aerobic exercise.

The writer might have mistakenly begun instead with this statement:

> This report examines low-impact aerobic exercise and communicates its findings to the general reader.

Notice how the first version sharpens the focus by expressing the precise subject of the analysis: not aerobics (a huge topic), but the alleged *therapeutic benefits* of aerobics.

Define your purpose by condensing your approach to a basic question: *Does low-impact aerobic exercise have therapeutic benefits?* or *Why have our sales dropped steadily for three months?* Then restate the question as a declarative sentence in your statement of purpose.

### A Report with No Bias

Interpret evidence impartially. Throughout your analysis, stick to your evidence. Do not force personal viewpoints on your material.

## Accurate and Adequate Data

Do not distort the original data by excluding vital points. Say you are asked to recommend the best chainsaw for a logging company. Reviewing test reports, you come across this information.

> Of all six brands tested, the Bomarc chainsaw proved easiest to operate. It also had the fewest safety features, however.

If you cite these data, present *both* findings, not simply the first—even though you may prefer the Bomarc brand. *Then* argue for the point you prefer.

As space permits, include the full text of interviews or questionnaires in appendixes.

## Fully Interpreted Data

Explain the significance of your data. Interpretation is the heart of the analytical report. You might interpret the chainsaw data in this way:

> Our cutting crews often work suspended by harness, high above the ground. And much work is in remote areas. Safety features therefore should be our first requirement in a chainsaw. Despite its ease of operation, the Bomarc saw does not meet our needs.

By saying "therefore . . ." you engage in analysis—not mere information-sharing. *Merely listing your findings is not enough.* Tell readers what your findings mean.

## Clear and Careful Reasoning

Reporting is not simply a mechanical collecting and recording of information. Each step of your analysis requires decisions about what to record, what to exclude, and where to go next. As you evaluate your data *(Is this reliable and important?)*, interpret your evidence *(What does it mean?)*, and make recommendations based on your conclusions *(What action is needed?)*, you might have to alter your original plan. You cannot know what you will find until you have searched. Remain flexible enough to revise when new evidence appears.

## Appropriate Length and Visuals

Depending on the problem or question, your analysis may range from a short memo to a long report with supplements. Make it just long enough to show readers how you arrived at your conclusions.

Use visuals generously (Chapter 12). Graphs are especially useful in an analysis of trends (rising or falling sales, radiation levels). Tables, charts, photographs, and diagrams work well in comparative analyses.

## Valid Conclusions and Recommendations

Throughout your analysis, you collect, evaluate, and interpret the best information on your topic or question. From this material, you decide on a *conclusion*—an overall judgment about what all your material means (that *X* is better than *Y*, that *B* failed because of *C*, that *A* is a good plan of action). Here is the interpretation that concludes a report on the feasibility of installing an active solar heating system in a large building:

1. Active solar space heating for our new research building is technically feasible because the site orientation will allow for a sloping roof facing due south, with plenty of unshaded space.

2. It is legally feasible because we are able to obtain an access easement on the adjoining property, to ensure that no buildings or trees will be permitted to shade the solar collectors once they are installed.

3. It is economically feasible because our sunny, cold climate means high fuel savings and a faster payback (fifteen years maximum) with solar heating. The long-term fuel savings justify our short-term installation costs (already minimal because the solar system can be incorporated during the building's construction—without renovations).

Having explained *what it all means,* you then recommend *what should be done.* Taking into account all possible alternatives, your recommendations urge specific action (to invest in *A* instead of *B*, to replace *C* immediately, to follow plan *A*, or the like). Here are the recommendations based on the previous interpretations:

1. I recommend that we install an active solar heating system in our new research building.

2. We should arrange an immediate meeting with our architect, building contractor, and solar-heating contractor. In this way, we can make all necessary design changes before construction begins in two weeks.

3. We should instruct our legal department to obtain the appropriate permits and easements immediately.

Your recommendations represent your ultimate decisions, the culmination of the whole report project. But keep in mind that a recommendation can be worth no more than the evidence and interpretations upon which it is based. A biased, incomplete, or otherwise faulty analysis is sure to produce worthless recommendations.

When you do achieve definite conclusions and recommendations, express them with assurance and authority. Unless you have reason to be unsure, avoid noncommittal statements ("It would seem that . . ." or "It looks as if . . ."). Be direct and assertive ("The earthquake danger at the reactor site is acute," or "I recommend an immediate investment"). Let readers know where you stand.

If, however, your analysis yields nothing definite, do not force a simplistic conclusion on your material. Instead, explain your position ("The contradictory responses to our consumer survey prevent me from reaching a definite conclusion. Before we make any decision about this product, I recommend a full-scale market analysis"). The wrong recommendation is far worse than no recommendation at all.

## EVALUATING AND INTERPRETING DATA

Researching (Chapter 6) is only part of your task. As you work with your information, you must evaluate its reliability as well as interpret it. Facts or statistics out of context can be interpreted in many ways, but you are ethically bound to find answers that stand the greatest chance of being truthful. To ensure the validity of your report, choose reliable sources, distinguish hard from soft evidence, and avoid specious reasoning.

### Choose Reliable Sources

Make sure each source is reputable, impartial, and authoritative. Say you are analyzing the alleged benefits of low-impact aerobics for stress reduction. You could expect claims in a reputable professional journal, such as the *New England Journal of Medicine,* to be reliable. Also, a reputable magazine, such as *Scientific American,* would be a reliable source. Obviously, claims in glamour or movie magazines are suspect. Even claims in monthly "digests," which offer simplified and mostly undocumented "wisdom" to mass audiences, should be verified.

Interview only people who have practiced aerobics extensively. Anyone who has practiced for only a few weeks could hardly assess bodily changes reliably. And to the extent you rely on personal experience reports, including those who have long practiced aerobics, you need a representative sample. Even reports from ten successful practitioners would be a small sample unless those reports were supported by laboratory data. On the other hand, with a hundred reports from people ranging from students to judges and doctors, you might not have "proved" anything, but the persuasiveness of your evidence would increase.

Your own experience, also, is not a valid base for generalizing. We cannot tell whether our experience is representative, regardless of how long we might have practiced meditation. If you have practiced aerobics with success, interpret your experience only within the broader context of your collected sources.

Some issues (the nuclear arms build-up or causes of inflation) are always controversial, and will never be resolved. Although we can get verifiable data and can reason persuasively on some subjects, no close reasoning by any expert and no supporting statistical analysis will "prove" anything about a controversial subject. Somebody's thesis that the balance of payments is at the root of the inflation problem cannot be proved. Likewise, one could only *argue* (more or less effectively) that federal funds will or will not alleviate poverty or unemployment. Some problems simply are more resistant to solution than others, no matter how reliable the sources.

Amid such difficulties, resist the temptation to report hasty but unverified answers. Better to report no answers than misleading ones. Take no claims for granted, and cross-check all data.

### Distinguish Hard from Soft Evidence

Hard evidence consists of facts. It can stand up under testing because it is verifiable. Soft evidence consists of uninformed opinion. It may collapse under testing unless the opinion is expert and unbiased.

Base your conclusions on hard evidence. Early in your analysis, you might read an article that makes positive claims about low-impact aerobics, but the article provides no data on measurements of pulse, blood pressure, or metabolic rates. Although your own experience and opinion might agree with the author's, do not hastily conclude that this form of exercise benefits everyone. So far, you have only two opinions — yours and the author's — without scientific support (e.g., tests of a cross-section under controlled conditions). Conclusions now would rest on soft evidence. Only after a full survey of reliable sources can you decide which conclusions are supported by the bulk of your evidence.

Until evidence proves itself hard, consider it soft. Say a local merchant claims that a nearby vacant shop is the best business location in town, but offers no facts to support his assertion. Before you rent the shop, cross-check other sources, and observe customer traffic in adjoining businesses. On the other hand, if the previous owner gave you the same judgment, even though he did not support it with fact, you might take it seriously — particularly if you know he is honest and has vacated the shop because he has just purchased a $325,000 retirement home at age fifty-five.

Be careful not to confuse probable, possible, and definite causes in a problem-solving analysis. Sometimes a definite cause can be identified easily (e.g., "The engine's overheating is caused by a faulty radiator cap"), but usually a good deal of searching and thought are needed to isolate a specific cause. Say you tackle this question: Why does our state college campus have no children's day-care facilities? A brainstorming session yields this list of possible causes:

- lack of need among students
- lack of interest among students, faculty, and staff
- high cost of liability insurance
- lack of space and facilities on campus
- lack of trained personnel
- prohibition by state law
- lack of legislative funding for such a project

Assume you proceed with interviews, questionnaires, and research into state laws, insurance rates, and local availability of qualified personnel. You gradually rule out some items, but others appear as probable causes. Specifically, you find a need among students, high campus interest, an abundance of qualified people for staffing, and no state laws prohibiting such a project. Three probable causes remain: lack of funding, high insurance rates, and lack of space. Further inquiry shows that lack of funding and high insurance rates *are* issues. These causes, however, can be eliminated by creating other sources of revenue: charging a fee for each child, holding fund-raising drives, or diverting funds from other campus organizations. Finally, after examining available campus space and speaking with school officials, you arrive at one definite cause: lack of space and facilities.[1]

Be sure readers can draw conclusions identical to your own on the basis of your evidence. A problem-solving analysis like this one requires that you discuss all possible causes, narrowing your focus to probable and then definite causes. The process might be diagrammed like this:

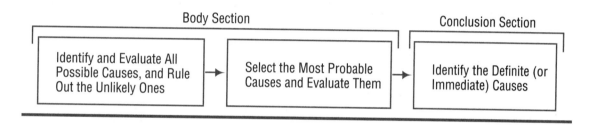

Early in your analysis you might have based your conclusions hastily on soft evidence (say, an opinion—buttressed by a newspaper editorial—that the campus was apathetic). Now you can reason from a basis of solid, factual evidence. You have moved from a wide range of possible causes to a few probable causes, to the most likely cause. Because you have covered your ground well, your recommendations will have credibility and authority.

Sometimes, finding a single cause is impossible, but this reasoning process can be tailored to most problem-solving analyses. Although few but the simplest effects have one cause, usually one or more principal causes become evident. By narrowing the field, you can focus on the real issues.

## Avoid Specious Reasoning

Specious reasoning is deceptive because it seems correct at first glance, but is faulty when scrutinized. Reasoning based on soft evidence is often specious. Conclusions you derive speciously fail under testing.

[1]Of course, one could argue that lack of space and facilities is somehow related to funding. And the college's being unable to find funds or space may be related to student need, which is not sufficiently acute or interest sufficiently high to exert real pressure. Lack of space and facilities, however, appear to be the *immediate* cause.

Assume you are an education consultant. Your community has asked you to analyze the accuracy of IQ testing as a measure of intelligence and as a predictor of students' performance. Reviewing your collected evidence, you find a positive correlation between low IQ scores and low achievers. You then verify your own statistics by examining a solid cross-section of reliable sources. You might now feel justified in concluding that IQ tests do measure intelligence and predict performance. This conclusion, however, would be specious unless you could show that

1. Parents, teachers, and the children tested had not seen individual test scores and had thus been unable to develop biased attitudes.

2. Children tested in all IQ categories had later been exposed to an identical curriculum at an identical pace. In other words, they were not channeled into programs on the basis of their scores.

Your total data could be interpreted only within the context of these two variables. Even hard evidence can be used to support specious reasoning, unless it is interpreted precisely, objectively, and within a context that accounts for all variables.

## WRITING THE REPORT

You might outline *before* or *after* writing a first draft. Whichever you choose to do, the finished report depends on a good outline. This model outline can be adapted to most analytical reports:

**I.** INTRODUCTION
  **A.** Definition, Description, and Background
  **B.** Purpose of the Report, and Intended Audience
  **C.** Sources of Information
  **D.** Working Definitions (here or in a glossary)
  **E.** Limitations of the Study
  **F.** Scope of the Inquiry (topics listed in order of importance)

**II.** COLLECTED DATA
  **A.** First Topic for Investigation
    **1.** Definition
    **2.** Findings
    **3.** Interpretation of findings
  **B.** Second Topic for Investigation
    **1.** First subtopic
      **a.** Definition
      **b.** Findings
      **c.** Interpretation of findings
    **2.** Second subtopic (and so on)

**III.** CONCLUSION
  **A.** Summary of Findings

**B.** Overall Interpretation of Findings (as needed)

**C.** Recommendations (as needed and appropriate)

This outline is only tentative. Modify if necessary.

Three sample reports in this text follow the model outline. The first report, "How SMU's Communication Program Can Best Serve Local Companies," is not reproduced entirely; however, parts of each major section (introduction, body, and conclusion) are illustrated and explained. The second report is entirely reproduced. The third is shown in Appendix C.

Each report responds to a slightly different question or problem. The first tackles the question, *How can X be improved?* The second tackles two questions: *Is X feasible?* and *Which version of X is better for our purposes?* The third: *Why does X happen, and what can we do about it?* At least one of these reports should serve as a model for your own analysis.

## Introduction

The introduction describes and defines the question or problem, and provides background. Identify your audience only in a report for your instructor, and only if you have not done so in an attached audience-and-use profile. Discuss briefly your sources of data, along with reasons for omitting data (e.g., key person not available for interview). List working definitions, unless you have so many that you need a glossary. If you do use a glossary and appendixes, refer to them here. Finally, preview the scope of your analysis by listing all major topics discussed in the body.

### HOW SMU'S COMMUNICATION PROGRAM
### CAN BEST SERVE LOCAL COMPANIES

#### INTRODUCTION

This analysis is part of an attempt to persuade SMU administrators of the need for additional courses in professional communication, beyond those now offered. *(Purpose and background)*

Primary data were collected from responses to a questionnaire (Appendix A [shown on pages 122–123]) mailed to presidents of local businesses. Of the 612 questionnaires sent, 103 were returned, for a return rate of 16.83 percent. Although a higher return rate would have been desirable, the responses do correlate with surveys done elsewhere. Moreover, secondary sources confirm this survey's findings, and show that the communication needs of local businesses resemble those elsewhere. *(Information sources / Limitations)*

Two major topics are covered in this report: a profile of communication in local companies, and an assessment of the needs and preferences expressed by survey respondents. *(Scope)*

## Body

The body divides a complex subject into related topics and subtopics, ranked in order of importance (ascending or descending). Carry your heading division as far as you can, to enable readers to follow the topic. In the sample that follows, the major topic is divided into several subtopics: "Preferred Courses," and "Preferred Course Content," and so on.

One major topic of the communications report follows. To save space, the first major topic is omitted.

### NEEDS AND PREFERENCES OF LOCAL COMPANIES

*Major topic*

Respondents identified their needs and preferences according to course offerings at SMU and willingness to participate in contract learning.

*Preferred Courses*

*First subtopic*

Companies were unanimous in their desire that SMU offer a greater range of communication courses. Preferences are ranked here, in decreasing order.

*Findings*

| Small companies | Large companies |
|---|---|
| 1. report writing | 1. report writing |
| 2. promotional writing | 2. interpersonal communication |
| 3. public speaking | 3. public speaking |
| 4. interpersonal communication | 4. promotional writing |

*Interpretation*

More than 60 percent of respondents preferred report writing as a sequel to professional communication. This local preference correlates with a nationwide survey by the *Harvard Business Review,* in which 72 percent of respondents (executives and managers) ranked report writing as "the best training for business" (Bennett 28). Another survey supports these findings, but with the qualification that "the informational report should receive stronger emphasis than the analytical report" (Bennett 32).

*Preferred Course Content*

*Second subtopic*

Both small and large companies rank topics in a writing program in this order:

*Findings*

1. organizing information
2. editing for clarity, conciseness, and fluency
3. summarizing information
4. adapting various messages for various audiences
5. grammar and punctuation

*Interpretation*

In a recent survey by Halpern, executives ranked these topics: organizing, polishing drafts, adapting writing to various audiences, clarifying purpose, and controlling the tone (39). These findings are consistent with recent rhetorical theory and with the emphasis in recent writing texts. For instance, Sigband's text emphasizes that "planning, organizing, writing, and editing are intimately associated" (69).

*Preferred Skills*

*Third subtopic*

Asked to rank communication skills in decreasing order of importance, company officers responded:

| Small companies | Large companies | Findings |
|---|---|---|
| 1. listening | 1. listening | |
| 2. writing | 2. reading | |
| 3. speaking face to face | 3. writing | |
| 4. reading | 4. speaking face to face | |
| 5. speaking to groups | 5. speaking to groups | |

Almost all companies surveyed consider ability to dictate of little importance.

These responses differ from those collected by Rochester and DiGaetani. In their survey, writing was ranked as most important, followed by listening, speaking, and reading (9). The conflicting rankings for writing and listening could result from differences in the groups surveyed. Rochester and DiGaetani questioned college faculty and chairpersons, whereas we surveyed business presidents.

*Interpretation*

While writing needs in business have long been recognized, listening skills have been ignored. DeGaetani writes in *Business Horizons:* "Listening is . . . the most often used yet least understood and researched of the communicating processes," even though "executives spend from 45 to 63 percent of their day listening" (622). But listening skills are attracting attention as "the newest area of communication interest" and "the key to growth and success" in business (Rochester and DiGaetani 10). One notable advocate of listening skills is the Sperry Corporation, which has a program to improve employees' listening skills, and which broadcasts this slogan: "When you know how to listen, opportunity only has to knock once" (Fortune 238).

### Desire for Seminars in Professional Communication

More than half the companies surveyed (small: 40 percent, large: 61 percent) indicate that SMU should offer Saturday seminars in professional communication. As part of a national trend, companies are sending more and more executives and managers to seminars (Main 234).

*Fourth subtopic*

As undergraduates, too few business, technical, and science majors develop the skills to communicate effectively in the workplace. Rochester and DiGaetani state that "the business world is in desperate need of personnel who can become total communicators" (10).

*Interpretation*

### Interest in Contract Learning

A limited number of companies are interested in contract learning, whereby SMU writing students work as communicators within the company. Among small companies surveyed, 15 percent are willing to participate. Among large companies, 18 percent would participate. Another 12 percent of large companies indicated that agreement would depend on the specific arrangement proposed. Other large companies support the concept of contract learning, but their companies prepare few specific documents.

*Fifth subtopic*

*Findings*

## Conclusion

The conclusion is likely to be of most interest to readers because it answers the questions that sparked the analysis in the first place. (Many workplace reports, therefore, are submitted with the conclusion *preceding* the introduction and body sections.)

Here you summarize, interpret, and recommend. Although you have interpreted evidence at each step in your analysis, your conclusion pulls the strands together in a broader interpretation and suggests action, where appropriate. This final section must be consistent in three ways:

1. Your summary must reflect accurately the body of your report.
2. Your overall interpretation must be consistent with the findings in your summary.
3. Your recommendations must be consistent with the purpose of the report, the evidence presented, and the interpretations given.

The summary and interpretations should lead logically to your recommendations. Here is the conclusion of the communications report:

## CONCLUSION

### Summary of Findings

Local business people typically write letters, memos, reports, and procedures. In small companies, most writing is done by presidents and vice presidents; in large companies, by managers. Few companies provide writing guidelines, and even fewer have in-house teaching programs. Yet many companies consider their writing only fair. Many large companies and some small ones offer tuition reimbursement.

All companies desire more communication courses at SMU, and more weekend seminars. Their first choice after professional communication is report writing, followed by interpersonal communication and public speaking. Small companies would also like a course in sales and promotional writing.

Both large and small companies want writing courses to emphasize organizing, editing, summarizing, audience analysis, and grammar. Respondents rate listening skills as vital, along with writing, reading, and speaking.

A limited number of companies are interested in contract-learning arrangements with SMU writing students.

### Recommendations

Based on my findings, I offer the following recommendations:

1. The major component in all professional communication courses should be either informational or analytical reports. Emphasis should also be given to letters, memos, and procedures.
2. Insofar as possible, these courses should involve students in the total communication process: writing, listening, speaking, and reading.
3. Because managers do most of the writing in the large companies surveyed, SMU should offer managerial communication as a sequel to professional communication.

4. SMU should offer weekend seminars in communication for the local business community.

5. Communication professors should explore the possibility of contract-learning arrangements with interested businesses (listed in Appendix B [not shown]).

6. All business and technical majors should take at least one course in professional communication.

## Supplements

Submit your completed report with these supporting documents, in order:

- title page
- letter of transmittal
- table of contents
- table of figures
- abstract
- **report text (introduction-body-conclusion)**
- glossary (as needed)
- appendixes (as needed)
- works-cited page (or alphabetical or numbered list of references)

Each of these supplements is discussed in Chapter 18. Forms of documentation are discussed in Chapter 6.

## APPLYING THE PRINCIPLES

The report in Figure 20.1, patterned after our model outline, combines a feasibility analysis with a comparative analysis.

### Audience-and-Use Profile

Richard Larkin, author of the following report, has a work-study job fifteen hours weekly in his school's placement office. Larkin knows that his boss, John Fitton (placement director), likes to keep abreast of trends in various fields. Larkin himself, an engineering major, has become interested in technical marketing and sales, a recent and rapidly growing choice of career for graduates in various specialized fields.

In need of a report topic for his writing course, Larkin offers to analyze the feasibility of a technical marketing and sales career, both for himself and for technical and science graduates in general. Fitton accepts Larkin's offer, looking forward to having the final report in his reference file for use by students choosing careers. Larkin wants his report to be useful in three ways: to satisfy a course requirement, to help him in choosing his own career, and to help other students with theirs.

With his topic approved, Larkin begins gathering his primary data, using interviews, letters of inquiry, telephone inquiries, and lecture notes. He supplements these primary sources with articles in recent publications. He will document his findings with a numbered list of references (discussed on pages 135–136).

As a guide for designing his final report, Larkin completes this audience-and-use profile (based on the worksheet, page 33).

### Audience Identity and Needs

My primary audience is composed of John Fitton, Placement Director, and the students who will be referring to my report as they choose careers. The secondary audience is my writing instructor. And, of course, the data I've uncovered will help me make my own choice of career.

Fitton is highly interested in this project, and he has promised to study my document carefully and to make copies available to interested students. Because he already knows something about the technical marketing field, Fitton will need very little background to understand my report. Many student readers, however, may know little or nothing about technical marketing, and so will need background, definitions, and detailed explanations. Here are the questions I can anticipate from my collective audience:

- What, exactly, is technical marketing and sales?
- What are the requirements for this career?
- What are pros and cons of this career?
- Could this be the right career for me?
- If so, how do I enter the field?
- Is there more than one option for entering the field? If so, which option would be best for me?

### Attitude and Personality

Readers likely to be most affected by my document are students who will be making career choices. And I would expect my readers' attitudes to vary widely:

1. Some readers will approach my document with great interest, especially those seeking a career that is more people-oriented than technology-oriented.
2. Some readers will be only casually interested in my document, as they investigate a range of possible careers.
3. Some readers are likely to feel overwhelmed by the variety and importance of the career choices they face. Members of this group might be looking for easy answers to the problem of choosing a career.
4. Other readers may approach my document skeptically, perhaps unwilling to reexamine their earlier decisions about a more traditional career.

To connect with this array of readers, I will need to persuade them that my conclusions are based on sound data and careful reasoning.

**Expectations about the Document**

Although I initiated this document, Fitton has been greatly supportive, even eager to read the final version. All my readers will expect me to spell things out. But I know that my readers are busy and impatient, and so I'll want to make this report concise enough to be read in no more than fifteen or twenty minutes.

Essential information will include an expanded definition of technical marketing and sales, the skills and attitudes needed for success, the career's advantages and drawbacks, and a description of various paths for entering the career. And throughout, I'll relate my material to many technical and science majors, not just engineers.

The body of this report combines a feasibility analysis with a comparative analysis. Therefore, I'll use a reasons-for and reasons-against structure in the feasibility section. In the comparison section, I'll use a block structure (page 196), followed by a table that presents a point-by-point comparison of the four entry paths. Because I want this report to lead to informed decisions, I will include concrete recommendations that are based solidly on my conclusions.

To address various readers who may not wish to read the entire report, I will include an informative abstract.

My tone throughout should be conversational. Because I am writing for a mixed audience (placement director, students, and writing instructor), I will use a third-person point of view.

---

This report's front matter (title page and so on) and end matter are shown and discussed in Chapter 18, pages 408–414.

(Figure 20.1 follows.)

FIGURE 20.1  A Feasibility Report

# FEASIBILITY ANALYSIS OF A CAREER IN TECHNICAL MARKETING

## INTRODUCTION

Technical marketing and sales is a most challenging and lucrative career for science and technology graduates. Specifically, this career involves identifying, reaching, and selling to customers a technical product or service.

Customer orientation is an ever-growing part of today's business and manufacturing climate. According to Richard Marks, sales manager for Global Industries, Inc., much of U.S. industry no longer is "manufacturing-driven" (where customers will buy any products that are available). Instead, companies are "market-driven" (customers demand that products be designed to exact specifications). These customer demands are the reason nearly 80 percent of top managers at Global Industries have sales and marketing experience beyond their technical background (1:84).

**FIGURE 1**  Top Managers with Sales and Marketing Backgrounds

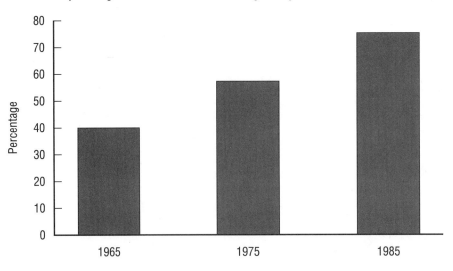

*Source:* Data from Campbell (1983).

**FIGURE 20.1** A Feasibility Report *Continued*

2

The emphasis on sales and marketing at Global Industries is part of a much larger trend. Nationwide, roughly 75 percent of today's top managers have worked their way up through the sales and marketing ranks (2:28). Figure 1 shows how marketing background has become part of the career track to top-management positions. In the product-oriented industries of 1965, technical marketing accounted for only 39 percent of top-management background. But the number had nearly doubled by 1985 because of growth in customer orientation.

Demand for technical salespeople is stimulating interest among technical graduates. Of the more than 40,000 engineering graduates in 1983, nearly 4 percent sought sales and marketing positions (2:20). Today, a growing number of high school seniors identify technical marketing as their career goal (1:85).

But the many undergraduates interested in this career need answers to these basic questions:

—Is this the right career for me?

—If so, how do I enter the field?

To help answer these questions, I have gathered information from professionals and from the literature.

In this report I analyze the feasibility of a technical marketing career, and compare the various paths for entering the field. My analysis includes these specific areas: definition of technical marketing, career requirements, advantages and disadvantages, and a comparison of four career options.

### COLLECTED DATA

#### *Feasibility of Technical Marketing as a Career*

Anyone considering technical marketing needs to assess whether this career is practical, considering the individual's interest and abilities.

#### *Definition of Technical Marketing*

Although the terms *marketing* and *sales* usually occur interchangeably, technical marketing involves far more than mere sales work. The process itself (identifying, reaching, and selling to customers) entails six major activities (3:43):

1. *Market research:* gathering information about the size and character of the market for a product or service.
2. *Product development and management:* producing the goods to fill a specific market need.

**FIGURE 20.1**  A Feasibility Report  *Continued*

3

3. *Cost determination and pricing:* measuring every expense in the production, distribution, advertising, and sale of the product, to determine a price for it.
4. *Advertising and promotion:* developing and implementing all strategies for reaching customers.
5. *Product distribution:* coordinating all elements of a technical product or service, from its conception through its final delivery to the customer.
6. *Sales and technical support:* creating and maintaining customer accounts, and servicing and upgrading products.

Fully engaged in all these activities, the marketing professional gains good understanding of the industry, the product, and the customer's needs.

*Career Requirements*

Besides a strong technical background, technical marketing requires a generous blend of motivation, skilled communication, and interpersonal skills, as summarized in Figure 2.

Motivation is essential for marketing work. Professionals must be energetic and able to function with minimal supervision. Ideal candidates are creative people who can plan and program their own tasks, who can manage their time for greatest efficiency, and who are not afraid of hard work (4). Leadership potential, as demonstrated by extracurricular activities, is an asset.

**FIGURE 2**  Requirements for a Career in Technical Marketing

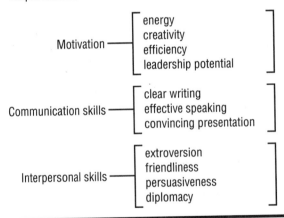

FIGURE 20.1  A Feasibility Report  *Continued*

4

But motivation alone is no guarantee of success. Marketing professionals are paid to communicate the virtues of their products or services to customers. This career, therefore, calls for skill in communication, both written and oral. Writing done for readers outside the organization includes advertising copy, product descriptions, sales proposals, sales letters, and instructions for customers. In-house writing includes recommendation reports, feasibility studies, progress reports, and memos to colleagues (5:89).

Skilled presentation is vital to any sales effort. Technical marketing professionals need to speak with enough confidence to be persuasive—to represent their products and services in the best light (6). And many of these sales presentations require public speaking before groups (at conventions, trade shows, and other functions).

No matter how strong one's motivation and ability in communication, interpersonal skills are the ultimate requirement for success in marketing (7:114). Consumers are more likely to buy a product or service if they *like* the person selling it. Marketing professionals are extroverted and friendly; they enjoy meeting new people. Because they understand diplomacy, they can motivate customers without alienating them.

*Advantages of the Career*

A technical marketing career offers the benefits of diversity and job mobility. Marketing personnel are exposed to every phase in a company's operation, from the design to the sale of a product. And many companies encourage employees to rotate periodically between marketing and their technical specialties (1:84). Because of the broad exposure it provides, a marketing position can provide a direct path to upper-management jobs.

Another benefit is the possibility of a high salary. Most marketing professionals receive base pay plus commission. According to John Turnbow, manager of National Electric's Technical Marketing Program, some of NE's new marketing engineers earn more than $60,000 in their first year. Many salaries reach six figures, sometimes higher than a company's executive salaries (1:87). Figure 3 compares the average salaries for engineers with five years of experience to those of marketing engineers. Compared to engineering salaries, the average salaries of marketing engineers have accelerated much more rapidly in recent years.

*Drawbacks of the Career*

Technical marketing is by no means a career for everyone. The job requires much time away from home. Depending on the region, personnel might spend 50 to 75 percent of work time traveling to meet with customers. And success requires long hours and occasional weekends (2:24). Above all, the job is highly stressful because of the pressure to sell and to meet quotas. Anyone considering this career should be able to thrive in a competitive environment.

**FIGURE 20.1** A Feasibility Report *Continued*

**FIGURE 3**  A Comparison of Average Salaries for Engineers and Marketing
Engineers with Five Years' Experience

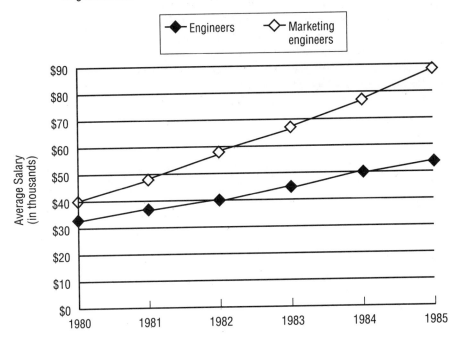

*Source:* Based on data from Basta (1984).

### *A Comparison of Entry Options*

Engineers and other technical graduates enter technical marketing through one of four options.
Some graduates join small companies and begin marketing work immediately. Others join
companies that offer formal training programs. Some begin by acquiring experience in their
specialty. And others earn a graduate degree beforehand. Each option is described below, and
then evaluated on the basis of specific criteria.

### *Option 1: Entry-Level Marketing Without Training*

Direct positions in marketing are available with smaller companies and with firms that
represent client manufacturers. Because this option offers no formal training, candidates must
be motivated and enterprising.

**FIGURE 20.1** A Feasibility Report *Continued*

6

Alan Carto, president of Abco Switch, believes that small companies can offer a unique opportunity: the entry-level salesperson can learn about all facets of an organization, and has the possibility for rapid advancement (6). Career counselor Phil Hawkins says, "It depends on whether you want to be a big fish in a small pond or a small fish in a big pond" (7).

Manufacturer's representatives constitute another entry-level position. These professionals represent products for manufacturers who have no marketing staff of their own. Manufacturers' representatives are, in effect, their own bosses; they can choose, from among many offers, the products they wish to represent (8).

An entry-level marketing position offers the notable advantage of immediate income and a chance for early promotion. A disadvantage, however, might be the gradual loss of any technical edge that one might have acquired in college.

*Option 2: A Marketing and Sales Training Program*
Formal training programs are the most popular route into marketing. Typically, large to mid-size companies offer these programs in two formats: (a) a product-specific program, in which trainees learn about a product or line of products, or (b) a rotational program, in which trainees learn about various products and work in various positions. Programs can last from several weeks to several months, teaching trainees about products, markets, competition, and customer relations.

Former trainees speak of the diversity and satisfaction offered by such training programs. Ralph Lang, a sales representative for Allied Products, enjoyed the constant interaction with personnel throughout his company (9). Bill Collins, a sales engineer with Intrex Computers, values his broad knowledge of Intrex's product line, instead of being narrowly focused on one technical area. Sarah Watts, also with Intrex, finds satisfaction in applying her training to a variety of sales challenges (2:29).

Like the direct-entry path, this option offers immediate income. A possible disadvantage, however, is that trainees may find their technical expertise compromised because they have had no chance to practice in their specialty.

*Option 3: Practical Experience in One's Specialty*
Instead of entering the marketing world directly, some candidates first gain practical experience in their specialty. This option provides direct exposure to the workplace, as well as a chance to sharpen technical ability in practical applications. Moreover, some companies offer marketing and sales positions to their engineers and other specialists who demonstrate interest and aptitude on the job.

FIGURE 20.1  A Feasibility Report  *Continued*

7

Many industry experts consider the practical experience option the best choice. Jane Doser, recruitment manager for Trans Electric, has this view of on-the-job experience: "Until you've done some work in your specialty, you really don't know yourself" (10:15).

This option ensures technical expertise, but not without a delay in entering technical marketing.

*Option 4: Graduate Program*
Instead of directly entering the workplace, some people choose graduate school, pursuing either an M.S. in their specialty, an M.B.A., or both. Master's degrees usually are not necessary for technical marketing unless a sale is highly complex (11). But the M.B.A. degree can be a real advantage because of the general business and marketing skills it provides. According to Kline Jorgenson, manager of college recruiting at Hayes-Proctor, a product-marketing specialist at HP would use much of the knowledge acquired in an M.B.A. program (10:15).

A motivated student might combine degrees. Dora Anson, president of Susimo Cosmic Systems, sees the M.S. and M.B.A. as the ideal combination for technical marketing (12). To employers, the advanced degree means greater technical competence, stronger motivation, and better performance (13).

One disadvantage of a full-time graduate program is the salary lost, compounded by school expenses. But these costs must be balanced against the prospect of promotion and monetary rewards later in one's career.

*An Overall Comparison of Advantages*
Table 1 compares the four entry options on the basis of three criteria: highest immediate income, fastest advancement through marketing ranks, and greatest potential for advancement. As Table 1 shows, choosing any entry can have major implications for one's career.

**Table 1** Relative Advantages Among Four Technical Marketing Entry Options

| Option | Relative Advantage | | |
| --- | --- | --- | --- |
| | Immediate income | Early advancement in marketing | Greatest long-term potential |
| Entry level, no training | yes | yes | no |
| Training program | yes | yes | no |
| Practical experience | yes | no | yes |
| Graduate program | no | no | yes |

**FIGURE 20.1** A Feasibility Report *Continued*

8

## CONCLUSION

### *Summary of Findings*

Technical marketing and sales involves identifying, reaching, and selling the customer a technical product or service. Besides a solid technical background, anyone entering the field will need motivation and communication and interpersonal skills. This career offers job diversity and excellent potential income, balanced against hard work and the pressure to perform.

College graduates who seek a career in technical marketing and sales have four entry options:

1. Direct entry with no formal training;
2. A formal training program;
3. Practical experience in a technical specialty prior to entry; or
4. Graduate programs.

Each option has advantages and disadvantages in immediacy of income, rapidity of advancement, and long-term potential.

### *Interpretation of Findings*

For graduates with a strong technical background and the right skills and motivation, technical marketing and sales offers great prospects for income and advancement. Anyone contemplating this field, however, needs to enjoy customer-contact work and to be able to thrive in a highly competitive environment.

Those who decide that technical marketing is for them can choose from among four entry options. Although any of the options might be adequate, one option or another might best suit a person's needs and inclinations:

—If immediate income is unimportant, graduate school is an attractive option.

—For hands-on experience, an entry-level job is the logical option.

—For sharpening technical skills, prior work in one's specialty is invaluable.

—For sophisticated sales training, a formal program with a large company is best.

### *Recommendations*

If your interests and abilities match the requirements outlined earlier, follow these suggestions in planning for a technical marketing career.

1. To get a firsthand view, seek the opinions and advice of people in the field.
2. Before settling on an entry option, consider all its advantages and disadvantages, and decide whether this option best coincides with your goals.

**FIGURE 20.1** A Feasibility Report *Continued*

9

3. Remember that you are never committed to one entry option only. For a fuller understanding of technical marketing, you might pursue any combination of options during your professional life.
4. When making any vital career decision, "listen to your brain and your heart" (13). Choose an option or options that offer not only professional advancement, but comfort and happiness as well.

## REFERENCES

1. Basta, Nicholas, "Take a good look at sales engineering." *Graduating Engineer* Sept, 1984: 84–87.

2. Campbell, Mary K. "Wanted: sales reps with EE degrees." *IEEE Potentials* Spring 1983: 28–29.

3. Cornelius, Hal, and William Lewis. *Career guide for sales and marketing.* New York: Monarch, 1983.

4. Lance, Norman, College Recruiting Director, Elco Electronic Systems. Personal interview, Placement Office, Southeastern Massachusetts University, North Dartmouth, 22 Feb. 1990.

5. *The job outlook in brief.* Washington: U.S. Department of Labor, Spring 1982.

6. Carto, Alan, President, Abco Switch, Inc. Personal interview, Boston, Mass., 8 Feb. 1990.

7. Splaver, Sarah. *Your personality and your career.* New York: Simon and Schuster, 1977.

8. Tolland, Mark, Marketing Vice President, Salford Associates. Telephone interview, 29 Mar. 1990, about the coming presentation on 25 Apr. at "Electro '90" in New York, "Alternate careers in marketing and sales."

9. Lang, Ralph, Sales Representative, Allied Products. Telephone interview, 10 Apr. 1989.

10. Schranke, Richard W. "EE and MBA: A winning combination?" *IEEE Potentials* Feb. 1985: 13–15.

11. McClone, Joan T. "The choice between an MS and an MBA." Lecture at Southeastern Mass. Univ., 8 Apr. 1989.

12. Anson, Dora. "The role of an engineer." Lecture at Southeastern Mass. Univ., 15 Jan. 1990.

13. Polgar, Warren. "How to plan a successful career." Lecture at Southeastern Mass. Univ., 10 Nov. 1989.

# REVISION CHECKLIST FOR ANALYTICAL REPORTS

Use this list to refine the content, arrangement, and style of your report. (Numbers in parentheses refer to the first page of discussion.)

## Content

☐ Does the report grow from a clear statement of purpose? (456)

☐ Is the report's length adequate and appropriate? (457)

☐ Are all limitations of the analysis spelled out? (463)

☐ Is each topic defined before it is discussed? (462)

☐ Are visuals used whenever possible to aid communication? (240)

☐ Is the analysis based on hard evidence? (460)

☐ Is the analysis free of specious reasoning? (461)

☐ Are all sources of data credible? (459)

☐ Are all data accurate? (457)

☐ Are all data unbiased? (456)

☐ Are all data complete? (457)

☐ Are all data fully interpreted? (457)

☐ Is the documentation adequate and correct? (127)

☐ Are recommendations based on accurate interpretations? (457)

## Arrangement

☐ Is there an introduction, body, and conclusion? (462)

☐ Are headings appropriate and adequate? (291)

☐ Have you built in enough transitions between related ideas? (518)

☐ Does the report have all needed front matter? (467)

☐ Does the report have all needed end matter? (467)

## Style

☐ Is the level of technicality appropriate for the stated audience? (468)

☐ Are all sentences clear, concise, and fluent? (199)

☐ Is the language convincing and precise? (221)

☐ Is the report written in correct English? (Appendix A)

Now revise all material that has not been checked off.

## EXERCISES

**1.** In the periodical section of your library, find examples of reports or articles that use each of these types of analysis:

  a. answering a practical question (Will *X* work for a specific purpose?)
  b. comparing two or more items (Is *X* or *Y* better for a specific purpose?)
  c. solving a problem (Why does *X* happen? *or* How can *X* be improved?)
  d. determining consequences (What are the effects of *X*?)
  e. assessing feasibility (Is *X* practical in a given situation?)

Provide full bibliographical information, along with a *descriptive* abstract (see page 148) of each article.

**2.** In the periodical and newspaper section of your library, compile a list of sources, by title, ranked in general order of reliability: (a) List five highly reliable sources. (b) List five sources that are less reliable. Briefly explain the reason for each choice by discussing your criteria for judgment.

**3.** These statements are followed by false or improbable conclusions. In order to prevent specious generalizations, what specific supporting data or evidence would be needed to justify each conclusion?

  a. Eighty percent of black voters in Mississippi voted for John Jones as governor. Therefore, he is not a racist.
  b. Fifty percent of last year's college graduates did not find desirable jobs. Therefore, college is a waste of time and money.
  c. Only 60 percent of incoming freshmen eventually graduate from this college. Therefore, the college is not doing its job.
  d. He never sees a doctor. Therefore, he is healthy.
  e. This house is expensive. Therefore, it must be well built.

**4.** Write an analytical report, using these guidelines (not necessarily in sequence):

  a. Choose a subject for analysis from the list at the end of this exercise, from your major, or from a subject that interests you.
  b. Identify the problem or question so that you will know exactly what you are looking for.
  c. Restate the main question as a declarative sentence in your statement of purpose.
  d. Identify an audience—other than your instructor—who will use your information for a specific purpose.
  e. Hold a private brainstorming session to generate major topics and subtopics.
  f. Use the topics to make an outline based on the model outline in this chapter. Divide as far as necessary to identify all points of discussion.
  g. Make a tentative list of all sources (primary and secondary) that you will investigate. Verify that adequate sources are available.
  h. Write your instructor a proposal memo (pages 423–425), describing the problem or question and your plan for analysis. Attach a tentative bibliography.
  i. Use your working outline as a guide to research and observation. Evaluate sources and evidence, and interpret all evidence fully. Modify your outline as needed.
  j. Submit a progress report to your instructor (pages 396–399), describing work completed, problems encountered, and work remaining.

*k.* Compose an audience-and-use profile. (Use the samples on pages 438 and 468 as models, along with the profile worksheet on page 33.)

*l.* Write the report for your stated audience. Work from a clear statement of purpose and be sure that your reasoning is shown clearly. Verify that your evidence, conclusions, and recommendations are consistent.

*m.* After writing your first draft, make any needed changes in the outline and revise your report according to the revision checklist. Include all necessary supplements.

*n.* Exchange reports with a classmate for further suggestions for revision.

*o.* Prepare an oral report of your findings for the class as a whole.

Here are some subjects for analysis:

- the effects of a vegetarian diet on physiological energy
- the causes of student disinterest in campus activities
- the student transportation problem to and from your college
- two or more brands of tools or equipment from your field
- noise pollution from nearby airport traffic
- the effects of toxic chemical dumping in your community
- the adequacy of veterans' benefits
- the effects of sun on skin
- the (causes, effects of) acid rain in your area
- the influence of a civic center or stadium on your community
- the best ways to lower cholesterol in your diet
- the best microcomputer to buy for a specific need
- the qualities employees seek in a job candidate
- the adequacy of security on your campus or in your dorm
- the adequacy of your college's remedial program
- the feasibility of opening a specific business
- the best location for a new business
- causes of the high dropout rate in your college
- the pros and cons of condominium ownership for you
- the feasibility of moving to a certain area of the country
- job opportunities in your field
- effects of the 200-mile limit on the fishing industry in your area
- the effects of budget cuts on public higher education in your state
- the best nonprescription cold remedy
- the adequacy of zoning laws in your town
- effect of population increase on your local water supply
- adequacy of your student group-health-insurance policy
- the feasibility of biological pest control as an alternative to pesticides
- the feasibility of large-scale desalination of sea water as a source of fresh water
- effective water conservation measures that can be used in your area
- the effects of legalizing gambling in your state
- effective measures for relieving the property-tax burden in your town
- the causes of low morale in the company where you work part-time
- causes of poor television reception in your area
- effects of thermal pollution from a local power plant on marine life
- the feasibility of using wood as a source of energy

- a comparison of two or more brands of wood-burning stove
- the feasibility of converting your home to solar heating
- adequacy of police protection in your town
- effective measures for improving the fire safety of your home
- the best energy-efficient, low-cost housing design for your area
- measures for improving productivity in your place of employment
- reasons for the success of a specific business in your area
- the feasibility of operating a campus food co-op
- the best investment (real estate, savings certificates, stocks and bonds, precious metals and stones) over the past ten years
- the problem of water supply and waste disposal for new subdivisions in your town
- the advisability of home birth (as opposed to hospital delivery)
- the adequacy of clean-up measures for toxic waste in your area
- the adequacy of the evacuation plan for your area in a nuclear emergency or other disaster
- the feasibility and cost of improving security in the campus dorms
- the advisability of pursuing a graduate degree in your field, instead of entering the work force with a bachelor's degree.

## TEAM PROJECTS

**1.** Divide into small groups. Choose a subject for group analysis—preferably, a campus issue—and partition the topic by group brainstorming. Next, select major topics from your list and classify as many items as possible under each major topic. Finally, draw up a working outline that could be used for an analytical report on this subject.

**2.** Prepare a questionnaire based on your work above, and administer it to members of your campus community. List the findings of your questionnaire and your conclusions in clear and logical form.

# Oral Reports

- Choosing the Best Delivery
- Preparing Extemporaneous Reports
- Giving Extemporaneous Reports

An oral report can be anything from a brief discussion to formal speeches or lectures. This chapter covers the more formal reporting situations, which require planning and preparation.

Compared to a written report, an oral report is more personal, may be more memorable, saves time, and elicits immediate response; a written report, though, is easier to refine and organize, can be more complex, can be studied at the reader's own pace, and can be easily reviewed.

Your formal oral reporting may include convention speeches, reports at national meetings, reports given through a teleconferencing network, and speeches to community groups. The higher your status, on the job or in the community, the more you will give formal talks.

## CHOOSING THE BEST DELIVERY

An oral report's effectiveness depends greatly on *how* it is delivered. Not all deliveries are equally effective.

### The Impromptu Delivery

The impromptu ("off-the-cuff") delivery generally is ineffective for formal reports. Being unified, coherent, fluent, and informative without preparation is nearly impossible. Don't fool yourself into thinking "It's all in my head." Get your plan down on paper.

### The Memorized Delivery

In the memorized delivery, you write your report and memorize it. The danger here is that you might sound like a parrot, and forgetting a line can be disastrous. Because your personality and body language influence audience interest, you can hardly expect to seem natural while reciting your lines.

## The Reading Delivery

Reading your report aloud is boring. Unless you maintain eye contact with your audience, and vary your gestures and tone, you might look like a robot. If you *do* plan to read the report, study it, and practice aloud to decrease reliance on the text. Otherwise, you might slur or mispronounce words, or keep your nose stuck to the text in fear of losing your place.

## The Extemporaneous Delivery

An extemporaneous delivery is carefully planned, practiced, and based on notes that keep you on track. This is the most popular speaking technique. By following notes in sentence-outline form, you maintain control of your material. Also, you speak in a natural, conversational style, with only brief glances at your notes. Because extemporaneous delivery is based on key ideas (represented by topic sentences to jog your memory) rather than fully developed paragraphs to be read or memorized, you can adjust your pace and diction as needed. Instead of coming across as a droning, mechanical voice, you are seen as a distinct personality. But the danger in extemporaneous delivery is that you may get off track, unless you are thoroughly prepared.

## PREPARING EXTEMPORANEOUS REPORTS

Plan your report step by step, to stay in control and build confidence. Some suggestions follow.

### Know Your Subject

A sure way to appear foolish is to speak on something you don't fully understand or know much about. Do your homework, *exhaustively*. Be prepared to explain and defend each assertion and statement of opinion with fact. Your audience expects to hear a knowledgeable speaker. Don't disappoint them.

### Identify Your Audience

Speakers who oversimplify and belabor obvious or trivial points are boring, but those who speak in generalities or fail to explain complex information are confusing. Adjust the amount of detail and the level of technicality to your audience.

Many audiences include people with varied technical backgrounds. Unless you know the background of each person, speak to a general audience (Chapter 2), as in a classroom of mixed majors.

As you plan the report, answer these questions:

1. What points do I want to make?
2. How can I make each point interesting and understandable?

If your subject is controversial, consider the average age, political views, education, and socioeconomic status of most audience members. Then decide how to speak candidly and persuasively without offending anyone.

## Plan Your Report

**Statement of Purpose.** Formulate, on paper, a statement of purpose, no more than two or three sentences long. Why are you speaking on this subject? Who is your audience? What effect do you wish to have?

**Sentence Outline.** If the oral report is simply a spoken version of your written report, you'll need substantially less preparation. Simply expand your outline for the written report into a sentence outline. For our limited purposes, we will assume that your oral report is based on a written report. Here is a typical sentence outline for a fifteen-minute oral report derived from a written report twenty pages long.

### ORAL REPORT OUTLINE

Arnold Borthwick

*Purpose:* By informing Cape Cod residents about the dangers to the Cape's freshwater supply posed by rapid population growth, this report is intended to increase local interest in the problem.

I. INTRODUCTION
   A. Do you know what you are drinking when you turn on the tap and fill a glass?
   B. The quality of our water is good, but not guaranteed to last forever.
   C. The somewhat unique natural storage facility for our water supply creates a dangerous situation.
   D. Cape Cod's rapidly increasing population could pollute our water.
   E. In fact, pollution in some towns already has begun.

II. BODY
   A. The groundwater is collected and held in an aquifer.
      1. This water-bearing rock formation forms a broad, continuous arch under the entire Cape (Visual #1).
      2. The lighter fresh water flows on top of the heavier salt water.
   B. With increasing population, sewage and solid waste from landfill dumps invade the aquifer.
      1. As wastes flow naturally toward the sea, they can invade the drawing radii of town wells (Visual #2).
      2. The Cape's sandy soil causes rapid seepage of wastes into the groundwater in the aquifer.

C. Increased population also causes overdraw on some town wells, resulting in saltwater intrusion.

D. Salt and calcium used in snow removal add to the problem by entering the aquifer from surface runoff.

E. The effects of continuing pollution of the Cape's water table will be far-reaching.

    1. Drinking water will have to be piped in more than 100 miles from Quabbin Reservoir.

    2. The Cape's beautiful freshwater ponds will be unfit for swimming.

    3. Aquatic and aviary marsh life will be threatened.

    4. The sensitive ecology of Cape Cod will be destroyed.

F. Such damage would, in turn, lead to economic disaster for Cape Cod's major industry—tourism.

### III. CONCLUSION

A. This problem is becoming more real than theoretical.

B. The conclusion is obvious: if the Cape is to survive ecologically and financially, immediate steps must be taken to preserve our *only* water supply.

C. These recommendations offer a starting point for effective action:

    1. Restrict population density in all Cape towns by creating larger building-lot requirements.

    2. Keep strict watch on proposed high-density apartment and condominium projects.

    3. Create a committee in each town to educate residents about conserving water in order to reduce the draw on town wells.

    4. Prohibit salt, calcium, and other additives in sand spread on snow-covered roads.

    5. Identify alternatives to landfill dumps for solid-waste disposal.

D. This crucial issue deserves the attention of every Cape resident.

E. Question-and-answer session.

Each sentence in this outline is a topic sentence for a paragraph the speaker will develop in detail. (Review outlining in Chapter 9.)

Identify areas in your outline where visuals are needed (Chapter 12). Flip charts and transparencies are especially effective for small (classroom-size) audiences.

Transfer your outline to notecards, which you can hold in one hand and shuffle as needed. Or insert the outline pages in a looseleaf binding for easy flipping. For either method, type or print clearly, leaving enough white space so that you can locate material at a glance.

**Visual Aids.** Well-chosen visuals increase audience interest. But visuals are no substitute for your report. Select visuals that will clarify and enhance your talk—without making you fade into the background.

Use visuals whenever possible to emphasize a point, when *showing* would be more effective than merely *telling*. The map in Figure 21.1 (3′ × 5′ as used in the actual talk) is designed to reinforce Arnold Borthwick's explanation of how solid wastes contaminate wells. Here, the *telling* is clarified greatly by the *showing*.

As you plan visuals for your report, ask yourself these questions:

1. What points do I want to highlight? What visual will provide the best emphasis in each case?

2. How large should the visual(s) be, to be seen by everyone?

3. What hardware is available (slide projector, opaque projector, overhead projector, film projector, videotape player, terminal with large-screen monitor)? What graphics programs are available? Which is best for my purpose and audience? How long in advance do I have to request this equipment?

FIGURE 21.1  Visual Aid for an Oral Report

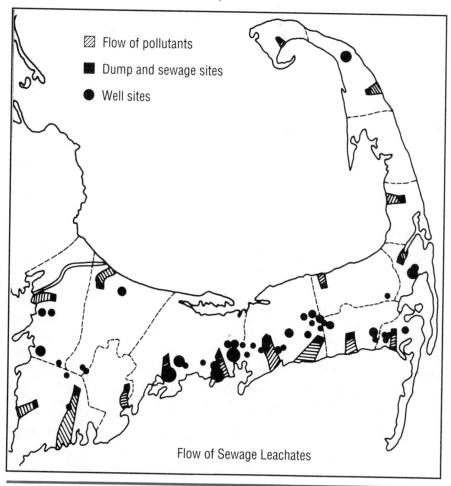

Flow of pollutants

Dump and sewage sites

Well sites

Flow of Sewage Leachates

4. Can I make (or have made) drawings, charts, graphs, or maps as needed? Can transparencies (for overhead projection) be made or slides collected? Do handouts have to be typed and reproduced?

Follow these suggestions for using visuals in your oral report:

- Keep visuals simple—for your unique audience. The audience will have no time to study each one. Adjust the amount and type of detail to your audience's level of technicality.

- Limit their number (no more than four or five in a twenty-minute talk). Otherwise, your presentation may seem like a media event. Be selective in choosing what to highlight with visuals.

- Try not to display your visuals until you refer to them during your presentation; otherwise, your audience may be distracted.

- Interpret each visual as you present it.

- Stand aside when discussing a visual, so that everyone can see it. Don't turn your back to the audience.

- After discussing the visual, remove it, to refocus attention on you.

- If you plan to do drawings on a blackboard, do them beforehand (in multicolored chalk). Otherwise, your audience will be twiddling their thumbs while you draw away.

- If you want the audience to remember material, prepare handouts for distribution *after* your talk.

- Check the room beforehand to make sure you have enough space, electrical outlets, furniture, and so on for your equipment. If you will be addressing a large audience by microphone, and plan to point to features on your visuals, be sure the microphone is movable. Don't forget a pointer if you need one.

**Delivery Time.** Unless you have reason to do otherwise, aim for a maximum delivery time of twenty minutes. Longer talks may cause your audience to lose attention. Time yourself in practice sessions, and trim as needed.

## Practice Your Delivery

Hold several practice sessions to learn the geography of your report. Then you won't fumble your actual delivery.

**Feedback.** Try to practice at least once before friends. Or use a full-length mirror and a tape recorder. Assess your delivery from your friends' comments or from your taped voice (which will sound high to you). Revise as needed.

**Organization.** Is your delivery unified, coherent, and logical? Does it hang together? Will the audience be able to follow your reasoning? Do you use enough transitional statements to reinforce the connections between ideas?

**Tone.** Maintain a conversational tone. Speaking as you do in the classroom should be appropriate.

## Anticipate Audience Questions

Consider the parts of your report that might elicit questions and challenges from your audience. You might need to clarify or justify information that is new, controversial, disappointing, or surprising. Be prepared to support all conclusions, recommendations, assertions, and opinions. If you doubt the validity of any statement, leave it unsaid. Try to predict your audience's responses so that you'll be ready to field questions. Repeat every question, to be sure that all audience members have heard it. If someone tries to engage you in a debate, offer to continue the discussion *after* the presentation.

## GIVING EXTEMPORANEOUS REPORTS

Present your report according to the guidelines:

### Use Natural Body Movements and Posture

If you move and gesture as you normally would in conversation, your audience will be more relaxed. Nothing seems more pretentious than a speaker with rehearsed moves and artificial gestures. Also, maintain a reasonable posture.

### Speak with Confidence, Conviction, and Authority

Show your audience you believe in what you say. Avoid qualifiers ("I suppose," "I'm not sure," "but . . . ," "maybe"). Also, clean up verbal tics ("er," "ah," "uuh," "mmm"). If you seem to be apologizing for your existence, you won't be impressive. But speaking with authority is not the same as speaking like an authoritarian; no sermons, please.

### Moderate Volume, Tone, Pronunciation, and Speed

When using a microphone, people often speak too loudly. Without a microphone, they may speak too softly. Be sure you can be heard clearly, without shattering eardrums. Ask your audience about the sound and speed of your delivery after a few sentences. Your tone should be confident, sincere, friendly, and conversational.

Because nervousness can cause too-rapid speech and unclear or slurred pronunciation, pay close attention to your pace and pronunciation. Usually, the rate you feel is a bit slow will be just about right for your audience.

### Maintain Eye Contact

Eye contact is vital in relating to your audience. Look directly into your listeners' eyes to hold their interest. With a small audience, eye contact is one of your best connectors. As you speak, establish eye contact with as many members of your audience as possible. With a large group, maintain eye contact with those in the first rows.

### Read Audience Feedback

Assess your audience's feedback continually and make adjustments as needed. If you are laboring through a long list of facts or figures and you notice people dozing or fidgeting, you might summarize the point you are making. Likewise, if frowns, raised eyebrows, or questioning looks indicate confusion, skepticism, or indignation, you can backtrack with a specific example or explanation. By tuning in to your audience's reactions, you can avoid leaving them confused, hostile, or simply bored.

### Be Concise

Say what you came to say, then summarize and close—politely and on time. Don't punctuate your speech with "clever" digressions that pop into your head. Unless a specific anecdote was part of your original plan to clarify a point or increase interest, avoid excursions. Each of us often finds what we have to say more interesting than our listeners do!

### Summarize

Before ending, summarize the major points and reemphasize anything of special importance. As you conclude, thank your listeners.

### Leave Time for Questions and Answers

At the very beginning, inform your audience that a question-and-answer period will follow. Announce a specific time limit (such as ten minutes), to avoid public debate. Then you can end the session gracefully, without making anyone feel cut off or excluded. Don't be afraid to admit ignorance. If you can't answer a question, say so, and move to the next question. End the session by saying, "We have time for one more question," or the like.

---

### REVISION CHECKLIST FOR ORAL REPORTS

Use this list to evaluate the content, arrangement, and style of the delivery. (Numbers in parentheses refer to the first page of discussion.)

### Content

☐ Is the report's content suited to this audience? (484)

☐ Does it begin with a clear statement of purpose? (485)

☐ Does the report achieve its stated purpose? (485)

☐ Does the report have adequate and appropriate visuals? (486)

☐ Is each assertion supported by facts? (490)

## Arrangement

☐ Is the line of reasoning clear? (488)

☐ Is the report coherent? (488)

☐ Is it free of needless digressions? (490)

☐ Does the speaker summarize effectively, before concluding? (490)

## Style

☐ Is the delivery relaxed and personable? (489)

☐ Does the speaker seem comfortable? (489)

☐ Does the speaker exhibit confidence and authority, without pretention? (489)

☐ Are all words pronounced distinctly? (489)

☐ Are the volume, tone, and speed of the delivery effective? (489)

☐ Does the speaker maintain good eye contact? (489)

☐ Is the delivery clear? (490)

## EXERCISES

**1.** In a memo to your instructor, identify and discuss the kinds of oral reporting duties you expect to encounter in your career.

**2.** Design an oral report for your writing class. (Base it on a written report.) Make a sentence outline, and include at least two visuals. Practice with a tape recorder or a friend. Use the revision checklist to evaluate your delivery.

**3.** Observe a lecture or speech, and evaluate it according to the revision checklist. Write a memo to your instructor (without naming the speaker), identifying strong and weak areas and suggesting improvements.

**4.** In an oral report to the class, present your findings, conclusions, and recommendations from the analytical report assignment in Chapter 20.

# Appendix A
# Review of Grammar, Usage, and Mechanics

No matter how vital and informative a message may be, its credibility is damaged by basic errors. Any of these errors—an illogical, fragmented, or run-on sentence; faulty punctuation; or a poorly chosen word—stands out and mars otherwise good writing. Not only do such errors confuse and annoy the reader, but they speak badly for the writer's attention to detail and precision. Your career will make the same demands for good writing that your English classes do. The difference is that evaluation (grades) in professional situations usually shows in promotions, reputation, and salary.

## COMMON SENTENCE ERRORS

Any piece of writing is only as good as each of its sentences. Here are common sentence errors, with suggestions for easy repairs.

### Sentence Fragment

A sentence is the expression of a logically complete idea. Any complete idea must have a subject and a verb and must not depend on another complete idea to make sense. Your sentence might have several complete ideas, but it must have at least one!

> Although he was nervous, he grabbed the line, and he saved the sailboat.
>    *(incomplete idea)*      *(complete idea)*      *(complete idea)*

However long or short your sentence, it should make sense to your reader. If the idea is not complete—if your reader is left wondering what you mean—you probably have left out something essential (the subject, the verb, or another complete idea). Such a piece of a sentence is called a *fragment*.

> Grabbed the line. *(a fragment because it lacks a subject)*
> Although he was nervous. *(a fragment because—although it has a subject and a verb—it needs to be joined with a complete idea)*

**TABLE A-1**  Correction Symbols

| Symbol | Meaning | Page | Symbol | Meaning | Page |
|--------|---------|------|--------|---------|------|
| *ab* | abbreviation | 521 | , / | comma | 508 |
| *agr p* | pronoun/referent agreement | 500 | – – / | dashes | 516 |
| *agr sv* | subject/verb agreement | 499 | . . . / | ellipses | 515 |
| *appr* | inappropriate diction | 232 | ! / | exclamation point | 507 |
| *bias* | biased tone | 235 | – / | hyphen | 517 |
| *ca* | pronoun case | 501 | *ital* | italics | 515 |
| *cap* | capitalization | 523 | ( ) / | parentheses | 516 |
| *chop* | choppy sentences | 496 | . / | period | 506 |
| *cl* | clutter word | 214 | ? / | question mark | 507 |
| *coh* | paragraph coherence | 191 | " / " | quotation | 514 |
| *cont* | contraction | 514 | ; / | semicolon | 507 |
| *coord* | coordination | 497 | *qual* | needless qualifier | 214 |
| *cs* | comma splice | 495 | *red* | redundancy | 209 |
| *dgl* | dangling modifier | 201 | *rep* | needless repetition | 210 |
| *euph* | euphemism | 224 | *ref* | faulty reference | 500 |
| *exact* | inexact word | 221 | *ro* | run-on sentence | 496 |
| *frag* | sentence fragment | 492 | *seq* | sequence of development in a paragraph | 192 |
| *gen* | generalization | 225 | *shift* | sentence shift | 505 |
| *jarg* | needless jargon | 223 | *st mod* | stacked modifiers | 201 |
| *len* | paragraph length | 196 | *str* | paragraph structure | 187 |
| *lev* | level of technicality | 18 | *sub* | subordination | 498 |
| *mng* | meaning ambiguous | 199 | *th op* | "th" sentence openers | 210 |
| *mod* | misplaced modifier | 502 | *trans* | transition | 518 |
| *noun ad* | noun addiction | 212 | *trite* | triteness | 224 |
| *om* | omitted word | 200 | *un* | paragraph unity | 191 |
| *over* | overstatement | 225 | *v* | voice | 203 |
| *par* | parallelism | 504 | *var* | sentence variety | 218 |
| *pct* | punctuation | 506 | *w* | wordiness | 208 |
| *ap/* | apostrophe | 513 | *wo* | word order | 202 |
| *[ ]/* | brackets | 516 | *ww* | wrong word | 221 |
| *: /* | colon | 508 | # | numbers | 524 |
| | | | ¶ | begin new paragraph | 187 |
| | | | *ts* | topic sentence | 188 |

The only exception to the sentence rule applies when we give a command (Run!) in which the subject (you) is understood. Because this is a logically complete statement, it is a sentence. And so is this one:

Sam is an electronics technician.

The statement is logically complete. Readers cannot miss your meaning: somewhere is a person; the person's name is Sam; the person is an electronics technician. Suppose instead we write:

Sam an electronics technician.

This statement is not logically complete, therefore not a sentence. The reader is left asking, "What about Sam the electronics technician?" The verb—the word that makes things happen—is missing. By adding a verb we can easily change this fragment to a complete sentence.

| | |
|---|---|
| Simple verb | **Sam is** an electronics technician. |
| Adverb | **Sam,** an electronics technician, **works hard**. |
| Dependent clause, verb, and subjective complement | **Although he is well paid,** Sam, an electronics technician, **is not happy**. |

Do not, however, mistake the following statement—which seems to have a verb—for a complete sentence:

Sam being an electronics technician.

Such "ing" forms do not function as verbs unless accompanied by such other verbs as *is, was,* and *will be.* Again, readers are left in a fog unless you complete your idea with an independent clause.

**Sam,** being an electronics technician, checked all **circuitry.**

Likewise, remember that the "to + verb" form does not function as a verb.

To become an electronics technician.

The meaning is unclear unless you complete the thought.

To become an electronics technician, **Sam had to pass an exam.**

Sometimes we can inadvertently create fragments by adding words—*because, since, if, although, while, unless, until, when, where,* and others—to an already complete sentence. We then change our independent clause (complete sentence) to a dependent clause.

**Although** Sam is an electronics technician.

Such words subordinate the words that follow them so that an additional idea is needed to make the first statement complete. That is, they make the statement dependent on an additional idea, which must itself have a subject and a verb and be a complete sentence. (See "Complex Sentences" and "Faulty Subordination.") Now round off the statement with a complete idea (independent clause).

Although Sam is an electronics technician, **he hopes to become an engineer.**

*Note:* Be careful not to use a semicolon or a period, instead of a comma, to separate elements in the preceding sentence. Because the dependent clause depends on the independent clause for its meaning, you need only a *pause* (symbolized by a comma), not a *break* (symbolized by a semicolon), between these ideas. Many fragments are created when the writer uses too strong a mark of punctuation (period or semicolon) between a dependent and an independent clause, severing the needed connection. (See the later discussion of punctuation.)

Here are some fragments. Each is repaired in several ways. Can you think of other ways of making these statements complete?

| | |
|---|---|
| Fragment | She spent her first week on the job as a researcher. **Compiling technical information from digests and journals.** |
| Correct | She spent her first week on the job as a researcher, compiling technical information from digests and journals. |
| | She spent her first week on the job as a researcher. She compiled technical information from digests and journals. |
| Fragment | **Because the operator was careless.** The new computer was damaged. |
| Correct | Because the operator was careless, the new computer was damaged. |
| | The operator's carelessness resulted in damage to the new computer. |

## Comma Splice

In a comma splice, two complete ideas (independent clauses), which should be *separated* by a period or a semicolon, are incorrectly *joined* by a comma:

**Comma splice**    Sarah did a great job, she was promoted.

You can choose among several possibilities for correcting this error:

1. Substitute a period followed by a capital letter:

    Sarah did a great job. She was promoted.

2. Substitute a semicolon to signal the relationship:

    Sarah did a great job; she was promoted.

3. Use a semicolon with a connecting adverb (a transitional word):

    Sarah did a great job; **consequently,** she was promoted.

4. Use a subordinating word to make the less important sentence incomplete, thereby dependent on the other:

**Because** Sarah did a great job, she was promoted.

5. Add a connecting word after the comma:

Sarah did a great job, **and** she was promoted.

Your choice of construction will depend, of course, on the exact meaning or tone you wish to convey. The following comma splice can be repaired in the ways mentioned above.

Comma splice    This is a fairly new technique, therefore, some people don't trust it.

Correct    This is a fairly new technique. Some people don't trust it.

This is a fairly new technique; therefore, some people don't trust it.

**Because** this is a fairly new technique, some people don't trust it.

This is a fairly new technique, **so** some people don't trust it.

## Run-on Sentence

The run-on sentence, a cousin to the comma splice, crams too many ideas without needed breaks or pauses.

Run-on    The hourglass is more accurate than the water clock for the water in a water clock must always be at the same temperature in order to flow with the same speed since water evaporates and must be replenished at regular intervals, thus not being as effective in measuring time as the hourglass.

Like a runaway train, this statement is out of control. Here is a corrected version:

Revised    The hourglasses more accurate than the water clock because water in a water clock must always be at the same temperature to flow at the same speed. Also, water evaporates and must be replenished at regular intervals. These temperature and volume problems make the water clock less effective than the hourglass for measuring time.

## Choppy Sentences

It is possible to write grammatically correct sentences that, nonetheless, read like Dick-and-Jane sentences. Short, choppy sentences make for tedious reading.

<dl>
<dt>Choppy</dt>
<dd>Brass-plated prongs are unacceptable. They prevent good contact in the outlet. They also rust or corrode. Some of the cheaper plugs also have no terminal screws. The conductors in the cord are soldered to the prongs. Sometimes they are just wrapped around them. These types often come as original equipment on small lamps and appliances. They are not worth repairing when a wire loosens.</dd>
</dl>

Correct this problem by combining related ideas within single sentences and by using transitions to increase coherence.

<dl>
<dt>Revised</dt>
<dd>Brass-plated prongs are unacceptable because they prevent good contact in the outlet and they rust and corrode. Furthermore, some of the cheaper plugs, which often come as original equipment on small lamps and appliances, have no terminal screws: the conductors in the cord are either soldered to the prongs or just wrapped around them. These plugs are not worth repairing when a wire loosens.</dd>
</dl>

The original eight sentences are replaced by three. (See also page 217.)

## Faulty Coordination

Give ideas of equal importance equal emphasis by joining them with coordinating conjunctions: *and, but, or, nor, for, so* and *yet.*

This course is difficult **but** worthwhile.
My horse is old **and** gray.
We must decide to support **or** reject the dean's proposal.

Do not coordinate excessively.

<dl>
<dt>Excessive coordination</dt>
<dd>The climax in jogging comes after a few miles **and** I can no longer feel stride after stride **and** it seems as if I am floating **and** jogging becomes almost a reflex **and** my arms **and** legs continue to move **and** my mind no longer has to control their actions.</dd>
<dt>Revised</dt>
<dd>The climax in jogging comes after a few miles when I can no longer feel stride after stride. By then I am jogging almost by reflex, nearly floating, my arms and legs still moving, my mind no longer having to control their actions.</dd>
</dl>

Avoid coordinating ideas that cannot be sensibly connected:

<dl>
<dt>Faulty</dt>
<dd>John had a weight problem and he dropped out of school.</dd>
<dt>Revised</dt>
<dd>John's weight problem depressed him so much that he couldn't study and so he quit school.</dd>
</dl>

|        |                                                                |
|--------|----------------------------------------------------------------|
| Faulty | I was late for work and wrecked my car.                        |
| Revised| Late for work, I backed out of the driveway too quickly, hit a truck, and wrecked my car. |

Don't use "try and." Use "try to" instead.

|        |                                          |
|--------|------------------------------------------|
| Faulty | I will try and help you.                 |
| Revised| I will try to help you.                  |
| Faulty | Bill promised to try and be on time.     |
| Revised| Bill promised to try to be on time.      |

## Faulty Subordination

When you subordinate, you make the less important idea dependent on the most important idea. With subordination you can combine related short sentences and also emphasize the most important idea. Consider these two ideas:

Joe studies hard. He has a learning disability.

Because these ideas are each expressed as a simple sentence, they appear to be coordinate (equal in importance). But if you wanted to express your opinion about Joe's chances of succeeding, you would need a third sentence: "His handicap probably will make success impossible"; or "His will power will help him succeed." An easier and more concise way to inject your opinion is by combining ideas and subordinating the one that deserves less emphasis:

Despite his learning disability *(subordinate idea)*, Joe studies hard *(independent idea)*.

This first version suggests that Joe will succeed. Below, subordination is used to suggest the opposite:

Despite his diligent studying *(subordinate idea)*, Joe has a learning disability *(independent idea)*.

Place the idea you want emphasized in the independent clause; don't write

Although Alfred is receiving excellent medical treatment, he is seriously ill.

if you mean to suggest that Alfred has a good chance of recovering.
Don't coordinate when you should subordinate:

|      |                                                                             |
|------|-----------------------------------------------------------------------------|
| Weak | Television viewers can relate to an athlete they idolize and they feel obliged to buy the product endorsed by their hero. |

Of the two ideas in the sentence above, one is the cause, the other the effect. Emphasize this relationship by subordination:

|         |                                                                          |
|---------|--------------------------------------------------------------------------|
| Revised | Because television viewers can relate to an athlete they idolize, they feel obliged to buy the product endorsed by their hero. |

When combining several ideas within a sentence, decide which is most important, and make the other ideas subordinate to it—don't simply coordinate:

> **Faulty**  This employee is often late for work, and he writes illogical reports, and he is a poor manager, and he should be fired.

> **Revised**  Because this employee is often late for work, writes illogical reports, and has poor management skills, **he should be fired.** *(The last clause is independent.)*

Do not overstuff sentences by subordinating excessively:

> **Overstuffed**  This job, which I took when I graduated from college, while I waited for a better one to come along, which is boring, where I've gained no useful experience, makes me anxious to quit.

> **Revised**  Upon college graduation, I took this job while waiting for a better one to come along. Because I find it boring, and have gained no useful experience, I am eager to quit.

## Faulty Agreement—Subject and Verb

Failure to make the subject of a sentence agree in number with the verb is a common error. Happily, it's an error easily corrected and avoided. We are not likely to use faulty agreement in short sentences, where subject and verb are not far apart. Thus we are not likely to say "Jack eat too much" instead of "Jack eats too much," but in more complicated sentences—in which the subject is separated from its verb by other words—we sometimes lose track of the subject-verb relationship.

> **Faulty**  The lion's **share** of diesels **are** sold in Europe.

Although "diesels" is closest to the verb, the subject is "share," a singular subject that must agree with a singular verb.

> **Correct**  The lion's **share** of diesels **is** sold in Europe.

Agreement errors are easy to correct when the subject and verb are identified.

> **Faulty**  A **system** of lines **extend** horizontally to form a grid.

> **Correct**  A **system** of lines **extends** horizontally to form a grid.

A second situation that causes trouble in subject-verb agreement occurs when we use indefinite pronouns such as *each, everyone, anybody,* and *somebody.* They function as subjects and usually take a singular verb.

> **Faulty**  **Each** of the crew members **were** injured.

> **Correct**  **Each** of the crew members **was** injured.

> **Faulty**  **Everyone** in the group **have** practiced long hours.

> **Correct**  **Everyone** in the group **has** practiced long hours.

Sometimes agreement problems can be caused by collective nouns such as *herd, family, union, group, army, team, committee,* and *board.* They can call for a singular or plural verb—depending on your intended meaning. When denoting the group as a whole, use a singular verb.

> Correct    The **committee meets** weekly to discuss new business.
>
> The editorial **board** of this magazine **has** high standards.

To denote individual members of the group use a plural verb.

> Correct    Not all of the editorial **board are** published authors.

Yet another problem occurs when two subjects are joined by *either . . . or* or *neither . . . nor.* Here, the verb is singular if both subjects are singular, and plural if both subjects are plural. If one subject is plural and one is singular, the verb agrees with the one that is closer.

> Correct    Neither **John** nor **Bill works** regularly.
>
> **Either apples** or **oranges are** good vitamin sources.
>
> Either Felix or his **friends are** crazy.
>
> Neither the boys nor their **father likes** the home team.

If two subjects (singular, plural, or mixed) are joined by *both . . . and,* the verb will be plural. *Or* suggests "one or the other," but *and* suggests both subjects, thereby requiring a plural verb.

> Correct    **Both** Joe and Bill **are** resigning.
>
> The **book and** the **briefcase appear** expensive.

A single *and* between singular subjects makes for a plural subject.

### Faulty Agreement—Pronoun and Referent

A pronoun can make sense only if it refers to a specific noun (its referent or antecedent), with which it must agree in gender and number.

> Correct    **Joe** lost **his** blueprints.
>
> The **workers** complained that **they** were treated unfairly.

Some instances, however, are less obvious. When an indefinite pronoun like *each, everyone, anybody, someone,* and *none* serves as the pronoun referent, the pronoun itself is singular.

> Correct    **Anyone** can get **his** degree from that college.
>
> **Anyone** can get **his or her** degree from that college.
>
> **Each** candidate described **her** plans in detail.

### Faulty or Vague Pronoun Reference

Whenever a pronoun is used, it must refer to one clearly identified referent; otherwise, your message will be confusing.

> Ambiguous    **Sally** told **Sarah** that **she** was obsessed with her job.

Does "she" refer to Sally or Sarah? Several interpretations are possible:

1. Sally is obsessed with her job.
2. Sally thinks Sarah is obsessed with Sally's job. (Sarah is envious.)
3. Sally thinks that Sarah is obsessed with her own job.
4. Sally is obsessed with Sarah's job.
5. Sally thinks that someone else is obsessed with Sally's job.
6. Sally thinks someone else is obsessed with Sarah's job.
7. Sally thinks someone else is obsessed with some other person's job.
8. Sally thinks someone else is obsessed with her own job.

> **Correct** Sally told Sarah, "I'm obsessed with my job."
> Sally told Sarah, "I'm obsessed with your job."
> Sally told Sarah, "You're obsessed with [your, my] job."
> Sally told Sarah, "She's obsessed with [her, my, your] job."

Avoid using *this, that,* or *it* — especially to begin a sentence — unless the pronoun refers to a specific antecedent (referent).

> **Vague** He drove way from his menial **job,** boring **lifestyle,** and damp **apartment**, happy to be leaving **it** behind.
>
> **Correct** He drove away, happy to be leaving behind his menial job, boring lifestyle, and damp apartment.
>
> **Vague** The problem with our **defective machinery** is compounded by the **operator's incompetence. That** annoys me!
>
> **Correct** I am annoyed by the problem with our defective machinery as well as by the new operator's incompetence.

## Faulty Pronoun Case

A pronoun's case (nominative, objective, or possessive) is determined by its role in a sentence: as subject, object, or indicator of possession.

If the pronoun serves as the subject of a sentence (*I, we, you, she, he, it, they, who*), its case is *nominative.*

> **She** completed her graduate program in record time.
> **Who** broke the chair?

When a pronoun follows a version of *to be* (a linking verb), it explains (complements) the subject, and so its case is nominative.

> It was **she.**
> The chemist who perfected our new distillation process is **he.**

If the pronoun serves as the object of a verb or a preposition (*me, us, you, her, him, it, them, whom*), its case is *objective.*

| Object of the verb | The employees gave **her** a parting gift. |
| Object of the preposition | Several colleagues left with **him**. |
| | To **whom** do you wish to complain? |

If a pronoun indicates possession (*my, mine, ours, your, yours, his, her, hers, its, their, theirs, whose*), its case is *possessive*.

The brown briefcase is **mine.**
**Her** offer was accepted.
**Whose** opinion do you value most?

Here are some of the most frequent errors made in pronoun case:

| Faulty | **Whom** is responsible to **who?** *(The subject should be nominative and the object should be objective.)* |
| Correct | **Who** is responsible to **whom?** |
| Faulty | The debate was between Marsha and **I.** *(As object of the preposition, the pronoun should be objective.)* |
| Correct | The debate was between Marsha and **me.** |
| Faulty | **Us** board members are accountable for our decisions. *(The pronoun accompanies the subject, "board members," and thus should be nominative.)* |
| Correct | **We** board members are accountable for our decisions. |
| Faulty | A group of **we** managers will fly to the convention. *(The pronoun accompanies the object of the preposition, "managers," and thus should be objective.)* |
| Correct | A group of **us** managers will fly to the convention. |

*Hint:* By deleting the accompanying noun from the two latter examples, we can easily identify the correct pronoun case ("We . . . are accountable . . ."; "A group of us . . . will fly . . .").

## Faulty Modification

The word order (syntax) of a sentence determines its effectiveness and meaning. Words or groups of words are modified (i.e., explained or defined) by adjectives, adverbs, phrases, or clauses. Modifiers add detail or otherwise help clarify an idea. Prepositional phrases, for example, usually define or limit adjacent words:

the foundation **with the cracked wall**
the repair job **on the old Ford**
the journey **to the moon**
the party **for our manager**

As do phrases with "-ing" verb forms:

the student **painting the portrait**
**Opening the door,** we entered quietly.

Or phrases with "to + verb" form:

**To succeed,** one must work hard.

Or certain clauses:

the man **who came to dinner**
the job **that I recently accepted**

When we use modifying phrases to begin sentences, we can get into trouble unless we read our sentences carefully.

> Dangling modifier    **Dialing the phone,** the cat ran out the open door.

Here, the introductory phrase signals the reader that the noun beginning the main clause (its subject) is what or who is answering the telephone. The absurd message occurs because the opening phrase has no proper subject to modify; it *dangles*.

> Correct    As Joe dialed the phone, the cat ran out the open door.

A dangling modifier can also obscure the meaning of your message.

> Dangling modifier    **After completing the student financial aid application form,** the Financial Aid Office will forward it to the appropriate state agency.

Who completes the form—the student or the financial aid office? Here are some other dangling modifiers that make the message confusing, inaccurate, or downright absurd:

> Dangling modifier    **While walking,** a cold chill ran through my body.

> Correct    While **I** walked, a cold chill ran through my body.

> Dangling modifier    Impurities have entered our bodies **by eating chemically processed foods.**

> Correct    Impurities have entered our bodies by **our** eating chemically processed foods.

> Dangling modifier    **By planting different varieties of crops,** the pests were unable to adapt.

> Correct    By planting different varieties of crops, **farmers** prevented the pests from adapting.

The position of adjectives and adverbs in a sentence is as important as the position of modifying phrases and clauses. Notice how changing word order affects the meaning of these sentences:

I **often** remind myself of the need to balance my checkbook.
I remind myself of the need to balance my checkbook **often.**

Position modifiers and the words they modify to reflect your meaning. Generally, they belong together. Notice below how the meaning changes as the modifier is moved.

| | |
|---|---|
| Misplaced modifier | Joe typed another memo on our computer **that was useless.** *(Was the typewriter or the memo useless?)* |
| Correct | Joe typed another useless memo on our computer. |
| | *or* |
| | Joe typed another memo on our useless computer. |
| Misplaced modifier | He read a report on the use of nonchemical pesticides **in our conference room.** *(Are the pesticides to be used in the conference room?)* |
| Correct | In our conference room, he read a report on the use of non-chemical pesticides. |
| Misplaced modifier | She volunteered **immediately** to deliver the radioactive shipment. *(Volunteering immediately, or delivering immediately?)* |
| Correct | She immediately volunteered to deliver . . . |
| | *or* |
| | She volunteered to deliver immediately . . . |

## Faulty Parallelism

Express items of the same importance in the same grammatical form:

| | |
|---|---|
| Correct | We here highly resolve . . . that government **of the people, by the people, for the people** shall not perish from the earth. |

The statement above describes the government with three modifiers of equal importance. Because the first modifier is a prepositional phrase, the others must be also. Otherwise, the message would be garbled, like this:

| | |
|---|---|
| Faulty | We here highly resolve . . . that government **of the people, which the people created and maintain, serving the people** shall not perish from the earth. |

If you begin the series with a noun, use nouns throughout the series; do the same for adjectives, adverbs, and specific types of clauses and phrases.

| | |
|---|---|
| Faulty | The new apprentice is **enthusiastic, skilled,** and **you can depend on her.** |
| Correct | The new apprentice is **enthusiastic, skilled,** and **dependable.** *(all subjective complements)* |
| Faulty | In his new job, he felt **lonely** and **without a friend.** |
| Correct | In his new job, he felt **lonely** and **friendless.** *(both adjectives)* |
| Faulty | She plans **to study** all this month and **on scoring well** in her licensing examination. |

|           |                                                                                              |
|-----------|----------------------------------------------------------------------------------------------|
| Correct   | She plans **to study** all this month and **to score well** in her licensing examination. *(both infinitive phrases)* |
| Faulty    | She **sleeps** well and **jogs** daily, **as well as eating** high-protein foods. |
| Correct   | She **sleeps** well, **jogs** daily, and **eats** high-protein foods. *(all verbs)* |

To improve coherence in long sentences, repeat words that introduce parallel expressions:

|           |                                                                                              |
|-----------|----------------------------------------------------------------------------------------------|
| Faulty    | Before buying this property, you should decide whether you will settle down and raise a family, travel for a few years, or pursue a graduate degree. |
| Correct   | Before buying this property, you should decide whether **to settle** down and raise a family, **to travel** for a few years, or **to pursue** a graduate degree. |

Make all headings parallel in your table of contents, your formal outline, and your report.

|         |                                |
|---------|--------------------------------|
| Faulty  | A. Picking the Fruit           |
|         |   1. When to pick    |
|         |   2. Packing         |
|         |   3. Suitable temperature |
|         |   4. Transport with care |

The logical connection between steps in that sequence is obscured because each heading is phrased in a different grammatical form.

|         |                                |
|---------|--------------------------------|
| Correct | A. Picking the Fruit           |
|         |   1. Choose the best time |
|         |   2. Pack the fruit loosely |
|         |   3. Store at a suitable temperature |
|         |   4. Transport with care |

Other forms of phrasing also would be correct here, as long as you express each item in a form parallel to that of all other items in the series.

## Sentence Shifts

Shifts in point of view damage coherence. If you begin a sentence or paragraph with one subject or person, don't shift to another.

|                  |                                                              |
|------------------|--------------------------------------------------------------|
| Shift in person  | When **you** finish the job, **one** will have a sense of pride. |
| Correct          | When **you** finish the job, **you** will have a sense of pride. |
| Shift in number  | **One** should sift the flour before **they** make the pie. |
| Correct          | **One** should sift the flour before **one** makes the pie. (Or better: Sift the flour before making the pie.) |

Don't begin a sentence in the active voice and then shift to passive.

**Shift in voice**    **He delivered** the plans for the apartment complex, and the building site **was also inspected by him.**

**Correct**    **He delivered** the plans for the apartment complex and also **inspected** the building site.

Don't shift tenses without good reason.

**Shift in tense**    She **delivered** the blueprints, **inspected** the foundation, **wrote** her report, and **takes** the afternoon off.

**Correct**    She **delivered** the blueprints, **inspected** the foundation, **wrote** her report, and **took** the afternoon off.

Don't shift from one mood to another (e.g., from imperative to indicative mood in a set of instructions).

**Shift in mood**    **Unscrew** the valve and then steel wool **should be used** to clean the fitting.

**Correct**    **Unscrew** the valve and then **use** steel wool to clean the fitting.

Don't shift from indirect to direct discourse within a sentence.

**Shift in discourse**    Jim wonders **if he will get the job** and **will he like it?**

**Correct**    Jim wonders **if he will get the job** and **if he will like it.**

Will Jim get the job, and will he like it?

## EFFECTIVE PUNCTUATION

Punctuation marks are like road signs and traffic signals. They govern reading speed and provide clues for navigation through a network of ideas; they mark intersections, detours, and road repairs; they draw attention to points of interest along the route; and they mark geographic boundaries. In short, punctuation marks give us a simple way of making ourselves understood.

### End Punctuation

The three marks of end punctuation—period, question mark, and exclamation point—work like a red traffic light by signaling a complete stop.

**Period.**   A period ends a sentence. Periods end some abbreviations.

| | | |
|---|---|---|
| Ms. | Assn. | N.Y. |
| M.D. | Inc. | B.A. |

Periods serve as decimal points for figures.

- $15.95
- 2.14%

**Question Mark.** A question mark follows a direct question.

> Where is the balance sheet?

Do not use a question mark to end an indirect question.

> Faulty   He asked if all students had failed the test?
>
> Correct   He asked if all students had failed the test.
> *or*
>             He asked, "Did all students fail the test?"

**Exclamation Point.** Because exclamation points suggest that you are excited or adamant, don't overuse them. Otherwise you might seem hysterical or insincere.

> Correct   Oh, no!
>             Pay up!

Use an exclamation point only when expression of strong feeling is appropriate.

## Semicolon

A semicolon usually works like a blinking red traffic light at an intersection by signaling a brief but definite stop.

**Semicolon Separating Independent Clauses.** Semicolons separate independent clauses (logically complete ideas), whose contents are closely related and are not connected by a coordinating conjunction.

> The project was finally completed; we had done a good week's work.

The semicolon can replace the conjunction-comma combination that joins two independent ideas.

> The project was finally completed, and we were elated.
> The project was finally completed; we were elated.

The second version emphasizes the sense of elation.

**Semicolons Used with Adverbs as Conjunctions and Other Transitional Expressions.** Semicolons must accompany conjunctive adverbs and other expressions that connect related independent ideas (*besides, otherwise, still, however, furthermore, consequently, therefore, in contrast, in fact,* and so on).

> The job is filled; however, we will keep your résumé on file.

> Your background is impressive; in fact, it is the best among our applicants.

**Semicolons Separating Items in a Series.** When items in a series contain internal commas, semicolons separate the items.

> We are opening branch offices in the following cities: Santa Fe, New Mexico; Albany, New York; Montgomery, Alabama; and Moscow, Idaho.

Members of the survey crew were John Jones, a geologist; Hector Lightweight, a draftsman; and Mary Shelley, a graduate student.

## Colon

A colon works like a flare in the road. It signals you to stop and then proceed, paying attention to the situation ahead, the details of which will be revealed as you move along. Usually, a colon follows an introductory statement that requires a follow-up explanation.

> We need the following equipment immediately: a voltmeter, a portable generator, and three pairs of insulated gloves.

> She is an ideal colleague: honest, reliable, and competent.

> Two candidates are clearly superior: John and Marsha.

In most cases—with the exception of *Dear Sir:* and other salutations in formal correspondence—colons follow independent statements (logically and grammatically complete.) Because colons, like end punctuation and semicolons, signal a full stop, they are never used to fragment a complete statement.

> **Faulty**   My plans include: finishing college, traveling for two years, and settling down in Boston.

No punctuation should follow "include."

Colons can introduce quotations.

> The supervisor's message was clear enough: "You're fired."

A colon normally replaces a semicolon in separating two related, complete statements when the second statement explains or amplifies the first.

> His reason for accepting the lowest-paying job offer was simple: he had always wanted to live in the Northwest.

The statement following the colon explains the "reason" mentioned in the statement preceding the colon.

## Comma

The comma is the most used—and abused—punctuation. Unlike the period, semicolon, and colon, which signal a full stop, the comma signals a *brief pause.* The comma works like a blinking yellow traffic light, for which you slow down without stopping. Never use a comma to signal a *break* between independent ideas; it is not strong enough.

**Comma as a Pause between Complete Ideas.**   In a compound sentence where a coordinating conjunction *(and, or, nor, for, but)* connects equal (independent) statements, a comma is usually placed immediately before the conjunction.

> This is a high-paying job, but the stress is high.

This vacant shop is just large enough for our boutique, and the location is excellent for walk-in customer traffic.

Without the conjunction, each of these statements would suffer from a comma splice, unless the comma were replaced by a semicolon or a period.

**Comma as a Pause between an Incomplete and a Complete Idea.** A comma is usually placed between a complete and an incomplete statement in a complex sentence to show that the incomplete statement depends for its meaning on the complete statement. (The incomplete statement cannot stand alone, separated by a break such as a semicolon, colon, or period.)

**Because he is a fat cat,** Jack diets often.

**When he eats too much,** Jack gains weight.

Above, the first idea is made incomplete by a subordinating conjunction *(since, when, because, although, where, while, if, until)*, which here connects a dependent with an independent statement. The first (incomplete) idea depends on the second (complete) for wholeness. When the order is reversed (complete idea followed by incomplete), the comma can usually be omitted.

Jack diets often **because he is a fat cat.**

Jack gains weight **when he eats too much.**

Because commas take the place of speech signals, reading a sentence aloud should tell you whether or not to pause (and use a comma).

**Commas Separating Items (Words, Phrases, or Clauses) in a Series.** Use a comma to separate items in a series.

**Sam, Joe, Marsha,** and **John** are joining us on the hydroelectric project.

The office was **yellow, orange,** and **red.**

He works hard **at home, on the job,** and even **during his vacation.**

The new employee complained **that the hours were long, that the pay was low, that the work was boring,** and **that the foreman was paranoid.**

Use no commas when *or* or *and* appear between all items in a series.

She is willing to work in San Francisco or Seattle or even in Anchorage.

Add a comma when *or* or *and* is used only before the final item in the series.

Our luncheon special for Thursday will be coffee, rolls, steak, beans, and ice cream.

Without the comma, the sentence might cause the reader to conclude that beans and ice cream are an exotic new dessert.

**Comma Setting Off Introductory Phrases.** Infinitive, prepositional, or verbal phrases introducing a sentence usually are set off by commas.

| | |
|---|---|
| Infinitive phrase | **To be or not to be,** that is the question. |
| Prepositional phrase | **In Rome,** do as the Romans do. |
| | **In other words,** you're fired. |
| | **In fact,** the finish work was superb. |
| Participial phrase | **Being fat,** Jack was a slow runner. |
| | **Moving quickly,** the army surrounded the enemy. |

When an interjection introduces a sentence, it is set off by a comma.

**Oh,** is that the final verdict?

When a noun in direct address introduces a sentence, it is set off by a comma.

**Mary,** you've done a great job.

**Commas Setting Off Nonrestrictive Elements.** A restrictive phrase or clause modifies the subject in such a way that deleting the modifier would change the meaning of the sentence.

All candidates **who have work experience** will receive preference.

The clause, "who have work experience," defines "candidates" and is essential to the meaning of the sentence. Without this clause, the meaning would be entirely different.

All candidates will receive preference.

This sentence also contains a restriction:

All candidates **with work experience** will receive preference.

The phrase, "with work experience," defines "candidates" and thus specifies the meaning of the sentence. Because this phrase *restricts* the subject by limiting the category, "candidates," it is essential to the sentence's meaning and so is not separated from the sentence by commas.

In contrast, a nonrestrictive phrase or clause does not limit or define the subject; such an element is nonessential because it could be deleted without changing the essential meaning of the sentence.

Our draftsperson, **who has little experience,** is highly competent.

This house, **riddled with carpenter ants,** is falling apart.

In each of those sentences, the modifying phrase or clause does not restrict the subject; each could be deleted:

Our new draftsperson is highly competent.
This house is falling apart.

Unlike a restrictive modifier, the nonrestrictive modifier does not supply the essential meaning to the sentence; any nonessential clause or phrase is set off by commas from the rest of the sentence.

To appreciate how commas can affect meaning, consider these statements:

**Restrictive**    Office workers **who drink martinis with lunch** have slow afternoons.

Because the restrictive clause limits the subject, "office workers," we interpret that statement as follows: some office workers drink martinis with lunch, and these have slow afternoons. In contrast, we could write

**Nonrestrictive**    Office workers, **who drink martinis with lunch,** have slow afternoons.

Here the subject, "office workers," is not limited or defined. Thus we interpret that all office workers drink martinis with lunch and have slow afternoons.

**Commas Setting Off Parenthetical Elements.**   Items that interrupt the flow of the sentence are called parenthetical and are enclosed by commas. Expressions such as *of course, as a result, as I recall,* and *however* are parenthetical and may denote emphasis, afterthought, clarification, or transition.

**Emphasis**    This deluxe model, **of course,** is more expensive.

**Afterthought**    Your report format, **by the way,** was impeccable.

**Clarification**    The loss of my job was, **in a way,** a blessing.

**Transition**    Our warranty, **however,** does not cover tire damage.

Direct address is parenthetical.

Listen, **my children,** and you shall hear . . .

A parenthetical expression at the beginning or the end of a sentence is set off by a comma.

**Naturally,** we will expect a full guarantee.

**My friends,** I think we have a problem.

You've done a good job, **Jim.**

**Yes,** you may use my name in your advertisement.

**Commas Setting Off Quoted Material.**   Quoted items included within a sentence are often set off by commas.

The customer said, **"I'll take it,"** as soon as he laid eyes on our new model.

**Commas Setting Off Appositives.**   An appositive, a word or words explaining a noun and placed immediately after it, is set off by commas.

Martha Jones, **our new president,** is overhauling all personnel policies.

Alpha waves, **the most prominent of the brain waves,** are typically recorded in a waking subject whose eyes are closed.

Please make all checks payable to Sam Sawbuck, **company treasurer.**

**Commas Used in Common Practice.** Commas are used to set off the day of the month from the year, in a date.

May 10, 1984

They are used to set off numbers in three-digit intervals.

11,215
6,463,657

They are also used to set off street, city, and state in an address.

The bill was sent to John Smith, 184 Sea Street, Albany, New York 01642.

When the address is written vertically, however, the commas that are omitted are those which would otherwise occur at the end of each address line.

John Smith
184 Sea Street
Albany, New York 01642

Use commas to set off an address or date in a sentence.

Room 3C, Margate Complex, is the site of our newest office.
December 15, 1977, is my retirement date.

Use them to set off degrees and titles from proper names.

Roger P. Cayer, M.D.
Gordon Browne, Jr.
Sandra Mello, Ph.D.

**Commas Used Erroneously.** Avoid sprinkling commas where they are not needed or simply do not belong. In fact, you are probably safer using too few commas than too many. Overuse of commas generally can be avoided if you read your sentences aloud.

**Faulty**    As I opened the door, he told me, that I was late. *(separates the indirect from the direct object)*

The universal symptom of the suicide impulse, is depression. *(separates the subject from its verb)*

This has been a long, difficult, project. *(separates the final adjective from its noun)*

John, Bill, and Sally, are joining us on the design phase of this project. *(separates the final subject from its verb)*

An employee, who expects rapid promotion, must quickly prove his worth. *(separates a modifier that should be restrictive)*

I spoke in a conference call with John, and Marsha. *(separates two words linked by a coordinating conjunction)*

The room was, eighteen feet long *(separates the linking verb from the subjective complement)*

We painted the room, red. *(separates the object from its complement)*

## Apostrophe

Apostrophes serve three purposes: to indicate the possessive, a contraction, and the plural of numbers, letters, and figures.

**Apostrophe Indicating the Possessive.** At the end of a singular word, or of a plural word that does not end in *s*, add an apostrophe plus *s* to indicate the possessive.

The people's candidate won.

The chain saw was Bill's.

The men's locker room burned.

The car's paint job was ruined by the hailstorm.

I borrowed Doris's book.

Have you heard Ray Charles's song?

For some words convention will require that you not add an *s*.

Correct    Moses' death

for conscience' sake

Do not use an apostrophe to indicate the possessive form of either singular or plural pronouns:

Correct    The book was hers.

Ours is the best sales record.

The fault was theirs.

At the end of a plural word that ends in *s*, add an apostrophe only.

Correct    the cows' water supply

the Jacksons' wine cellar

At the end of a compound noun, add an apostrophe plus an *s*.

Correct    my father-in-law's false teeth

At the end of the last word in nouns of joint possession, add an apostrophe plus *s* if both own one item.

Correct    Joe and Sam's lakefront cottage

Add an apostrophe plus *s* to both nouns if each owns specific items.

Correct    Joe's and Sam's passports

**Apostrophe Indicating a Contraction.** An apostrophe shows that you have omitted one or more letters in a phrase that is usually a combination of a pronoun and a verb.

| Correct | | |
|---|---|---|
| | I'm | they're |
| | he's | you'd |
| | you're | who's |

Don't confuse *they're* with *their* or *there*.

| Faulty | |
|---|---|
| | there books |
| | Their now leaving. |
| | living their |

| Correct | |
|---|---|
| | their books |
| | They're now leaving. |
| | living there |

Remember the distinction in this way:

Correct    Their boss knows they're there.

Don't confuse *it's* and *its*. *It's* means "it is." *Its* is the possessive.

Correct    It's watching its reflection in the pond.

Don't confuse *who's* and *whose*. *Who's* means "who is," whereas *whose* indicates the possessive.

Correct    Who's interrupting whose work?

Other contractions are formed from the verb and the negative.

| Correct | | |
|---|---|---|
| | isn't | can't |
| | don't | haven't |
| | won't | wasn't |

**Apostrophe Indicating the Plural of Numbers, Letters, and Figures.** The 6's on this new typewriter look like smudged G's, the 9's are illegible, and the %'s are unclear.

## Quotation Marks

Quotation marks set off exact words borrowed from another speaker or writer. At the end of a quotation the period or comma is placed within quotation marks.

Correct    "Hurry up," he whispered.
             She told me, "I'm depressed."

The colon or semicolon is always placed outside quotation marks:

Correct    Our contract clearly defines "middle-management personnel"; however, it does not state salary range.

             You know what to expect when Honest John offers you a "bargain": a piece of junk.

Sometimes a question mark is used within a quotation that is part of a larger sentence.

> **Correct**   "Can we stop the flooding?" inquired the foreman.

When the question mark or exclamation point is part of the quotation, it belongs within quotation marks, replacing the comma or period.

> **Correct**   "Help!" he screamed.
> He asked John, "Can't we agree about anything?"

If, however, the question mark or exclamation point is meant to denote the attitude not of the quoter but of the quotee, it is placed outside the quotation mark.

> **Correct**   Why did he wink and tell me, "It's a big secret"?
> He actually accused me of being an "elitist"!

When quoting a passage of fifty words or longer, indent the entire passage five spaces and single-space between its lines to set it off from the text. Do not enclose the indented passage in quotation marks.

Use quotation marks around titles of articles, book chapters, poems, and unpublished reports.

> **Correct**   The enclosed article, "The Job Market for College Graduates," should provide some helpful insights.

The title of a published work—book, journal, newspaper, brochure, or pamphlet—should be underlined to represent italics.

Finally, use quotation marks (with restraint) to indicate your ironic use of a word.

> **Correct**   He is some "friend"!

## Ellipses

Use three dots in a row ( . . . ) to indicate that you have left some material out of a quotation. If the omitted words come at the end of the original sentence, a fourth dot indicates the period. Use several dots centered in a line to indicate that a paragraph or more has been left out. Ellipses help you save time and zero in on the important material within a quotation.

> **Correct**   "Three dots . . . indicate that you have left some material out. A fourth dot indicates the period. . . . Several dots centered in a line . . . indicate a paragraph or more. . . . Ellipses help you . . . zero in on the important material. . . . "

## Italics

In typing or longhand writing, indicate italics by <u>underlining.</u> Use italics for titles of books, periodicals, films, newspapers, and plays; for the names of ships; for foreign words or scientific names; for emphasizing a word (used sparingly); for indicating the special use of a word.

*The Oxford English Dictionary* is a handy reference tool.
The *Lusitania* sank rapidly.
He reads *The Boston Globe* often.
My only advice is *caveat emptor*.
*Bacillus anthracis* is a highly virulent organism.
Do *not* inhale these spores, under any circumstances!
Our contract defines a *full-time employee* as one who works a minimum of thirty-five hours weekly.

## Parentheses

Use commas normally to set off parenthetical elements, dashes to give some emphasis to the material that is set off, and parentheses to enclose material that defines or explains the preceding statements.

> This organism requires an anaerobic (oxygenless) environment.
> The cost of manufacturing our Beta II transistors has increased by 10 percent in one year. (See Appendix A for full cost breakdown.)
> This new three-colored model (made by Ilco Corporation) is selling well.

Notice that material within parentheses, like all other parenthetical material discussed earlier, can be deleted without harming the logical and grammatical structure of the sentence.

Also, use parentheses to enclose numbers or letters that segment items of information in a series.

> The three basic steps in this procedure are (1) . . . , (2) . . . , and (3). . . .

## Brackets

Use brackets within a quotation to add material which was not in the original quotation but which is needed for clarification. Sometimes a bracketed word will provide an antecedent (or referent) for a pronoun.

> "She [Jones] was the outstanding candidate for the job."

Brackets can enclose information from some other location within the context of the quotation.

> "It was in early spring [April 2, to be exact] that the tornado hit."

Use brackets to correct a quotation.

> "His report was [full] of mistakes."

Use *sic* ("thus so") when quoting a mistake in spelling, usage, or logic.

> His secretary's comment was clear: "He don't [*sic*] want any of these."

## Dashes

Dashes are effective—as long as they are not overused. Make dashes on your typewriter by placing two hyphens side by side, with a space before and after. Although parentheses deemphasize the enclosed material, dashes emphasize it.

Used selectively, dashes provide emphasis but they are not a substitute for all other forms of punctuation. When in doubt, do not use a dash!

Dashes can be used to denote an afterthought.

Have a good vacation—but don't get sunstroke.

or to enclose an interruption in the middle of a sentence.

This building's designer—I think it was Wright—was, above all, an artist.

Our new team—Jones, Smith, and Brown—is already compiling outstanding statistics.

Although they can often be used interchangeably with commas, dashes dramatize a parenthetical statement more than commas do.

Mary, a true friend, spent hours helping me rehearse for my interview.

Mary—a true friend—spent hours helping me rehearse for my interview.

Notice the added emphasis in the second version.

## Hyphen

Use a hyphen to divide a word at the right-hand margin. Consult your dictionary for the correct syllable breakdown:

    com-puter
    comput-er

Actually, it is best to avoid altogether this practice of dividing words at the ends of lines in a typewritten text.

Use a hyphen to join compound modifiers (two or more words preceding the noun as an adjective).

    the rough-hewn wood
    the well-written report
    the all-too-human error
    a three-part report

Do not hyphenate these same words if they *follow* the noun.

    The wood was rough hewn.
    The report is well written.
    The error was all too human.

Hyphenate an adverb-participle compound preceding a noun.

    the high-flying glider

Do not hyphenate compound modifiers if the adverb ends in *-ly*.

    the finely tuned engine

Hyphenate all words that begin with the prefix *self-*.

    self-reliance

self-discipline
self-actualizing

Hyphenate to avoid ambiguity.

re-creation *(a new creation)*
recreation *(leisure activity)*

Hyphenate words that begin with *ex-* only if *ex-* means "past."

ex-foreman
expectant

Hyphenate all fractions, along with ratios which are used as adjectives and which precede the noun.

a two-thirds majority
In a four-to-one ratio they defeated the proposal.

Do not hyphenate ratios if they do not immediately precede the noun.

The proposal was voted down four to one.

Hyphenate compound numbers from twenty-one through ninety-nine.

Thirty-eight windows were broken.

Hyphenate a series of compound adjectives preceding a noun.

The subjects for the motivation experiment were fourteen-, fifteen-, and sixteen-year-old students.

## TRANSITIONS AND OTHER CONNECTORS

Transitions help make meaning clear, signaling readers that we are in a specific time or place, that we are giving an example, showing a contrast, shifting gears, concluding our discussion, or the like. Common transitions and the relationships they signal:

| | |
|---|---|
| **Addition** | I am majoring in naval architecture; **furthermore,** I spent three years crewing on a racing yawl. |

moreover      and
in addition   again
also          as well as

| | |
|---|---|
| **Place** | Here is the switch that turns on the stage lights. **To the right** is the switch that dims them. |

beyond        to the left
over          nearby
under         adjacent to
opposite to   next to
beneath       where

| | |
|---|---|
| **Time** | The crew will mow the ball field this morning; **immediately afterward** we will clean the dugouts. |

| | |
|---|---|
| first | the next day |
| next | in the meantime |
| second | in turn |
| then | subsequently |
| meanwhile | while |
| at length | since |
| later | before |
| now | after |

| | |
|---|---|
| **Comparison** | Our reservoir is drying up because of the drought; **similarly,** water supplies in neighboring towns are dangerously low. |

likewise
in the same way
in comparison

| | |
|---|---|
| **Contrast or alternative** | Felix worked hard; **however,** he received poor grades. **Although** Jack worked hard, he was never promoted. |

| | |
|---|---|
| however | but |
| nevertheless | on the other hand |
| yet | to the contrary |
| still | notwithstanding |
| in contrast | conversely |

| | |
|---|---|
| **Results** | Jack fooled around; **consequently** he was shot by a jealous husband. **Because** he was fat, he was jolly. |

| | |
|---|---|
| thus | therupon |
| hence | as a result |
| therefore | so |
| accordingly | as a consequence |

| | |
|---|---|
| **Example** | Competition for part-time jobs is fierce; **for example,** eighty students applied for the clerk's job at Sears. |

| | |
|---|---|
| for instance | namely |
| to illustrate | specifically |

| | |
|---|---|
| **Explanation** | She had a terrible semester; **that is,** she flunked four courses. |

| | |
|---|---|
| in other words | in fact |
| simply stated | put another way |

| | |
|---|---|
| **Summary or conclusion** | Our credit is destroyed, our bank account is overdrawn, and our debts are piling up; **in short,** we are bankrupt. |

| | |
|---|---|
| in closing | to sum up |
| to conclude | all in all |

|  |  |
|---|---|
| to summarize | on the whole |
| in brief | in retrospect |
| in summary | in conclusion |

Pronouns serve as connectors, because a pronoun refers back to a noun in a preceding clause or sentence.

As the **crew** neared the end of the project, **they** were all willing to work overtime to get the job done.

**Low employee morale** is damaging our productivity. **This** problem needs immediate attention.

Repetition of key words or phrases is another good connecting device—as long as it is not overdone. Here is another example of effective repetition:

Overuse and drought have depleted our water supply to critical level. Because of our **depleted supply**, we need to enforce strict **water**-conservation measures.

Here, the repetition also emphasizes a critical problem.

Because this paragraph lacks transitions, sentences seem choppy and awkward:

> **Choppy**  Technical writing is a difficult but important skill to master. It requires long hours of work and concentration. This time and effort are well spent. Writing is indispensable for success. Good writers derive pride and satisfaction from their effort. A highly disciplined writing course should be part of every student's curriculum.

Here is the same paragraph rewritten to improve coherence:

> **Revised**  Technical writing is a difficult but important skill to master. It requires long hours of work and concentration. This time and effort, **however**, are well spent **because** writing is an indispensable tool for success. **Moreover,** good writers derive pride and satisfaction from their effort. A highly disciplined writing course, **therefore,** should be a part of every student's curriculum.

Besides increasing coherence *within* a paragraph, transitions and other connectors can emphasize relationships *between* paragraphs by linking related groups of ideas. Here are two transitional sentences that could serve as the conclusion for some paragraphs or as topic sentences for paragraphs that would follow; or they could stand alone for emphasis as single-sentence paragraphs:

Because the A-12 filter has decreased overall engine wear by 15 percent, it should be included as a standard item in all our new models.

With the camera activated and the watertight cover sealed, the diving bell is ready to be submerged.

These sentences look both ahead and back, providing clear direction for continuing discussion.

Topic headings, like those used in this book, are another device for increasing coherence. A topic heading is both a link and a separation between related yet distinct groups of ideas.

Finally, a whole paragraph can serve as a connector between major sections of your report. Assume that you have just completed a section in a report on the advantages of a new oil filter and are now moving to a section on selling the idea to the buying public. Here is a paragraph you might write to link the two sections:

> Because the A-12 filter has decreased overall engine wear by 15 percent, it should be included as a standard item in all our new models. Tooling and installation adjustments will add roughly $100 to the list price of each model, however, and so we have to explain the filter's long-range advantages to the customer. Let's look at ways of explaining these advantages.

Effective use of transitional devices in your revisions is one way of transforming a piece of writing from adequate to excellent.

## EFFECTIVE MECHANICS

Correctness in abbreviation, capitalization, use of numbers, and spelling is an important sign of your attention to detail. Don't ignore these conventions. (See Chapter 13 for format conventions.)

### Abbreviations

(For correct abbreviations in documentation, see pages 129–136.)

In using abbreviations, consider your audience; never use one that might confuse your reader. Often, abbreviations are not appropriate in formal writing. When in doubt, write the word out in full.

Abbreviate some words and titles when they precede or immediately follow a proper name.

| Correct | Mr. Jones | Raymond Dumont, Jr. |
|---------|-----------|---------------------|
|         | Dr. Jekyll | Warren Weary, Ph. D. |
|         | St. Simeon |                     |

Do not, however, write abbreviations such as these:

| Faulty | Mary is a Dr. |
|--------|---------------|
|        | Pray, and you might become a St. |

In general, do not appreviate military, religious, and political titles.

| Correct | Reverend Ormsby |
|---------|-----------------|
|         | Captain Hook |
|         | President Bush |

Abbreviate time designations only when they are used with actual times.

**Correct**    A.D. 576
400 B.C.
5:15 A.M.

Do not abbreviate these designations when they are used alone.

**Faulty**    Plato lived sometime in the B.C. period.
She arrived in the A.M.

In formal writing, do not abbreviate days of the week, months, words such as *street* and *road*, or names of disciplines such as *English*. Avoid abbreviating states, such as *Me.* for *Maine*; countries, such as *U.S.* for *United States;* and book parts such as *Chap.* for *Chapter, pg.* for *page*, and *fig.* for *figure*.

Use *no.* for *number* only when the actual number is given.

**Correct**    Check switch No. 3.

Abbreviate units of measurement only when they are used often in your report and are written out in full on first use. Use only those abbreviations which you are sure your reader will understand. Abbreviate items in a visual aid only if you need to save space.

Here are common technical abbreviations for units of measurement:

| | | | |
|---|---|---|---|
| ac | alternating current | freq | frequency |
| amp | ampere | ft | foot |
| Å | angstrom | g | gram |
| az | azimuth | gal | gallon |
| bbl | barrel | gpm | gallons per minute |
| BTU | British Thermal Unit | gr | gram |
| C | Celsius | hp | horsepower |
| cal | calorie | hr | hour |
| cc | cubic centimeter | in. | inch |
| circ | circumference | iu | international unit |
| cm | centimeter | j | joule |
| cps | cycles per second | ke | kinetic energy |
| cu ft | cubic foot | kg | kilogram |
| db | decibel | km | kilometer |
| dc | direct current | kw | kilowatt |
| dm | decimeter | kwh | kilowatt hour |
| doz | dozen | l | liter |
| dp | dewpoint | lat | latitude |
| F | Fahrenheit | lb | pound |
| f | farad | lin | linear |
| fbm | foot board measure | long | longitude |
| fl oz | fluid ounce | log | logarithm |
| FM | frequency modulation | m | meter |
| fp | foot pound | max | maximum |

| | | | | |
|---|---|---|---|---|
| mg | milligram | sec | second |
| min | minute | sp gr | specific gravity |
| ml | milliliter | sq | square |
| mm | millimeter | T | ton |
| mo | month | temp | temperature |
| mph | miles per hour | tol | tolerance |
| oct | octane | ts | tensile strength |
| oz | ounce | v | volt |
| psf | pounds per square foot | va | volt ampere |
| | | w | watt |
| psi | pounds per square inch | wk | week |
| | | wl | wavelength |
| qt | quart | yd | yard |
| r | roentgen | yr | year |
| rpm | revolutions per minute | | |

Here are some common abbreviations for reference in manuscripts:

| | | | |
|---|---|---|---|
| anon. | anonymous | fig. | figure |
| app. | appendix | i.e. | that is |
| b. | born | illus. | illustrated |
| © | copyright | jour. | journal |
| c., ca. | about (c. 1988) | l. ll. | line(s) |
| cf. | compare | ms., mss. | manuscript(s) |
| ch. | chapter | no. | number |
| col. | column | p., pp. | page(s) |
| d. | died | pt., pts. | part(s) |
| ed. | editor | rev. | revised or review |
| e.g. | for example | rep. | reprint |
| esp. | especially | sec. | section |
| et al. | and others | sic | thus, so (to cite an |
| etc. | and so on | | error in the |
| ex. | example | | quotation) |
| f. or ff. | the following page or pages | trans. | translation |
| | | vol. | volume |

For abbreviations of other words, consult your dictionary. Most dictionaries list abbreviations at the front or back or alphabetically with the word entry.

## Capitalization

Capitalize these: proper nouns, titles of people, books and chapters, languages, days of the week, the months, holidays, names of organizations or groups, races and nationalities, historical events, important documents, and names of structures or vehicles. In titles of books, films, and so on, capitalize the first word and all those following, except articles or prepositions.

> *A Tale of Two Cities*
> Protestant
> Wednesday
> the *Queen Elizabeth II*
> the Statue of Liberty
> April
> Chicago
> the Bill of Rights
> the Chevrolet Corvette
> Russian
> Labor Day
> Dupont Chemical Company
> Senator John Pasteur
> France
> the War of 1812
> Daughters of the American Revolution
> the Emancipation Proclamation

Do not capitalize the seasons, names of college classes (*freshman, junior*), or general groups (*the younger generation,* or *the leisure class*).

Capitalize adjectives that are derived from proper nouns.

> Chaucerian English

Capitalize titles preceding a proper noun but not those following:

> State Senator Marsha Smith
> Marsha Smith, state senator

Capitalize words like *street, road, corporation, college* only when they accompany a proper noun.

> Bob Jones University
> High Street
> the Rand Corporation

Capitalize *north, south, east,* and *west* when they denote specific locations, not when they are simple directions.

> the South
> the Northwest
> Turn east at the next set of lights.

Begin all sentences with capitals.

## Use of Numbers

As a rule, write out numbers that can be expressed in two words or fewer.

                    fourteen                      five
                    eighty-one                    two million
                    ninety-nine

Use numerals for all others.

                    4,364                         2,800,357
                    543                           200 million
                    3¼

Use numerals to express decimals, precise technical figures, or any other exact measurements. Numerals are more easily read and better remembered than numbers that are spelled out.

                    50 kilowatts                  15 pounds of pressure
                    14.3 milligrams               4,000 rpm

Express these in numerals: dates, census figures, addresses, page numbers, exact units of measurement, percentages, ages, times with A.M. or P.M. designations, and monetary and mileage figures.

                    page 14                       1:15 P.M.
                    18.4 pounds                   9 feet
                    115 miles                     12 gallons
                    the 9-year-old tractor        $15
                    15.1 percent

Do not begin a sentence with a numeral.

    Six hundred students applied for the 102 available jobs.

If the numeral takes more than two words, revise word order.

    The 102 available jobs brought 780 applicants.

Do not use numerals to express approximate figures, time not stated with A.M. or P.M. designations, or streets named by numbers less than 100.

    about seven hundred fifty
    four fifteen
    108 East Forty-second Street

If one number immediately precedes another, spell out the first and use a numeral to express the second:

    Please deliver twelve 18-foot rafters.

Only in contracts and other legal documents should a number be stated both in numerals and in words:

    The tenant agrees to pay a rental fee of three hundred seventy-five dollars ($375.00) monthly.

# Appendix B
# Writers and Audiences
# on the Job

This appendix contains the full (and mostly unedited) text of interviews with four working professionals who write daily. The first two can be classified as part-time writers; the latter two, full-time. The varied reporting tasks of these four respondents typify the range of primary audiences and information needs that writers face on the job:

- Blair Cordasco's audience needs instructions for improving their job performance.
- James North's audience needs to make major investment and marketing decisions.
- Bill Trippe's audience needs to make major government policy decisions.
- Pam Herbert's audience needs clear documentation to understand and use her company's mainframe computer.

As the interviews show, each of these writers has secondary audiences as well.

Each respondent was asked eight questions about amount and types of writing, audiences, writing challenges, deadlines, individual writing processes, and advice to students. The collected responses contain much useful advice for anyone whose career will depend in some way on good writing.

## FIRST RESPONDENT—WRITING FOR COLLEAGUES

Blair Cordasco is a training specialist for an international bank. Her main job is to develop instructional programs and procedures that help improve employees' performance at all levels.

*Question 1:* What percentage of your job is spent writing, editing, and dealing with written communication?
*Answer:* Roughly 75 percent.

*Question 2:* What types of writing or editing do you do?
*Answer:* I do a few types:

1. I write memos recommending types of training needed in various divisions. I also write status (periodic) reports on the types and evaluation of training that has been given. Basically, I keep management informed about the activities of the training department.

2. I write materials to be used for training (lectures, procedures, and manuals). Also, I rewrite technical manuals that accompany high-tech equipment so that nonspecialized managers can understand the operating principles of the equipment, to better supervise the staff using the equipment.

3. I edit my staff's material for clarity, conciseness, fluency, and basic correctness. Reading their material from the viewpoint of the person who has to learn from it, I adjust the level of technicality and detail for the intended audience. Then the edited material goes to the content experts, who review the simplified versions for technical accuracy. Organization and format are *very* important. Word processors make the formatting much easier, and we use them for our training manuals.

*Question 3:* Who are the audiences for your writing, and what are their needs?
*Answer:* I write for two audiences:

1. Management staff (B.A.'s in finance, economics, et cetera, and M.B.A.'s), who range from college trainees all the way to upper management, with varying degrees of technical and business experience. Senior members of the audience need executive briefings (overviews, summaries), not a lot of detail, with technical material as appropriate. The junior members of the audience need training materials either for skills (interviewing, quality-control techniques, et cetera) or for knowledge (overview of international banking policy, or fundamentals of data processing, et cetera).

2. Clerical staff (high school education), such as secretaries and terminal operators, who need to learn specific skills (word processing, key punching, et cetera) they can apply directly to the job. The material has to be at their level of comprehension.

*Question 4:* What does your audience expect from your documents?
*Answer:* Upper management asks, "How does it affect the bottom line? How can I use this information to increase productivity and quality?" Lower management and trainees ask, "What's in it for me? Can it help me do *my* job better?"

I spend much of my time trying to convince people that the information we offer can help them. Managers have to be persuaded that your recommendations are worthwhile and the best way to go.

Clerical staff want to know, "Will this help me keep my job?" Their lifeline is the information and training we offer. An underlying assumption in training clericals is that the training will increase their self-esteem and sense of belonging: "Will this make me feel like a jerk or help me feel I can accomplish something?" Will the tone be condescending, or the material too technical? Or will the tone be friendly and engaging, and the material at the right level of technicality?

*Question 5:* What is your biggest writing challenge on the job?
*Answer:* Aside from the persuasive challenge I just described, the biggest challenge is taking a body of information and packaging it *accurately* in different ways for different audiences.

For instance, a new piece of equipment arrives, and the vendor's documentation is too technical for operators (how to) and managers (overview). We have to take the vendor's information and rewrite (how to) for operators. Then, for managers, we have to explain the operating principles, point out potential problems and solutions, and identify major problem areas. So, from one technical manual, we have to extract instructions for the operators, and explanations and overviews for the managers.

A trainer really is dealing with all kinds of information and all kinds of vehicles, and trying to reduce them to the lowest common denominator—to make the meaning absolutely clear.

*Question 6:* What about deadlines?
*Answer:* Deadlines are a way of life. Nothing is more essential than getting information to the right people at the right time.

*Question 7:* Do you follow a standard and predictable process when writing?
*Answer:* Deadlines affect how we approach the writing process. With plenty of time, we can afford the luxury of the whole process: the careful decisions about our audience, purpose, content, organization, and style—and plenty of revisions. At times, we have to take shortcuts.

Each project presents its own problem, so we have no pat way of writing training materials. With limited time, we have limited revisions—so you have to think on your feet.

Say we receive word November 1 that a new check-processing system will be installed and operating by December 1. Operators will begin training on this equipment one week before the system is implemented (last week of November). In this case, we might write backward: the basic "How to" comes first, and the more elaborate documentation (for management) would come whenever we could complete it

*Question 8:* What advice do you have for students?
*Answer:* Make sure that *whatever* you're writing is clear *to you* first, and that the way you've organized it makes sense *to you*. Then take that material and become more objective. Try to understand how your audience thinks: "How

can I make this logical to my audience? Will they understand what I want them to understand?"

Organizing is the key. Develop the type of outlining or listing or brainstorming tool that works best for you, but *find* one that works, and use it consistently. Then you'll be comfortable with that general strategy whenever you sit down to write, especially under rigid deadline.

## SECOND RESPONDENT—WRITING FOR CLIENTS

James North is a senior project manager for a market-research firm. Most of his writing is for clients who will make investment decisions based on feasibility and strategy for marketing new products.

*Question 1:* What percentage of your job is spent writing, editing, and dealing with written communication?
*Answer:* From 50 to 60 percent.

*Question 2:* What types of writing and editing do you do?
*Answer:* Aside from an occasional internal memo, my writing is designed for two purposes: (1) to gather, analyze, and interpret information from respondents, and (2) to report our findings to clients.

For respondents, I have to translate the market-research (or information) needs of clients into precise questions that *cannot* be misinterpreted. The questions have to be so precise and unambiguous that the respondent knows *exactly* what we're asking.

Reports, on the other hand (of findings gathered from questionnaires), have to allow the clients to make marketing decisions based on their understanding of complex data analyses presented in a concise, *nontechnical* way. These people are not in any way experts. They want to know, "What do I do next?" Our reports interpret the findings ("What do they mean?"), and offer recommendations for marketing strategy ("What should I do?").

*Question 3:* Who are the audiences for your writing, and what are their needs?
*Answer:* Among questionnaire respondents, the range is vast: from scientists making a breakthrough in robotics, to homemakers choosing a brand of coffee off a supermarket shelf.

Clients range from an express-mail carrier who's spending $3 million to assess the feasibility of expanding into international express service, to an individual who wants to market a new brand of salad dressing or low-calorie chocolate.

Both respondents and clients need material that is concise, clear, and unambiguous.

*Question 4:* What does your audience expect from your documents?

*Answer:* Respondents expect questions they can interpret accurately—that is, in only *one* way.

Clients expect enough information to make basic investment and marketing decisions. Some clients need answers to very specific statistical questions: "How many overnight letters are sent daily from England to the United States? How many potentially could be sent daily—classified by type of industry (pharmaceutical, aerospace, and so on) and size of company?" Other clients have more general questions:

- "How does one break into the salad-dressing market?"
- "What are your recommendations for establishing a new brand of salad dressing?"
- "How can I make any money with my low-calorie chocolate?"
- "I have a million, and want to market this chocolate. What can you do for me?"

For the latter group of clients, we might research the taste, packaging, and price-point (say, $4 a pound for chocolate) that is most popular for that product.

Our audience's expectations range from those of real experts who know exactly what they want—down to the *n*th decimal point—to somebody who calls with a vague idea and asks, "What can you do for me?"

*Question 5:* What is your biggest writing challenge on the job?
*Answer:* My writing has to have *one* interpretation *only*. I have to be certain that respondents are answering *exactly* the question I had in mind, not inventing their own version of the question. Then I have to take these data and translate them into accurate interpretations and recommendations for our clients.

*Question 6:* What about deadlines?
*Answer:* The priority is to *meet* the deadline, regardless of the quality of the report. You never have the time you need, and so the people who survive are those who *are able to write it*—no matter what—with enough quality to satisfy clients' expectations.

The difference between real-world and college writing is that, in the real world, the deadlines are *never* extended—never any excuses for not getting the report done on time.

Whether you keep the client depends on how well you can write under impossible time pressures.

*Question 7:* Do you follow a standard and predictable process when writing?
*Answer:* I look at data and move to some kind of prose that gets revised as often as time allows. For long reports, I rely on outlines. The final product is a recommendation, but the process begins with my looking at a computer printout.

Even when we devise questions that we're sure are unambiguous to our intended respondents, we pretest them—often finding that the questions have been misinterpreted. Then we write them and test them again, until we get them right.

The basic element in my writing process is checking and rechecking for ambiguous messages, and revising as often as time allows.

*Question 8:* What advice do you have for students?

*Answer:* Learn to write as quickly, accurately, and unambiguously as you can in a given amount of time. The issue is to have it there when it's expected — maybe not perfect, but to develop a style so that the document is out and it gets the job done.

Organize your time in such a way that you can always meet the deadline. Prepare to labor under Murphy's Law: there will always be some crisis. The survivors are those who anticipate the crises, and are able to deliver nonetheless.

## THIRD RESPONDENT—WRITING FOR GOVERNMENT DECISION MAKERS

Bill Trippe is a communications specialist with a military contract company. He provides all-around communications support for a large group of engineers who perform technical analyses for the government.

*Question 1:* What percentage of your job is spent writing, editing, and dealing with written communication?

*Answer:* Writing is probably 30 to 40 percent of my job, with editing tasks taking up another 50 percent, and training and tutoring accounting for the remaining 10 to 20 percent.

*Question 2:* What types of writing and editing do you do?

*Answer:* The writing I do falls into four categories:

1. Semitechnical reports from my managers to upper managers and to our military sponsors. These are usually progress reports on our larger projects, and run from 1 to 20 pages.

2. Some sections of highly technical (engineer-to-engineer) reports. The engineers and scientists usually write the "body" of highly technical reports, and I write the abstract, introduction, executive summary, acknowledgments, conclusions, lists of references, and bibliography. I also fill out sections that need more material — add figures and tables, expand explanations for equations and other raw material.

3. Miscellaneous reports and articles for a general audience. These include articles and announcements for the company newspapers, project updates for the board of directors, and short papers about our work, to be used for recruiting new employees.

4. Training materials for engineers. I'm currently writing short essays on grammar, audience analysis, and techniques for oral presentation. I've

also put together my own editorial style sheet that I constantly update and expand.

*Question 3:* Who are the audiences for your writing, and what are their needs?
*Answer:* The typical primary audience is either our military sponsors or our management.

Military. Specifically, the military sponsor is the project officer: a Colonel or Lieutenant Colonel who is the commanding officer in charge of procuring a military system. He usually has help from other commissioned officers and civilian engineers, who coordinate testing, management and technical reviews, quality assurance, and other specific areas of the program. The project officer is the decision maker—the primary reader—and his supporting staff are the secondary readers.

Their needs? Usually, they've asked our company to perform a specific technical analysis. Our formal products to them are a technical report, regular progress reports, and related briefings. They want thorough analysis that relates the state of the art and our company's expertise to their specific program. If they want to build a radar to operate in extreme cold, we'll tell them if it can be done, how reliable the radar would be, and how much it would probably cost.

Their uses for the papers? As a basis for making decisions, or—once the decision has been made—as a means of persuading their superiors (right on up through the Pentagon and Congress) to go along with their recommendations.

Company Management. The progress reports that I ghostwrite for my bosses are usually written for what I would call "middle managers"—my bosses are the low-level managers and their superiors (the readers) are just above them and just below vice presidents. They are all ex-engineers-turned-managers, and so the technical content can be anywhere from general to complex. They are looking at how the project team is handling potential problem and risk areas—difficult technical tasks, areas where cost overruns are possible, and any problems with scheduling. I write weekly reports about every project, and other reports as necessary.

Their needs? They are a fairly easy audience to write for, mainly because they can read at almost any level. They are, however, a critical audience—almost to the point of hostility. They are tough on their underlings probably because their superiors are then tough on them, and the stakes get higher up there. But, as I said above, they are looking for highlights about management concerns: cost, schedule, and technical risk. They are also more concerned with the quality of the writing and presentation because they are conscious of company image and how the customer and the public perceive us.

Their use? They use all the information given them as a means for making sure the individual projects and tasks are running smoothly. They put together a picture of their area that they can then present to their superiors and to the customers.

For the other writing I do, the audience is everyone from the readers of the company newspaper to the board of directors: general readers, I guess. Everything in this category is really written for information purposes, for entertainment, or as a bit of fluff.

*Question 4:* What does your audience expect from your documents?
*Answer:* The military audience expects a technically complete report with a rigorous organization, a good balance of detail, strong graphics and tables, and a clear and readable writing style. Because not all the military people are engineers, certain sections of the report (the introduction, executive summary, and conclusions) have to be semitechnical. On the other hand, to satisfy the various interests of the specialists in the project office, many areas have to be covered in great detail, but without bogging down the basic paper. For this, we resort to appendixes—lots of them. Occasionally we will publish completely separate reports: an executive summary, the analysis, and supporting appendixes; this type of report usually goes over very well.

I've covered the expectations of our management audience in my response to question 3. But I should add that these readers really expect the writing to be tailored to their needs. When they ask for a one-page memo, they expect a one-page memo. The same is true for briefings; they ask for and get a three-minute presentation.

*Question 5:* What is your biggest writing challenge on the job?
*Answer:* All my writing tasks are a big challenge because they are highly visible and fall under the tightest deadlines.

The greatest pressure comes from the writing tasks because they are my product alone, whereas the editing I do is the author's product I've helped along. If the author's paper is unsatisfactory, I don't hear about it (unless there's an editorial problem, but that hasn't happened). If my reports are late or contain a mistake, however, then my bosses come directly to me. Therefore I watch what I write very closely, and rewrite and proofread everything. I also work very closely with the typists and artists, to see that everything is exactly as I want it.

In a corporation, you are invisible unless you do something spectacular or make a mistake. Because the spectacular rarely happens, management tends not to notice you until you make a mistake.

*Question 6:* What about deadlines?
*Answer:* I rarely miss a deadline.

I know this isn't my own rule, but I use it anyway: when I need to estimate how much time it will take me to do something, I make a reasonable estimate and then double it. Then when I know something is due, I use my doubled estimate and go to work on it, leaving everything else aside. This technique usually works, unless the computer goes down and I lose an afternoon's writing to that great electronic void, or a secretary is out sick, or everyone I need to see is in a meeting.

The important thing is juggling jobs and deadlines. I work for six projects and really have eight bosses. If everyone is screaming for work, then I go to everyone, explain my schedule, and see if they can come up with a new deadline. If conflicts persist, then I explain the problem to the big boss, and he tells me what to work on first. My negotiations are usually enough; I've had to involve the big boss only once or twice.

I've brought work home, stayed late, gotten extra help from secretaries and co-op students, and done jobs faster than I've wanted to. But by reasonably estimating how long a job will take and by being assertive about how to handle conflicting deadlines, I haven't had to do any of these other things to the point of discomfort.

*Question 7:* Do you follow a standard and predictable process when writing?
*Answer:* The process I follow depends on the length of the piece.

For short pieces (say, less than three pages) I outline in my head, draft, and revise. On my first draft of a short piece, I spend 40 to 50 percent of the time on the first one or two paragraphs, and crank out the rest quickly. Then I revise two or three times, and tinker with the mechanics and format right up until printing. I use a word processor for everything I write and edit, and so revising is easy: I'm never afraid to change something because of the extra work mechanical typing would create. I usually don't bother to make a printout until I'm pretty close to a final product, and so it's easy for me to revise something short several times a day, tinker with it a half-dozen times, and then make a printout and tinker some more once I see it on paper.

For longer pieces, I try to follow a rigorous pattern:

a. Interview knowledgeable people first for ideas and inspiration
b. Write a paragraph-long purpose statement and sentence outline
c. Review the outline and thesis with the boss of the project
d. Write a first draft, paying particular attention to the introduction
e. Have the draft reviewed for technical accuracy
f. Revise for accuracy and begin revising for editorial quality
g. Revise (two to five revisions, depending on deadline)
h. Correct the mechanics and format

*Question 8:* What advice do you have for students?
*Answer:* Whew! I could say a million things here.

The following paragraphs are for non-engineers interested in technical writing as a career.

First of all, technical writing is as tough a writing job as newspaper writing, advertising writing, creative writing, and any other job where you work primarily with words. And so to enter this field you have to be fully competent as a writer: you need to have an inquisitive mind and critical abilities, have eagerness to interview and do research, be a wide and thoughtful reader, have a college background that includes many writing courses beyond freshman English, have experience writing against deadlines, have at least a strong interest in and an ability to learn about technical areas (courses in math, science, and computers would give you a great advantage), have the social ability to deal with many kinds of people of all ages and backgrounds, and have the instincts and composure to deal with corporate life.

Get practical experience before graduation. I combined my undergraduate writing program with assorted part-time writing and publication jobs. I was on the staffs of the college newspaper, literary magazine, yearbook, and radio station. I wrote for a local daily newspaper and did other freelance writing for newspapers and magazines. I packed in as much practical experience as possible—and learned a great deal about writing, editing, production, graphic design, and photography. Despite the pressure you might feel to learn as much about the technical areas as possible, keep in mind that technical writers and editors are expected to be publications specialists, not engineers.

I also had a strong academic program. I had an excellent writing program in college, minored in philosophy, and took an assortment of interesting and challenging courses. In this job, I have to think on my feet and learn complex things quickly. But I've never felt at a loss—and I've gained the respect of my peers and bosses—by being able to tackle a subject and get things down on paper. My liberal arts background—with its emphasis on reading, and writing—is an asset. On the other hand, you need some math and science knowledge and ability. My only college science courses were in biology and chemistry, but my high school program included physics and trigonometry. In interviews for technical writing jobs, I had to convince interviewers that I had an aptitude for technical subjects. I would recommend getting a few technical courses under your belt, and you won't need anything else after that.

Another point: Learn what the established industry procedures are for writing, editing, production, art, and printing. Specifically:

- Learn how to use the editorial style books. The *U.S. Government Printing Office Style Manual* and the *Chicago Manual of Style* are the two most widely used.

- Learn everything you can about production; a little book published by the International Paper Company, *Pocket Pal,* is stuffed with information about production and printing. A part-time job on the school newspaper will get you experience in basic paste-up. Learn typing, word processing, and typesetting; these may be your biggest practical skills on the job. If you know what can be done and how long it takes to do it, then you will be able to reduce your production problems by at least half. Learn about how artwork is prepared, because illustrations are as much as 50 percent of a technical document; I work closely with two illustrators. Learn also what makes a good graphic and what doesn't. Learn something about how maps are made.

- Learn photography (including darkroom work—not just snapshots). I work with photographs all the time, and though I don't have to take pictures myself, I assign photographers to take pictures, work with them when they are taking pictures, choose pictures for papers and for audiovisual presentations, and work them into page layouts, presentations, and three-dimensional displays.

- Learn printing. Read what you can about it in *Pocket Pal,* and then look for a large printer in your area who offers tours and try to arrange one for a group from your school. If you know printing, then you will know a great deal about production—what looks good and what doesn't, what works and what doesn't. I deal with the print shop every day. I often have to rush work through there against the shop's normal schedule; if I weren't reasonable in my demands on these guys, they'd have my head.

And a word or two for engineers. According to most surveys I've read, you'll spend about 25 percent of your time on the job writing, and another 25 percent giving formal and informal oral presentations on your work. After your technical knowledge, your most important skill will be your ability to communicate. Your career will soar or die based on your ability to communicate your technical knowledge to other people (especially your bosses); I see this happening all the time.

## FOURTH RESPONDENT—WRITING FOR COMPUTER USERS

Pamela Herbert is a technical writing for a nationwide distributor. She writes user documentation (manuals for using the company's mainframe computer in various applications in various company locations), and upgrades existing manuals.

*Question 1:* What percentage of your time is spent writing, editing, and dealing with written communication?
*Answer:* The actual physical process of writing constitutes maybe 30 percent of my work time. I spend more time coordinating research (I have to catch the field-support people, who are my information sources, when they are in town), collecting information, organizing notes, planning a manual's general structure, preparing visuals to go with the text, and keeping track of which reviewer has which draft. Because I rely on others' feedback, I circulate materials often. And so I write memos fairly often, about three to five times per week.

*Question 2:* What types of writing and editing do you do?
*Answer:* My writing consists of

- user documentation: procedure manuals written for the Hewlett-Packard 3000, a mainframe computer. (Our plants use computer applications such as Accounts Payable, Accounts Receivable, Inventory Control, and Vending. The manuals for these applications are called "control procedures." They are written according to a very structured format.)

- supplements or updates to existing documents. (I'm currently writing a user guide for a "report-writer" program to be used in our plants. The guide provides examples specific to our business, so that users can have models. Also, it defines terms and summarizes important information from the original manual for the software.)

- three to five memos per week, requesting information or reporting on my work.

My editing consists of work on

- revisions of articles written for our company newsletter.

- revisions of procedures written by programmers to be used by computer operators. These have less detail and less supplementary information than the user documentation. They are "skeleton" procedures that simply need rewriting for clarity.

*Question 3:* Who are the audiences for your writing, and what are their needs? *Answer:* My audience varies according to the type of writing involved. User documentation is called "control procedures"; the procedures are written both for upper management (so that they can understand the system in general) and for data-entry clerks (so that they can use the system). Therefore, control procedures are written as *tasks* required to perform a function with the computer.

System users in our branch offices are not, for the most part, data-processing people. They are clerical workers using the computer to do the tasks they once did manually; they see the computer as a "black box," and so they need very specific, nontechnical instructions. Managers and other administrators simply can read the task overviews and outlines for a picture of the whole process.

When researching a topic, I try to anticipate my audience's needs, and I ask the technical source person (usually a programmer or systems analyst) specific questions keyed to my audience's needs: Who performs the task? What materials are required? What does the task accomplish? What can go wrong? Otherwise, I would waste time soaking up like a sponge any and all information the source person feels like rattling off, whether it's important or not.

For memos, I try to keep the style consistent regardless of who will be reading it. Anything worth writing as a memo (instead of a scribbled note or phone call) is always in my plainest English—even though my boss initially balked at my writing "here it is . . ." instead of "Enclosed please find . . ." I have a hard time with the business-ese that so many people use. If they ask me to proofread memos or letters, I can't resist striking all the heretofore's and please-be-advised-that-pursuant-to-your-request's. People dislike surrendering those, though, so I'm rarely asked to proofread those things anymore!

*Question 4:* What does your audience expect from your documents?

*Answer:* As I mentioned earlier, my manuals have to be specific and straightforward.

Also, because I'm a member of the Quality Assurance group, one function of my job is to detect potential problems with the computer system: inconsistencies or "user-unfriendliness" that could result in a service call from a branch office to our systems support staff here. I am expected to document a system as it is; then, while reviewing my document, the department managers decide whether something in the system should be changed.

Also, I am expected to submit written reports of any Quality Assurance issues I encounter while researching the applications. Readers expect a crystal-clear description of the problem. Considered a "naive user," I am a fairly good test of how understandable a procedure will be to average data-entry persons. And so I am expected to clarify any procedure that strikes me as unclear.

*Question 5:* What is your biggest writing challenge on the job?

*Answer:* The biggest challenge in my job is staying on top of a project. It is too easy to procrastinate, put off sifting through those notes, sitting down with that programmer, or making those phone calls to Jacksonville. My boss doesn't watch over me to make sure I'm working, but if a project is unusually delayed, he wants to know why.

The biggest *writing* challenge is getting my reviewers to *review!* They hang on to a draft for weeks, and only after I hound them do they drag the thing out and review it. Then they hound *me* to finish the thing and get it into the field. I have to be diplomatic, of course, but it is frustrating to work hard and meet a deadline and then have the draft stagnating on someone's desk. All I can do is document the delays, so that I'm not held responsible for the lengthy production time on a document.

*Question 6:* What about deadlines?

*Answer:* Although I construct project timetables with deadlines, I always end up revising the charts and extending the deadlines. I know pretty much how long it will take to research, organize, and draft a manual. What I can't predict are reviewers who drop my work when they are suddenly called out of town, "emergency" projects that cut into my time, source people who are on the road for weeks, and downtime on our system here. Sometimes I spend a week at a branch office myself to do research, rather than wait for a technical person to become available at the home office. At any rate, I try to meet a project's deadline, but I do not lose sleep over it if I can't do it.

*Question 7:* Do you follow a standard and predictable process when writing?

*Answer:* The process involved in producing a typical manual is

  a. Once the project is assigned, I set up research time with the technical person.

b. I compile notes, collect report samples, and roughly sketch any data-entry screens I'll have to document.

c. I analyze the notes, organizing information into tasks, procedure steps, knowledge topics, nice-to-knows, rules, cautions, and so on.

d. I write a draft (longhand) and enter it into my word processor.

e. I assemble a complete draft, which is sent around on its in-house review. Major problems are ironed out, and I revise the first draft.

f. The revised draft is mailed to a user who is familiar enough with the system to review the document competently.

g. I usually assemble "dummy data" to run the procedure on the system myself, as a final test of the accuracy of what I've written.

h. After "cosmetic" editing, the document is issued.

Deadlines do affect the process. If I'm really strapped for time, I'll scrap the field review (step f). I usually make sure the draft is as accurate as possible before it goes out on its first review in-house. If time gets really tight, though, I won't bother to double-check every issue with the first draft. I'll rap it out quickly and let any errors be caught in the review process. I try not to do this often; shortcuts never save much time, and the glaring gaps in the text make me look like a sloppy writer.

*Question 8:* What advice do you have for students?
*Answer:* DON'T let yourself slide into the jargon and business-ese trap. Hold fast to the way you learned to write—correctly.

DO be flexible, though, when it comes to other people's writing. You can't preach English grammar to adults if they don't want to hear it.

DO contribute your ideas. Many times I've found that the only reason something hadn't been done in a better way was that no one had bothered to suggest it.

DO keep lists. They help you to keep tabs on which document is at what stage, who is reviewing what, and so on. It's much easier to concentrate on writing when you have everything written down neatly and don't have to commit it all to memory.

DO try to get hands-on experience with what you're writing about, if possible. If you're writing a procedure, perform it yourself. If you're describing the parts of a device, get your hands on one. Knowing your subject thoroughly can make the writing 100 percent easier.

DO keep up to date with the technical writing field. Subscribe to documentation newsletters (I gets lots of info from the Society for Technical Communication, as well as a neat newsletter called *Simply Stated*, from the Document Design Center). Try to keep learning more about different documentation developments and techniques. You'll be more useful in your present job, and more marketable if you decide to look for a new one.

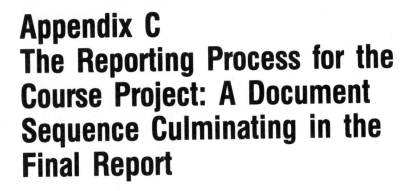

# Appendix C
# The Reporting Process for the Course Project: A Document Sequence Culminating in the Final Report

**P**rofessionals in the workplace engage in one common and continual activity; they struggle with big decisions like these:

- Should we promote Jones to general manager, or should we bring in someone from outside the company?
- Why are we losing customers, and what can we do about it?
- How can we decrease work-related injuries
- Should we encourage foreign investment in our corporation?
- Will this employee-monitoring program ultimately increase productivity or alienate our workers?

Research projects are routinely undertaken to answer questions like those above—to provide decision makers the information they need.

A course research project, like any in the workplace, is designed to fill a specific information need: to answer a question, to solve a problem, to recommend a course of action. And so students of technical communication often spend much of the semester preparing a research project.

Virtually any major project in the workplace requires that all vital information at various stages (the plan, the action taken, the results) be recorded *in writing*. The proposal-progress report-final report sequence keeps readers informed at each stage of a project. This appendix presents a unified and sequential view of the documents that record project information.

# A SAMPLE RESEARCH PROJECT[1]

**M**ike Cabral, communications major, works part-time as assistant to the production manager for *Megacrunch*, a computer magazine specializing in small-business applications. (*Production* is the transforming of manuscripts into a published form.) Mike's writing instructor assigns a course project, and encourages students to select topics from the workplace, when possible. Mike asks *Megacrunch* production manager Marcia White to suggest a research topic that might be useful to the magazine.

White outlines a problem she thinks needs careful attention: Now six years old, *Megacrunch* has enjoyed steady growth in sales volume and advertising revenue—until recently. In the past year alone, *Megacrunch* has lost $150,000 in subscriptions and one advertising account worth $60,000. White knows that most of these losses are caused by increasing competition (three competing magazines have emerged in 18 months). In response to these pressures, White and the executive staff have been exploring ways of reinvigorating the magazine—through added coverage and "hot" features, more appealing layouts and page design, and creative marketing. But White is concerned about another problem that seems partially responsible for the fall in revenues: too many errors are appearing in recent issues.

When *Megacrunch* first hit the shelves, the occasional error in grammar or accuracy seemed unimportant. But as the magazine grew in volume and complexity, the errors increased. Errors in recent issues include misspellings, inaccurate technical details, unintelligible sentences and paragraphs, and scrambled source code (in sample programs).

White asks Mike if he's interested in researching the error problem and looking into quality-control measures. She feels that his three years' experience with the magazine qualifies him for the task. Mike accepts the assignment.

White cautions Mike that this topic is politically sensitive, especially to the editorial staff and to the investors. She wants to be sure that Mike's investigation doesn't merely turn up a lot of dirty laundry. Above all, White wants to preserve the confidence of investors—not to mention the morale of the editorial staff, who do a good job in a tough environment, plagued by impossible deadlines and constant pressure. White knows that offending people—even unintentionally—can be disastrous.

She therefore insists that all project documents express a supportive rather than critical point of view: "What could we be doing better?" instead of "What are we doing wrong?" Before agreeing to release the information from company files (complaint letters, notes from irate phone calls, and so on) White asks Mike to submit a proposal, in which the intent of this project is made absolutely clear.

---

[1]In the interest of privacy, the names of all people, publications, companies, and products in Mike's documents have been changed.

# THE PROJECT DOCUMENTS

Three types of documents lend shape and sequence to the research project: **the proposal,** which spells out the plan; **the progress report,** which keeps track of the investigation; and **the final report,** which analyzes the findings. This appendix shows how these documents function together in Mike Cabral's reporting process.

## The Proposal Stage

Proposals offer plans for meeting needs. A proposal's primary readers are those who will decide whether to approve, fund, or otherwise support the project. Readers of a research proposal usually begin with questions like these:

- *What, exactly, do you intend to find out?*
- *Why is the question worth answering, or the problem worth solving?*
- *What benefits can we expect from this project?*

Once readers agree the project is worthwhile, they will want to know all about the plan:

- *How, exactly, do you plan to do it?*
- *Is the plan realistic?*
- *Is the plan acceptable?*

Besides these questions, readers may have others: *How much will it cost? How long will it take? What makes you qualified to do it?* and so on. See Chapter 19 for more discussion and examples.

Mike Cabral knows his proposal will have only Marcia White (and possibly some executive board members) as primary readers. The secondary reader is Mike's writing instructor, who must approve the topic as well. And at some point, Mike's documents could find their way to his coworkers.

White already knows the background and she needs no persuading that the project is worthwhile, but she does expect a realistic plan before she will approve the project. (For his instructor, Mike attaches a short appendix [not shown here] outlining the background and his qualifications). Also, Mike concentrates on his emphasis: he wants the proposal to be positive rather than critical, so as not to offend anyone. He therefore focuses on achieving *greater accuracy* rather than *fewer errors*. So that his instructor can approve the project, Mike submits the proposal in Figure C.1 by the semester's fourth week.

**FIGURE C.1** A Proposal for a Research Project

## Rangeley Publishing Company

TO:    Marcia White, Production Manager        September 26, 19xx

FROM:   Mike Cabral, Production Assistant  *MC*

SUBJECT: <u>Proposal for Studying Ways of Improving Quality Control at *Megacrunch* Magazine</u>

### Introduction

*Tells what the problem is, and what it means to the company*

The growing number of grammatical, informational, and technical errors in each monthly issue of *Megacrunch* is raising complaints from authors, advertisers, and readers. Beyond compromising the magazine's reputation for accurate and dependable information, these errors— almost all of which seem avoidable—endanger our subscription and advertising revenues.

### Summary of the Problem

*Further defines the problem and its effects*

More and more authors are complaining of errors in published versions of their articles. Software developers assert that errors in reviews and misinformation about products have damaged reputations and sales. For instance, Osco Scientific, Inc. claims to have lost $150,000 in software sales because of an erroneous review in *Megacrunch*.

Although we continue to receive a good deal of "fan mail," we also receive letters speculating about whether *Megacrunch* has lost its edge as the leading resource for small-business users.

### Proposed Study

*Describes the project and tells who will carry it out*

I propose to examine the errors that most frequently recur in our publication, to analyze their causes, and to search for ways of improving the quality of the magazine.

### Methods and Sources

*Tells how the writer plans to carry out the project*

In addition to close examination of recent *Megacrunch* issues and competing magazines, my primary data sources will include correspondence and other feedback now on file from authors, developers, and readers. I also plan telephone interviews with some of the above sources. In addition, interviews with our editorial staff should yield valuable insights and suggestions. As secondary material, books and articles on editing and writing can provide sources of theory and technique.

**FIGURE C.1** A Proposal for a Research Project *Continued*

**Encourages reader support by telling *why* the project will be beneficial**

**Conclusion**

We should not allow clearly avoidable errors to eclipse the hard work that has made *Megacrunch* the leading Cosmo resource for small-business users. I hope that my research project will help resolve many such errors. With your approval, I will begin immediately.

## The Progress-Report Stage

The progress report keeps readers up-to-date on the project's activities, new developments, accomplishments or setbacks, and timetable. Depending on the size and length of the particular project, the number of progress reports will vary. (Mike's course project will require only one.) Readers approach any progress report with two big questions:

- *Is the project moving ahead according to plan and schedule?*
- *If not, why not?*

Readers may have a host of subordinate questions as well. See pages 396–399 for more discussion and examples.

Mike Cabral designs his progress report in Figure C.2 for his boss *and* his instructor, and turns it in by the semester's tenth week. Each of these readers will want to know what Mike has accomplished so far.

**FIGURE C.2**  A Progress Report for a Research Project

# Rangeley Publishing Company

TO:      Marcia White, Production Manager                    November 6, 19xx

FROM:    Mike Cabral, Production Assistant ꟽℓ

SUBJECT: Report of Progress on My Research Project: A Study of Ways for
         Improving Quality Control at *Megacrunch* Magazine

**Work Completed**

*Tells what has been done so far, and what is now being done*

My topic was approved on September 28, and I immediately began both primary and secondary research. I have since reviewed file letters from contributors and readers, along with notes from phone conversations with various clients and from interviews with *Megacrunch's* editing staff. I have also surveyed the types and frequency of errors in recent issues. Recent books and articles on writing and editing have rounded out my study. The project has moved ahead without complications. With my research virtually completed, I have begun to interpret the findings.

**Preliminary Interpretation of Findings**

*Tells what has been found so far, and what it seems to mean*

From my primary research and my own editing experience, I am developing a focused idea of where some of the most avoidable problems lie and how they might be solved. My secondary sources offer support for the solutions I expect to recommend, and they suggest further ideas for implementing the recommendations. With a realistic and efficient plan, I think we can go a long way toward improving our accuracy.

**Work Remaining**

*Comments on the project schedule; describes what remains to be done; and gives a date for completion*

So far, the project is on schedule. I plan to complete the interpretation of all findings by the week of November 29, and then to organize, draft, and revise my final report in time for the December 14 submission deadline.

## The Final-Report Stage

The final report presents the results of the research project: findings, interpretations, and recommendations. Readers will expect this document to answer questions like these:

- *What did you find?*
- *What does it all mean?*
- *What should we do?*

Depending on the topic and situation, of course, readers will have specific questions as well. See Chapter 20 for discussion and examples.

During his research, Mike Cabral discovered problems over and above the published errors he had been assigned to investigate. For instance, after looking at competing magazines he decided that *Megacrunch* needed improved page design, along with a higher quality stock (the paper the magazine is printed on). He also concluded that a monthly section on business applications would help. But despite their usefulness, none of these findings or ideas was part of Mike's *original* assignment. White expected him to answer these questions, specifically:

- *Which errors recur most frequently in our publication?*
- *Where are these errors coming from?*
- *What can we do to prevent them?*

Mike therefore decides to focus exclusively on the error problem. (He might later discuss those other issues with White—if the opportunity arises. But if *this* report were to include material that exceeds the assignment *and* the reader's expectations, Mike could end up appearing arrogant or presumptuous.)

Mike tries to give White only what she requested. He analyzes the problem and the causes, and then recommends a solution. Mike adapts the general outline on page 462 to shape the three major sections of his report: *introduction-findings and conclusions-recommendations.* For the reader's added convenience and orientation, he includes the report supplements discussed in Chapter 18: *front matter* (title page, transmittal letter, tables of contents, and informative abstract) and *end matter*, as needed (works-cited page and appendixes [not shown here]).

After several revisions, Mike submits copies of the report in Figure C.3 to his boss and to his writing instructor.

FIGURE C.3  The Final Report for a Research Project

Virtually any long
report has a title page

The title promises
what the report will
deliver

# Quality-Control Recommendations
## for *Megacrunch* Magazine

Prepared for

**Marcia S. White**
**Production Manager**
**Rangeley Publications**

The primary reader's
name, title, and
organization

by

**Michael T. Cabral**

Author's name

Submission date

December 14, 19xx

**FIGURE C.3** The Final Report for a Research Project *Continued*

**A letter of transmittal usually accompanies a long report**

82 Stephens Road
Boca Grande, FL 08754
December 14, 19xx

**Addressed to the primary reader**

Marcia S. White, Production Manager and Vice President
Rangeley Publications, Inc.
167 Dolphin Ave.
Englewood, FL 08567

Dear Ms. White:

**Identifies the subject of the report, and tells what is covered**

**A confident tone throughout**

**Gives an overview of the entire project, and its findings**

I am happy to submit my report recommending quality-control measures for *Megacrunch* magazine. The report briefly discusses the history of our quality-control problem, identifies the types of errors we are up against, analyzes possible causes, and recommends three realistic solutions.

My research confirmed exactly what you had feared. The problem is big and deeply rooted: our authors have legitimate complaints; our readers justifiably want information they can put to work; and developers and advertisers have the right to demand fair and complete representation. As a result of client dissatisfaction, competing magazines are gaining readers and authors at our expense.

**Request for action**

**Invites follow-up investigations**

To have an immediate impact on our quality-control problem, we should act now. Because of our limited budget, I have tried to recommend low-cost, high-return solutions. If you have other solutions in mind, I would be happy to research them for projected effectiveness and feasibility.

Sincerely,

*Michael T. Cabral*

Michael T. Cabral
Production Assistant
Rangeley Publications

**FIGURE C.3** The Final Report for a Research Project *Continued*

Table of Contents
(Reports with
numerous visuals
also have a Table of
Figures)

Front matter (items
that precede the
report)

Headings and
subheadings from the
report itself

All headings in the
table of contents
follow the exact
phrasing of those in
the report text

The various typefaces
and indentations
reflect the respective
rank of various
headings in the report

Each heading listed in
the table of contents
is assigned a page
number

End matter (items
that follow the report)

iii

# CONTENTS

**FIGURE C.3** The Final Report for a Research Project *Continued*

iv

# INFORMATIVE ABSTRACT

An investigation of the quality-control problem at *Megacrunch* magazine identifies not only the types of errors and their causes, but also recommends a plan.

*Megacrunch* suffers from the following types of avoidable errors:

- *Grammatical errors* are most frequent: misspellings, fragmented and jumbled sentences, misplaced punctuation, etc.

- *Informational errors:* incorrect prices, products attributed to wrong companies, mismarked visuals, etc.

- *Technical errors* are less frequent, but the most dangerous: garbled source code, mismarked diagrams, misused technical terms, etc.

- *Distortions of the author's original meaning:* introduced by editors who attempt to improve clarity and style

The above errors seem to have the following causes:

- *Poor initial submissions from contributors* ignore basic rules of grammar, clarity, and organization

- *Lack of structure in the editing cycle* allows for unrestrained and often excessive editing at all stages

- *Lack of diversity in the editorial staff* leaves language specialists responsible for catching technical and informational errors

- *Lack of communication with authors and advertisers* leaves the primary sources out of the production process

On the basis of my findings, I offer three recommendations for improving quality control during the production process:

- *Expanded author's guide* that includes guidelines for effective use of active voice, visuals, direct address, audience analysis, etc.

- *Five-stage editing cycle* that specifies everyone's duties at each stage. The cycle would require two additional staff members: a technical editor and a fact checker/typist for editorial changes.

- *More communication with contributors* by exchanging galley proofs and increasing our use of the electronic network.

---

**Sidebar (left margin):**

For busy readers, the informative abstract summarizes the report's essential message (findings, conclusions, recommendations). It is the one part of a long report read by most readers.

The summary stands alone in meaning—a kind of mini-report written for the most general reader.

Busy readers need to know quickly what is most important. A summary gives them enough information to decide whether they should read the whole report, parts of it, or none of it.

**FIGURE C.3** The Final Report for a Research Project *Continued*

This section tells what the report is about, why it was written, and how much it covers

An overview of the problem and its effects on the magazine's revenues

Readers are referred to appendixes for details that would interrupt the report flow

Request for action

Purpose and scope of the report; overview of research methods and data sources

Because his primary reader knows the background, Mike keeps the introduction brief

1

# INTRODUCTION

The reputation of *Megacrunch* magazine is jeopardized by grammatical, technical, and other errors appearing in each issue.

*Megacrunch* has begun to lose some long-time readers, advertisers, and authors. Although many readers continue to praise the usefulness of our information, complaints about errors are increasing and subscriptions are falling. Advertisers and authors increasingly point to articles or layouts in which excessive editing has been introduced, and some have taken their business and articles to competing magazines. One disgruntled reader sums up our problem by asking that we devote "more effort to publishing a magazine without the kinds of elementary errors that distract readers from the content" (Grendel 2). This kind of complaint is typical of the sample letters in Appendix A.

Granted, complaints are inevitable—as can be seen in a quick review of "Letters to the Editor" in virtually any publication. But if *Megacrunch* is to withstand the competition and uphold its reputation as the leading resource for Cosmo applications in small business, we must minimize such complaints.

This report identifies the major errors that recur in our magazine, and investigates their causes. My data is compiled from interviews with our editorial staff, a review of complaint letters from authors and readers, and a spot-check for errors in the magazine itself. Books and articles on writing and editing provide theory and technique. The report concludes by recommending a three-part solution to our error problem.

**FIGURE C.3** The Final Report for a Research Project *Continued*

This section tells what the writer found and what it means

The first subsection analyzes the problem; the second will examine causes

Introduction and justification for an analysis of the problem

A lead-in to the visual

A visual that illustrates various parts of the problem

To further segment the report, each of the three major sections (INTRODUCTION, FINDINGS AND CONCLUSIONS, RECOMMENDATIONS) begins on its own separate page

Discussion of the visual, and lead-in to the analysis

Overview of the next subsection, so that readers know what to expect

One part of the problem defined, with examples

Effects of the problem

The writer cites authorities, to clarify and support his position

2

# FINDINGS AND CONCLUSIONS

## ELEMENTS OF THE PROBLEM

Errors in *Megacrunch* are limited to no single category. For example, some errors are tied to technical slip-ups, while others result from editors changing the author's intended meaning. My spot-check of *Megacrunch* 8.10, our most recent issue, revealed errors of the types listed in Table 1.

**TABLE 1 Sample Errors**

| Spot-check of *Megacrunch* 8.10 | | |
|---|---|---|
| **Error type** | **As Published** | **Corrected Version** |
| mechanical | <u>varity</u> of software | <u>variety</u> of software |
| technical | <u>Dos</u> | <u>DOS</u> |
| informational | <u>Deluxe Panel</u> | <u>DeluxePanel</u> |
| grammatical | <u>This</u> will help . . . . | <u>Editing</u> will help . . . . |
| grammatical | . . . everyone helps <u>for of</u> a program's release date approaches . . . . | . . . everyone helps <u>as</u> a program's release date approaches . . . . |
| technical | <u>ram</u> | <u>RAM</u> |
| informational | <u>cosmo</u> | <u>Cosmo</u> |
| mechanical | We're back<u>,</u> now we will | We're back<u>;</u> now we will |

As Table 1 illustrates, errors in our recent issue were plentiful and diverse. My random analysis of only six pages identified errors in four categories: grammatical/mechanical, technical, informational, and distortions of intended meaning.

### Errors in Grammar and Mechanics

Basic correctness is a "given"—and a problem—for any publication. Sentence fragments, confused punctuation, and poor spelling cause readers to "question the professionalism or diligence of both author and editor" (Chang 39)—an assertion borne out by the sampling of complaints in Appendix A.

In a recent survey of college and workplace writing, Haswell (168) found that the average writer suffers from the following basic problems:

**FIGURE C.3** The Final Report for a Research Project *Continued*

Findings

The writer interprets and relates the material to his readers, who always want to know what something means to them personally

Request for action

3

- Three-to-four words are misspelled in a memo-length piece.
- One of every ten sentences is a run-on or an "attachable sentence fragment."
- Every fifth possessive is incorrectly formed.

Given these findings, we should not be surprised to receive imperfect manuscripts. However, we must eliminate the slips of the so-called average writer before final copy goes to press.

Another part of the problem defined

Examples

Effects of the problem

Long-term effects

Other examples

The need for action

**Errors in Information Accuracy and Access**

Beyond basic errors, we have published some inaccurate information. For instance, we sometimes attribute products to the wrong companies or we list incorrect prices. Inaccuracies of this kind infuriate readers, product developers, and suppliers alike. And a retraction printed in the magazine's subsequent issue has little impact once the damage has been done.

Besides inaccurate information, *Megacrunch* too often presents inaccessible information. Mismarked visuals, misplaced headings, and misnumbered page references make the magazine hard to follow and use selectively. For readers to find what they need, a magazine must offer easy access.

Another part of the problem defined

Effects of the problem

Examples

**Technical Errors**

Technical errors seem one of our biggest problems. While some readers might raise a proverbial eyebrow over grammatical errors or skim over informational errors, technical errors are more frustrating and incapacitating. On a page of text, a misplaced comma or a missing bracket can be irritating, but in a program listing, these same errors can render the program useless. Even worse, a misnamed or misnumbered pin or socket in a hardware diagram might cause users to inadvertently destroy their data or damage their hardware.

Some technical slips in *Megacrunch* have veered close to disaster. Consider, for example, the flawed diagram in Figure 1, from our 7.12 issue.

A vivid example, illustrated in a visual

**FIGURE 1** Partial View of the Port Panel on the AXL-100

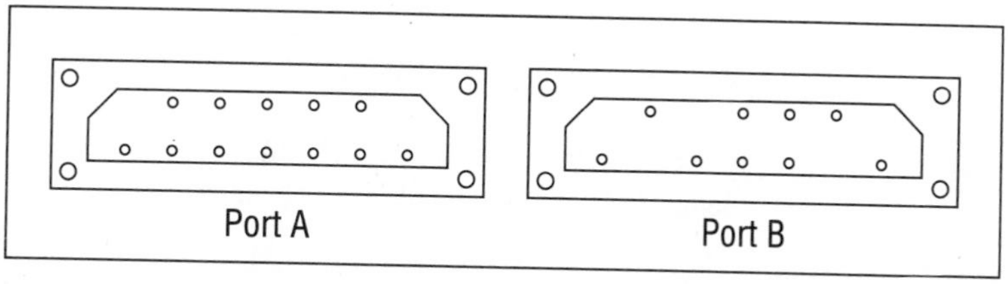


*To connect the external drive, plug Cable Y into Port A*

FIGURE C.3  The Final Report for a Research Project  *Continued*

4

Our published diagram instructed users to plug a 9-pin external-drive cable into Port A, a 12-pin modem port; the correct connection was to have been made to Port B. Ron Catabia, author of the article and respected tech wizard, explained the flaw in our reproduction of his diagram: "Had any users followed the instructions as printed, the read/write head on their external drive could have suffered permanent damage" (Catabia 1). Our lengthy correction printed one month later was in no way a sufficient response to an error of this importance. Nor could our belated correction placate an enraged and discredited author.

Such errors do little to encourage reader's perception of *Megacrunch* as the serious user's resource for the latest technical information.

### Distortions of the Author's Original Meaning

Authors routinely complain that, in our efforts to increase clarity and readability, we distort their original, intended meaning. After reading the edited version of her article, one author insisted that "too often, edits actually changed what I had said to something I hadn't said—sometimes to the point of altering the facts" (Dimmersdale 1).

Editorial liberties inevitably alienate authors. Overzealous editors who set out to shorten a sentence or fine-tune a clause—while knowing nothing about the program being discussed—can distort the author's meaning. As a recent study confirms, "When reviewers [editors] criticize in areas outside their expertise, their misguided reviews are seen as an intrusion" (Barker 37).

Following is an excerpt that typifies the distortions in recent issues of *Megacrunch*. Here, a seemingly minor editorial change (from "but" to "even") radically changes the meaning.

**As submitted:** A user can complete the Filibond program without ever having typed but a single command.

**As published:** A user can complete the Filibond program without ever having typed even a single command.

As the irate author later pointed out, "My intent was to indicate that a single command must be typed during the program run" (Klause 2).

This type of wholesale editing (of which more examples are shown in Appendix B) is a disservice to all parties: author, reader, and magazine.

### CAUSES OF OUR EDITORIAL INACCURACY

Before devising a plan for dealing with our editing difficulties, we have to answer questions such as these:

---

Discussion of the visual

Effects of the error

Interpretation—what it all means to the magazine

Another part of the problem defined

Example

Interpretation

The writer cites an authority, to support his interpretation

Another example

Effect of the error

Second subsection

Justification for an analysis of causes

**FIGURE C.3** The Final Report for a Research Project *Continued*

5

• Where are these errors coming from?
• Can they be prevented?

The scope of this subsection, so that readers will know what to expect

Interviews with our editing staff along with analysis of our editing practices and review of letters on file uncovered the following causes: (1) poor initial submissions from contributors, (2) lack of structure in our editing cycle, (3) lack of diversity in our editing staff, and (4) lack of communication with authors and advertisers. Each cause is discussed below.

**First cause defined**

**Findings**

**Conclusion**

### Poor Initial Submissions from Contributors

Some contributors submit poorly written manuscripts. And so we edit heavily whenever "a submission otherwise deserves flatout rejection," as one editor argues. Our editors claim that printing poor writing would be more damaging than the occasional editing excesses that now occur. Although editors can improve clarity and readability without in-depth knowledge of the subject we often misinterpret the author. Clearer writing guidelines for authors would result in manuscripts needing less editing to begin with. Our single-page author's guide is inadequate.

**Second cause defined**

### Lack of Structure in Our Editing Cycle

In our current editing cycle, the most thorough editors see an article repeatedly, as often as time allows. Various editors are free to edit heavily at all stages. And these editors are entirely responsible for judgments about grammatical, informational, and technical accuracy.

**Findings**

**No citation here, to protect in-house sources and to allow employees to speak candidly without fear of reprisal**

**Interpretation—what it means**

Although "having your best give their best" throughout the cycle seems a good idea, this approach leads to inconsistent editing and/or overediting. Some editors do a light editing job, choosing to preserve the original writing. Others prefer to "overhaul" the original. With light-versus-heavy editing styles entering the cycle randomly, errors slip by. As one editor noted, "Sometimes an article doesn't get a tough edit until the third or fourth reading. At that point, we have no time to review these last-minute changes."

Any article heavily edited and rewritten in the final stages stands a chance of containing typographical and mechanical errors, some questionable sentence structures, inadvertent technical changes, and other problems that result from a "rough-and-tumble edit."

**Findings**

**Interpretation**

While some articles are edited inconsistently, others are overedited. Our editors tend to be vigilant in pursuit of clarity, conciseness, and tone. Unfortunately, they seem less vigilant about technical accuracy.

FIGURE C.3  The Final Report for a Research Project  *Continued*

6

**Conclusion**

Instead of full-scale editing at all stages, we need a cycle that makes a manuscript progress from inadequate (or adequate) to excellent, through different levels of editorial attention. For example, a first edit should be thorough, while a final proofreading should be merely a fine-combing for typographical and mechanical errors.

**Third cause defined and interpreted**

**Conclusion**

### Lack of Diversity in Our Editing Staff

The many types of errors suggest that our present staff alone cannot spot all problems. Strong writing backgrounds have not prepared our editors to recognize a jumbled line of programming code or a misquoted price. To snag all errors, we must hire technical specialists. We need both a technical editor and a fact checker, to pick up where current editors leave off.

**Fourth cause defined and interpreted**

**Conclusion**

### Lack of Communication with Authors and Advertisers

Many of our editing troubles emerge from a gap between the meaning intended by contributors and the interpretation by editors. In the present system, contributors submit manuscripts without seeing any editorial changes until the published version appears. Along with an expanded author's guide, regular communication throughout the editing process (and perhaps the writing process as well) would involve contributors in developing the published piece, and thus make authors more responsible for their work.

**FIGURE C.3** The Final Report for a Research Project *Continued*

This section tells what should be done

The scope of this section, so that readers will know what to expect

Lead-in to the first recommendation

The recommendation

Lead-in to the second recommendation

Lead-in to the third, and final recommendation

First part of the final recommendation

Second part of the final recommendation

7

# RECOMMENDATIONS

To eliminate published inaccuracies, I recommend: (1) an expanded author's guide, (2) a five-stage editing cycle, and (3) improved communication with contributors.

## EXPANDED AUTHOR'S GUIDE

The obvious way to limit editing changes would be to accept only near-perfect submissions. But as a technical resource we cannot afford to reject poorly written articles that are nonetheless technically valuable.

To reduce editing required on submissions, I recommend we expand our author's guide to include topics like these: audience analysis, use of direct address and active voice, principles of outlining and formatting, and use of visuals.

## FIVE-STAGE EDITING CYCLE

In place of haphazard editing, I propose a progressive, five-stage cycle: Stages one through three would refine grammar, clarity, and readability. Two additional staff members, a fact checker and a technical editor, would check facts and technical accuracy in the final two stages. Figure 2 outlines responsibilities at each stage.

## IMPROVED COMMUNICATION WITH CONTRIBUTORS

The following measures would reduce errors caused by misunderstandings between contributors and editors.

### Author-Client Verification of Galley Proofs

Two weeks before our deadline, we could send authors pre-publication galley proofs [which show the text as it will appear in published form]. Authors could check for technical errors or changes in meaning, and return proofs within five days.

### Regular Use of Our Electronic-Mail Network

At any time during production, authors and editors could communicate through CompuServe by leaving questions and messages in one another's electronic mailboxes. (Virtually all our regular authors subscribe to CompuServe.) In addition, all parties could log onto CompuServe at one or more scheduled times daily to discuss the manuscript. The E-mail alternative is cheaper than the telephone, eliminates "telephone tag," and could serve as a "hot line" for authors while they prepare a manuscript for submission.

**FIGURE C.3**  The Final Report for a Research Project  *Continued*

8

The visual illustrates
and summarizes the
process being
recommended

**FIGURE 2**  The Five-Stage Editing Cycle

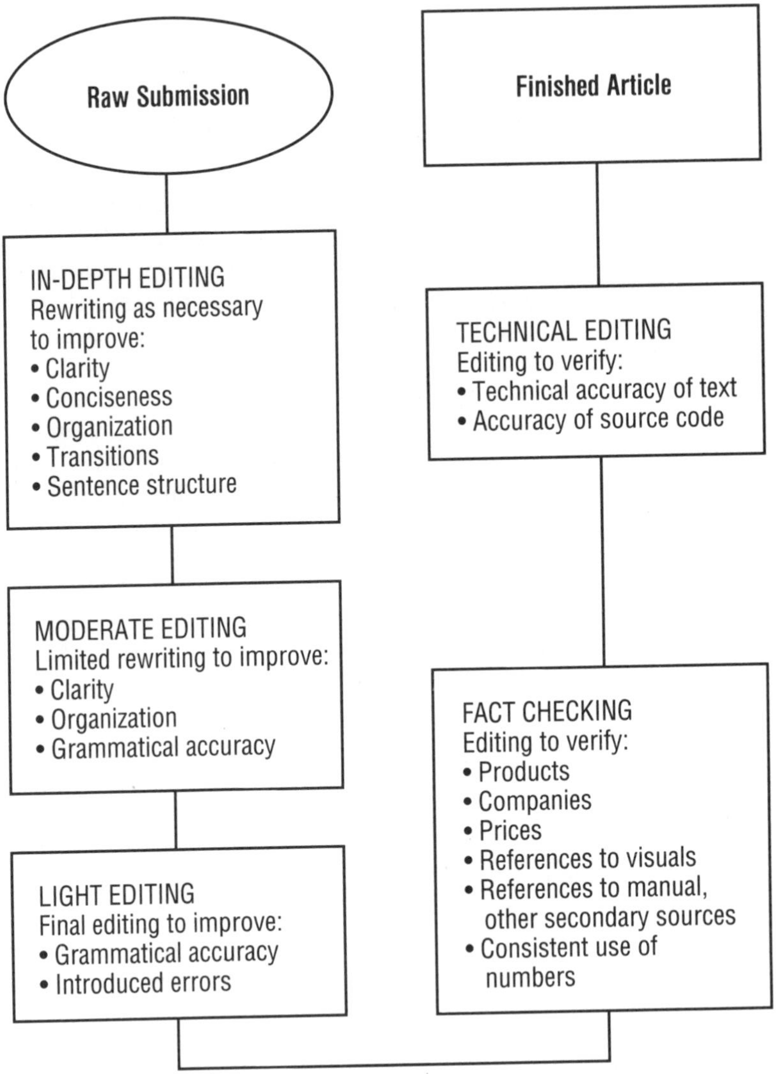

**Raw Submission**

**Finished Article**

**IN-DEPTH EDITING**
Rewriting as necessary
to improve:
• Clarity
• Conciseness
• Organization
• Transitions
• Sentence structure

**TECHNICAL EDITING**
Editing to verify:
• Technical accuracy of text
• Accuracy of source code

**MODERATE EDITING**
Limited rewriting to improve:
• Clarity
• Organization
• Grammatical accuracy

**FACT CHECKING**
Editing to verify:
• Products
• Companies
• Prices
• References to visuals
• References to manual,
  other secondary sources
• Consistent use of
  numbers

**LIGHT EDITING**
Final editing to improve:
• Grammatical accuracy
• Introduced errors

Data sources are
listed right below the
visual

*Source:* Adapted from Rainey, Kenneth T. "Technical Editing at the Oak Ridge National
Laboratory." *Journal of Technical Writing and Communication.* 18 (1988): 175-181 and
from Unikel, George. "The Two-Level Concept of Editing." *The Technical Writing
Teacher.* XV (1988): 49-54.

FIGURE C.3 The Final Report for a Research Project *Continued*

The list of references, or works cited, clearly identifies each source cited in the report

This writer documents his sources according to the format given in Chapter 6

9

# WORKS CITED

Barker, Thomas T. "Feedback in Hightech Writing." *Journal of Technical Writing and Communication.* 1(1988): 35–51.

Catabia, Ronald X. Notes from author Catabia's phone conversation with the Managing Editor, 10 March 1988.

Chang, Frederick R. "Revise: A Computer-Based Writing Assistant." *Journal of Technical Writing and Communication.* 1(1987): 25–41.

Dimmersdale, Olga B. Author's letter to the Managing Editor. 6 July 1987.

Grendel, M. L. Reader's letter to the Managing Editor. 14 May 1987.

Haswell, Richard H. "Toward Competent Writing in the Workplace." *Journal of Technical Writing and Communication.* 2(1988): 161–172.

Klause, Marcia C. Author's letter to the Managing Editor. 11 May 1988.

*Megacrunch* editing staff. Interviews. 12–14 April 1989.

[** Pages 10–21 contain the appendixes referred to in the report text. To save space, these pages are omitted here.]

# Works Cited

## CHAPTER 1

Barnum, Carol, and Robert Fisher. "Engineering Technologists as Writers: Results of a Survey." *Technical Communication* 31.2 (1984): 9–11.

"Calming Words for Computer Anxiety." *Changing Times* Apr. 1986: 99.

Dumont, R. A. "Writing, Research and Computing." *Critical Thinking: An SMU Dialogue,* 1988.

Elbow, Peter. *Writing Without Teachers.* New York: Oxford University Press, 1973.

Florman, Samuel, quoting Donald A. Rikard. "Toward Liberal Learning for Engineers." *Technology Review* Mar. 1986: 18–25.

Franke, Earnest A. "The Value of the Retrievable Technical Memorandum System to an Engineering Company." *IEEE Transactions on Professional Communication* 32.1 (March 1989): 12–16.

"Gassee Steps into the Limelight." *MacWorld* Aug. 1986: 20–21.

Gubernat, Susan. "Template for Success." *Publish!* Sept./Oct. 1986: 70–75.

Hauser, Gerald. *Introduction to Rhetorical Theory.* New York: Harper, 1986.

Ilkovics, D. "Information Systems Technologies: An Introduction." *Electrical Communication* 60.1 (1986): 9.

International Data Corporation. "White Paper." *Fortune* 10 Dec. 1984.

Keyes, Langley Carleton. "Profits in Prose." *Harvard Business Review* 1 (1961): 105–12.

Max, Robert R. "Wording It Correctly." *Training and Development Journal* Mar. 1985: 5–6.

Reade, Nathaniel. "Hay Yo! Take Dis Lettah, Willyah?" *New England Monthly* Apr. 1986: 20–21.

Richards, Thomas O., and Ralph A. Richards. "Technical Writing." Paper presented at the University of Michigan, 11 July 1941. Published by the Society for Technical Communication.

Spruell, Geraldine. "Teaching People Who Already Learned How To Write, To Write." *Training and Development Journal* October 1986: 32–35.

# CHAPTER 3

Gilsdorf, Jeanette W. "Executives' and Academics' Perception on the Need for Instruction in Written Persuasion." *The Journal of Business Communication* 23.4 (Fall 1986): 55–68.

Goodall, H. Lloyd, Jr., and Christopher L. Waagen. *The Persuasive Presentation.* New York: Harper, 1986.

Hauser, Gerald A. *Introduction to Rhetorical Theory.* New York: Harper, 1986.

Hays, Robert. "Political Realities in Reader/Situation Analysis." *Technical Communication* 31.1 (1984): 16–20.

Kelman, Herbert C. "Compliance, Identification, and Internalization: Three Processes of Attitude Change." *Journal of Conflict Resolution* 2 (1958): 51–60.

Kipnis, David, and Stuart Schmidt. "The Language of Persuasion." *Psychology Today* (April 1985): 40–46, in Ross, Raymond S. *Understanding Persuasion.* 3rd. ed. Englewood Cliffs, New Jersey: Prentice, 1990: 101–102.

Littlejohn, Stephen W. *Theories of Human Communication.* 2nd. ed. Belmont, California: Wadsworth, 1983.

Littlejohn, Stephen W., and David M. Jabusch. *Persuasive Transactions.* Glenview, Illinois: Scott, 1987.

Rokeach, Milton. *The Nature of Human Values.* New York: Free Press, 1973.

Ross, Raymond S. *Understanding Persuasion.* 3rd Edition. Englewood Cliffs, N.J.: Prentice, 1990.

Rottenberg, Annette T. *Elements of Argument.* 2nd ed. New York: St. Martin's, 1988.

Sherif, Muzafer, et al. *Attitude and Attitude Change: The Social Judgment-Involvement Approach.* Philadelphia: W.B. Saunders, 1965.

Stonecipher, Harry. *Editorial and Persuasive Writing.* New York: Hastings House, 1979.

Varner, Iris I., and Carson H. Varner. "Legal Issues in Business Communications." *ABCA Bulletin* (September 1983): 31–40.

# CHAPTER 4

Christians, Clifford G. et al. *Media Ethics: Cases and Moral Reasoning.* 2nd Edition. White Plains, NY: Longman, 1978.

Clark, Gregory. "Ethics in Technical Communication: A Rhetorical Perspective." *IEEE Transactions on Professional Communication* 30.3 (1987): 190–195.

Girill, T. R. "Technical Communication and Ethics." *Technical Communication* 34.3 (1987): 178–179.

Golen, Steven, et al. "How to Teach Ethics in a Basic Business Communication Class." *The Journal of Business Communication* 22.1 (1985): 75–84.

Gouran, Dennis, et al. "A Critical Analysis of Factors Related to Decisional Processes Involved in the Challenger Disaster." *Central States Speech Journal* 37.3 (1986): 119–135.

Hauser, Gerald A. *Introduction to Rhetorical Theory.* New York: Harper, 1986.

Janis, Irving L. *Victims of Groupthink: A Psychological Study of Foreign Policy Decisions and Fiascos.* Boston: Houghton, 1972.

Johannesen, Richard L. *Ethics in Human Communication.* 2nd Edition. Prospect Heights, Illinois: Waveland, 1983.

Lewis, Philip L. and N.L. Reinsch, "The Ethics of Business Communication." *Proceedings of the 1981 American Business Communication Conference.* Champaign, Illinois. In *Technical Communication and Ethics.* Ed. John R. Brockmann and Fern Rook. Washington, DC: Society for Technical Communication, 1989: 29–44.

Littlejohn, Stephen W., and David M. Jabusch. *Persuasive Transactions.* Glenview, Illinois: Scott, 1987.

Marx, Gary T., and Sanford Sherizen, "Monitoring on the Job." *Technology Review* (Nov./Dec. 1986): 63–72.

Mokhiber, Russell. "Crime in the Suites." *Greenpeace* 14.5 (1989): 14–16.

Pace, Roger C. "Technical Communication, Group Differentiation, and the Decision to Launch the Space Shuttle Challenger." *Journal of Technical Writing and Communication* 18.3 (1988): 207–220.

Porter, James E. "Truth in Technical Advertising: A Case Study." *Transactions in Professional Communication* 30.3 (1987): 182–189.

Presidential Commission, *Report to the President on the Space Shuttle Challenger Accident.* Volume I. Washington, DC: U.S. Government Printing Office, 1986.

Rowland, Robert C. "The Relationship between the Public and the Technical Spheres of Argument: A Case Study of the Challenger Seven Disaster." *Central States Speech Journal* 37.3 (1986): 136–146.

Rubens, Philip M. "Reinventing the Wheel?: Ethics for Technical Communicators." *Journal of Technical Writing and Communication* 11.4 (1981): 329–339.

Stevenson, Richard W. "Workers Who Turn in Bosses Use Law to Seek Big Rewards." *New York Times.* 10 July 1989: 1.

Unger, Stephen H. Controlling Technology: *Ethics and the Responsible Engineer.* New York: Holt, 1982.

Walter, Charles, and Thomas F. Marsteller. "Liability for the Dissemination of Defective Information." *IEEE Transactions on Professional Communication* 30.3 (1987): 164–167.

Wicclair, Mark R., and David K. Farkas. "Ethical Reasoning in Technical Communication: A Practical Framework." *Technical Communication* 31.2 (1984): 15–19.

Winsor, D. A. "Communication Failures Contributing to the Challenger Accident: An Example for Technical Communicators." *IEEE Transactions on Professional Communication* 31.3 (1988): 101–107.

Yoos, George. "A Revision of the Concept of Ethical Appeal. *Philosophy and Rhetoric* 12 (Winter 1979): 41–58.

## CHAPTER 8

"Earthquake Hazard Analysis for Nuclear Power Plants." *Energy and Technology Review* June 1984: 8.

Gartaganis, Arthur. "Lasers." *Occupational Outlook Quarterly* Winter 1984: 22–26.

Stanton, Mike. "Fiber Optics." *Occupational Outlook Quarterly* Winter 1984: 27–30.

## CHAPTER 9

Felker, Daniel B., et al. *Guidelines for Document Designers*. Washington: American Institutes for Research, 1981.

Halpern, Jean W. "An Electronic Odyssey." *Writing in Nonacademic Settings*. Ed. Dixie Goswami and Lee Odell. New York: Guilford, 1985. 157–201.

Redish, Janice C., et al. "Making Information Accessible to Readers." *Writing in Nonacademic Settings*. Ed. Dixie Goswami and Lee Odell. New York: Guilford, 1985. 129–53.

Roundy, N., and D. Mair, "The Composing Process of Technical Writers: A Preliminary Study." *Journal of Advanced Composition* 3 (1982): 89–101.

## CHAPTER 10

Consumer Product Safety Commission. *Fact Sheet No. 65*. Washington: GPO, 1979.

Kidder, Tracy, *The Soul of a New Machine*. Boston: Little, 1981.

U.S. Air Force Academy. *Executive Writing Course*. Washington: GPO, 1981.

## CHAPTER 11

Bailey, Edward P. *Writing Clearly: A Contemporary Approach*. Columbus, Ohio: Merrill, 1984.

Felker, Daniel B., et al. *Guidelines for Document Designers*. Washington: American Institutes for Research, 1981.

Kremers, Marshall. "Teaching Ethical Thinking in a Technical Writing Course." *IEEE Transactions on Professional Communication* 32.2 (1989): 58–61.

Williams, Joseph. *Style*. Glenview, Ill.: Scott, 1981.

Zinsser, William. *On Writing Well*. New York: Harper, 1980.

## CHAPTER 12

Cochran, Jeffrey K., et al. "Guidelines for Evaluating Graphical Designs." *Technical Communication* 36.1 (1989): 25–32.

Felker, Daniel B., et al. *Guidelines for Document Designers*. Washington: American Institutes for Research, 1981.

Girill, T. R. "Technical Communication and Art." *Technical Communication* 31.2 (1984): 35.

Hartley, James. *Designing Instructional Text*. 2nd ed. London: Kogan Page, 1985.

Journet, Debra. Unpublished review of *Technical Writing*, 3rd ed.

Lambert, Steve. *Presentation Graphics on the Apple® Macintosh*. Bellevue, Wash.: Microsoft Press, 1984.

Murch, Gerald R. "Using Color Effectively: Designing to Human Specifications." *Technical Communication* 32.4 (1985): 14–20.

Schmeupe, Ken D. "Upgrading Your Business Graphics." *Popular Computing* Dec. 1985: 51–56.

*The Aldus Guide to Basic Design.* Aldus Corporation, 1988.

"Using icons as communication." *Simply Stated.* 75 (Sept./Oct. 1987): 1, 3.

Van Pelt, William. Unpublished review of *Technical Writing,* 3rd ed.

Williams, Robert I. "Playing with Format, Style, and Reader Assumptions." *Technical Communication* 30.3 (1983): 11–13.

## CHAPTER 13

Benson, Phillipa J. "Visual Design Considerations in Technical Publications." *Technical Communication* 32.4 (1985): 35–39.

Felker, Daniel B., et al. *Guidelines for Document Designers.* Washington: American Institutes for Research, 1981.

Hartley, James. *Designing Instructional Text.* 2nd ed. London: Kogan Page, 1985.

Kirsh, Lawrence. "Take It from the Top." *MacWorld* Apr. 1986: 112–15.

Pinelli, Thomas E., et al. "A Survey of Typography, Graphic Design, and Physical Media in Technical Reports." *Technical Communication* 33.2 (1986): 75–80.

Redish, Janice C., et al. "Making Information Accessible to Readers." *Writing in Nonacademic Settings.* Ed. Lee Odell and Dixie Goswami. New York: Guilford, 1985.

*The Aldus Guide to Basic Design.* Aldus Corporation, 1988.

White, Jan. *Visual Design for the Electronic Age.* New York: Watson Guptill, 1988.

Wight, Eleanor. "How Creativity Turns Facts into Usable Information." *Technical Communication* 32.1 (1985): 9–12.

Williams, Robert I. "Playing with Format, Style, and Reader Assumptions." *Technical Communication* 30.3 (1983): 11–13.

## CHAPTER 14

Glidden, H. K. *Reports, Technical Writing, and Specifications.* New York: McGraw, 1964.

Riney, Larry A. *Technical Writing for Industry.* Englewood Cliffs: Prentice, 1989.

## CHAPTER 15

Bedford, Marilyn S. and F. Cole Stearns. "The Technical Writer's Responsibility for Safety." *IEEE Transactions on Professional Communication* 30.3 (1987): 127–132.

Clement, David E. "Human Factors, Instructions and Warnings, and Product Liability." *IEEE Transactions on Professional Communication* 30.3 (1987): 149–156.

Girill, T. R. "Technical Communication and Law." *Technical Communication* 32.3 (1985): 37.

Meyer, Benjamin D. "The ABCs of New-Look Publications." *Technical Communication* 33.1 (1986): 16–20.

Van Pelt, William. Unpublished review of *Technical Writing,* 3rd ed.

Walter Charles and Thomas F. Marsteller. "Liability for the Dissemination of Defective Information." *IEEE Transactions on Professional Communication* 30.3 (1987): 164–167.

Weiss, Edmond H. *How to Write a Usable User Manual.* Philadelphia: ISI Press, 1985.

# Index